Thomas Schupp
Hazardous Substances

Also of interest

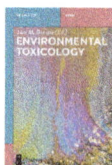

Environmental Toxicology
Luis M. Botana (Ed.), 2018
ISBN 978-3-11-044203-8, e-ISBN 978-3-11-044204-5

Nanomaterials Safety.
Toxicity And Health Hazards
Shyamasree Ghosh; 2019
ISBN 978-3-11-057808-9, e-ISBN 978-3-11-057909-3

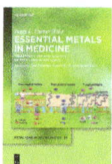

Essential Metals in Medicine.
Therapeutic Use and Toxicity of Metal Ions in the Clinic
Metal Ions in Life Sciences 19
Peggy L. Carver (Ed.) 2019
ISBN 978-3-11-052691-2, e-ISBN 978-3-11-052787-2

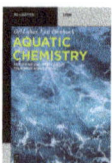

Aquatic Chemistry.
For Water and Wastewater Treatment Applications
Ori Lahav, Liat Birnhack, 2019
ISBN 978-3-11-060392-7, e-ISBN 978-3-11-060395-8

Thomas Schupp

Hazardous Substances

Risks and Regulations

DE GRUYTER

Author
Dr. Thomas Schupp
Muenster University of Applied Science
Chemical Engineering
Stegerwaldstr. 39
48565 Steinfurt
Thomas.schupp@fh-muenster.de

ISBN 978-3-11-061805-1
e-ISBN (PDF) 978-3-11-061895-2
e-ISBN (EPUB) 978-3-11-061979-9

Library of Congress Control Number: 2020941920

Bibliographic information published by the Deutsche Nationalbibliothek
The Deutsche Nationalbibliothek lists this publication in the Deutsche Nationalbibliografie;
detailed bibliographic data are available on the Internet at http://dnb.dnb.de.

© 2020 Walter de Gruyter GmbH, Berlin/Boston
Cover image: Thomas Schupp
Typesetting: Integra Software Services Pvt. Ltd.
Printing and binding: CPI books GmbH, Leck

www.degruyter.com

Preface

Giving courses in hazardous substance-related risks and regulations for master's degree in chemical engineering revealed soon the lack of a concise textbook that could be recommended to the students. There are already numerous excellent textbooks about toxicology on the market, and also for environmental behavior and ecotoxicology, although not that much as in toxicology. However, all these textbooks that I am aware of do not match the requirements for my master's degree courses, and frequently the content goes far beyond than what is required for passing the course. As a result, I had generated some documents as "supporting information" for the students, and soon the idea came up to summarize these documents as textbook tailored for our master's degree course "Hazardous Substances – Risks and Regulations."

Besides studying chemistry and working for BASF, I had spent much time in doing services at the German Red Cross, and in total about 20 weeks holiday were invested to attend lectures in toxicology, time that then was not available for my family. Nevertheless, my family accepted the way I proceeded, and I am very grateful to my wife Maria and our children for their patience and support.

I would also like to thank Dr. Gitta Egbers, Dr. Ines Bruening and Dr. Torsten Jeschke (BASF Polyurethanes GmbH) and Bjoern Hidding (BASF SE) for their review.

Steinfurt, March 2020
Thomas Schupp

https://doi.org/10.1515/9783110618952-202

Contents

1 Introduction

Chemical engineers typically find employment in chemical industries and also in other industries where the knowledge about chemical reactions and chemical behavior and large-scale treatment of chemicals can be transformed into customer benefit. They are responsible for production, distribution and quality control of chemicals, for marketing and also for employees who handle chemicals.

Regulations in the European Union (EU) and in many other regions charge chemical engineers with responsibility for product safety, workplace safety, plant safety and regulatory compliance. As the text in the following chapters refers to EU legislation, a few words shall be dedicated to the basic meaning of the EU laws.

In the EU, legislative acts are enacted as "Regulation," "Directive," "Decision," "Recommendation" or "Opinion".[1] "Regulations" are directly and literally binding for the EU member states; member states have only that freedom for shaping the implementation that is explicitly granted by the regulation.

"Directives" are setting a certain framework and define common goals. Within this framework, the EU member states have some freedom how these "common denominators" are achieved and to be put into force. Member states are granted a certain time slot to integrate the directive into national laws.

A "Decision" addresses relevant parties directly; the addressed can be certain member states, companies, institutions or others. To those addressed, the decision is directly binding.

A "Recommendation" is an instrument to allow EU institutions to present their view on a kind of best or common practice for a "how to do," without imposing any legal obligation to anybody.

An "Opinion" is a kind of an "expert judgement" of an EU institution; it is not legally binding.

EU Regulation is a legislation that is directly and literally binding to the EU member states. Freedom for national implementations is only possible as far as it is defined in the regulation.
EU Directives define common goals. After a certain time slot, EU member states have to implement the EU directive into national legislation.

The flow of product and safety-relevant information is illustrated by the simplified organigram shown in Fig. 1. Together with the product purchased, safety-relevant information is provided to the Health, Safety and Environment (HSE) department. The consequences for plant safety and workplace safety are evaluated and supported by the occupational medicine and expert departments. These latter departments are also to be involved in the evaluation of own products. The expert departments may be

---

1 https://europa.eu/european-union/eu-law/legal-acts_en

https://doi.org/10.1515/9783110618952-001

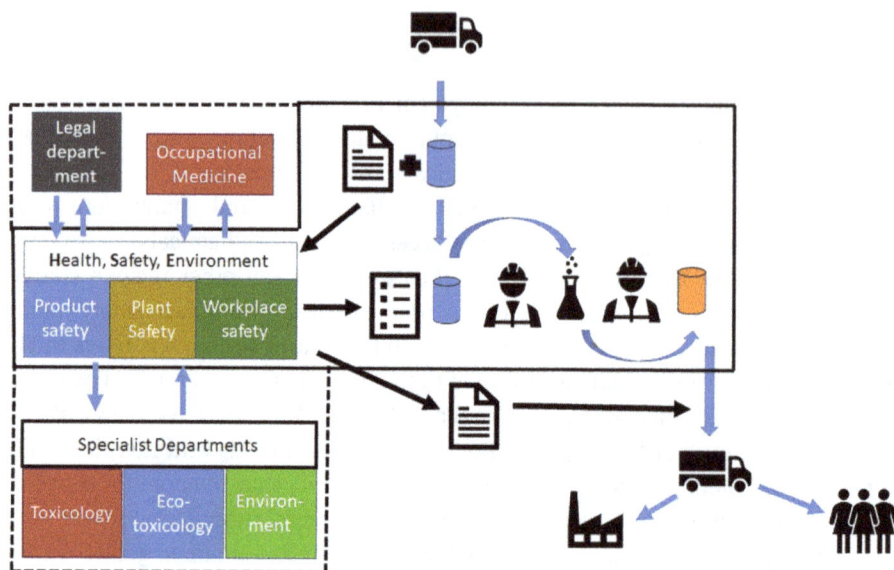

Fig. 1: Simplified organigram of a company purchasing and producing chemicals. The lines indicate the company boundary. Expert departments may be incorporated in the company or bought-in as services. Occupational medicine is compulsory and may be part of the company or a service from a competent service provider. Products are supplied to other customers and/or directly to consumers. Safety instructions for customers have to be made available by the HSE department.

part of the company, but services can be bought in by competent service providers. To involve expertise in occupational medicine is mandatory, and the involvement of expert departments is mandatory under certain circumstances, but always highly recommended for product liability reasons. For certain legal issues like export controls or high-liability applications (e.g., medicinal products, food contact materials), special legal advice might be necessary. Safety instructions are to be generated for the workers inside the company as well as for customers, workplace-related risk assessments and occupational hygiene measures, and the evaluation in terms of transport regulations (expert for dangerous goods) has to be organized by the HSE department. HSE is a "must have" in a company handling and dealing with dangerous substances, and chemical engineers may find employment in the HSE department. However, even if not employed in the HSE department, chemical engineers need to understand risks associated with chemicals to assure they neither do pose themselves, others or the environment to unacceptable risks nor run into product liability claims.

As a result, chemical engineers need to understand the behavior of hazardous substances, the associated risks and regulatory consequences. For that reason, lectures should be offered at least in master courses. For responsible management of chemicals associated risks, there is a need to understand basic principles of toxicology, ecotoxicology, environmental behavior as well as classification and labelling, regulations for the marketing of hazardous products and workplace safety. This textbook is intended as a guide accompanying students of chemical engineering through the respective areas of risks posed by chemicals and to provide the necessary background information.

In some parts, this textbook goes beyond the subjects needed to fulfill regulatory requirements. There are certain areas where chemical engineers should feel comparatively comfortable due to their education background. For example, toxicokinetics has many aspects in common with mass transfer in chemical reactors, and metabolism as well as environmental behavior can be matched against knowledge in reaction engineering. More in-depth information in these areas shall raise interest of chemical engineers in toxicology and ecotoxicology, enable them to actively participate in discussions with toxicologists and ecotoxicologists, and also make them ambassadors to the general public in discussing chemical risks on a rational basis.

Biology and physiology, on the contrary, are areas where students of chemical engineering are not familiar with. These subjects will be addressed on a limited scale, just sufficient to provide the necessary orientation in the area of risks posed by chemicals.

2 Physical and chemical properties of substances

Understanding physical and chemical properties of substances is a prerequisite for proper understanding of not only physical–chemical hazards, but also for the understanding of substance behavior in the environment and in organisms.

Environmental behavior covers the emission, distribution and fate of a substance in the environment, and this perspective shows similarities to the adsorption, distribution, metabolism and excretion (ADME) of a substance in an organism as discussed in Chapter 3, if an ecosystem (or nature) is regarded as an organism.

Physical–chemical properties of substances are part of their identification, and in the field of hazardous substances and risk assessment, proper identification and characterization is crucial for the success of toxicological and environmental investigations. This chapter starts with the identification of substances.

2.1 Identification of substances

Identification of a substance starts with simple physical properties that can be checked quickly and without sophisticated equipment. For solids, it is the melting point and visual appearance; for liquids, it is the refraction index, perhaps extended to boiling point, viscosity and density. These methods may provide a hint on purity, already, and it may occur that some limited data on identification are sufficient to check the specification of a product agreed upon between supplier and customer. Nevertheless, a more in-depth check on purity or identification of impurities can be very important. Just as one example: the substance aniline was the parent for the name "aniline-cancer," describing bladder cancer that was comparatively frequently detectable in workforces in aniline production plants. Later, it turned out that impurities in the technical aniline – namely benzidine and 2-naphthylamine – were the causative agents. To detect and quantify such by-products and impurities, spectrometric and chromatographic methods are required. The European chemicals regulation actually requires the full characterization of a "mono-constituent" substance (see Chapter 8) submitted for registration by UV, IR, NMR and MS spectra, as far as appropriate, and to make use of chromatographic methods [gas chromatography (GC), high-performance liquid chromatography (HPLC) and thin-layer chromatography (TLC)] to identify every component that contributes to at least 0.1% to the technical substance. Lower detection limits may be required if the presence of specific critical impurities cannot be excluded.

https://doi.org/10.1515/9783110618952-002

2.1.1 Solid substances: melting point, granulometry, density

These are properties to characterize solids. Melting points are taken in melting point apparatus using capillaries, but also a heating plate equipped with a microscope may be used for the investigation of small quantities. Melting points shall also be identified for substances which are liquids at room temperature. Differential scanning calorimetry (DSC) is a precise and comfortable method, but it requires access to the appropriate technical equipment. A less accurate but easy to run method is to pour the liquid into a test tube, place a thermometer into the substance and cool the tube in an ice–water mix (273 K), or place it in a fridge (about 255 K), or in an ice–sodium chloride mix (253 K) or in a dry-ice-acetone bath (195 K) and to check whether or not it solidifies. At least a crude detection of the melting point as " >" or " <" is possible. When the substance is allowed to warm up slowly and the temperature is taken over time, around the melting point the increase in temperature per unit time is slowed down (Fig. 2).

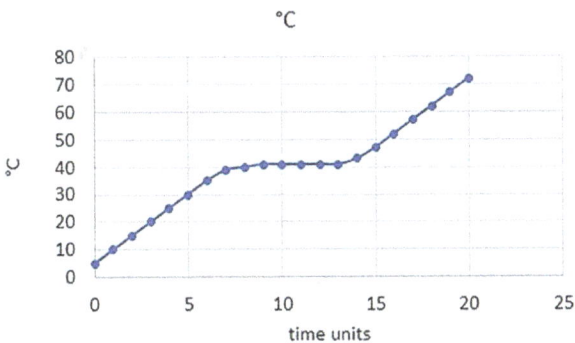

Fig. 2: Melting point detection by temperature–time curve (arbitrary sample).

Granulometry characterizes the particle sizes and particle size distribution of granular/powdered material. This value is an important indication for potential dust exposure. Depending on the handling of the material, aerosols (dust) can be formed and inhaled. Particles smaller than 10 µm in aerodynamic diameter can enter the alveoli of human lung, which is critical in terms of inhalation toxicity.

Density of solids is also characteristic and has to be distinguished between solid substance block density and the bulk density of piled up granules. The density is used for the conversion of volume into mass and vice versa, and it tells us where the substance is to be found if spilled into a lake or river.

2.1.2 Liquid substances: boiling point, refraction index, density

Density mentioned in the previous section is a property than can/has to be taken for liquids. Further, boiling points have to be taken which can be done simply by distillation (and, thereby, purifying the substance) or by methods consuming only small amounts like the Siwoloboff method (look up 440/2008/EC or lab-books on organic chemistry).

The refraction is taken from transparent liquids, and it is the quotient of the speed of light in vacuum divided by the speed of light in the liquid. The propagation of the electromagnetic wave is influenced by the electron distribution and density in the compound investigated. Therefore, the refraction index allows a quick and easy control of substance identity and purity for liquid products.

The density of liquids is taken with pycnometers, or by detection of weight loss of a solid body with defined volume and mass in air dived into the liquid. The density is used for the conversion of volume into mass and vice versa, and it tells us where the substance is to be found if spilled into a lake or river.

2.1.3 Formula, structure

Of the pure substance, spectral data like UV-Vis spectra, IR spectra, ^{1}H and ^{13}C-NMR spectra and mass spectra help to identify the structure of the substance, especially for organic molecules, and allow to distinguish them from isomers. For inorganic molecules in crystallized form, X-ray structure analysis is a kind of a gold-standard for the identification. Please refer to textbooks for organic and inorganic chemistry.

2.2 Characterization

Characterization means establishment of physical–chemical data which have a more direct impact on the environmental behavior and toxicity of the substance. For example, the vapor pressure provides insight into how easily the substance evaporates, and how far airborne exposure plays a role. Water solubility indicates whether the aquatic compartment in the environment may be affected, and together with the octanol–water partitioning (K_{OW}) whether a substance is more likely to be found in aquatic or in lipophilic phase in the environment or an organism.

2.2.1 Vapor pressure

The relation between temperature, vapor pressure and heat of evaporation is given by the so-called Clausius–Clapeyron equation and its simplified form, the August formula (see textbooks of physical chemistry). The vapor pressure can be extrapolated from boiling temperatures at different pressures in the distillation apparatus. Having at least two data points allows to derive the parameters of the August formula:

$$\log(p) = A - \frac{B}{T}$$

with $\log(p)$ being the decadic logarithm of the vapor pressure at temperature T; A and B are substance-specific constants. This method (called "dynamic method") allows an access to an approximate vapor pressure with the help of standard lab equipment. Vapor pressures between 1,000 and 100,000 Pa can be established with satisfying accuracy.

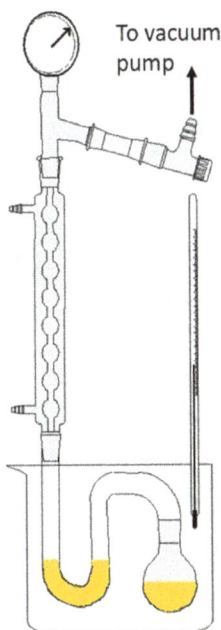

Fig. 3: Equipment for vapor pressure detection: static method.

In the static method, a vessel with the liquid sample is connected to a U-tube and placed in a heat bath (Fig. 3). The U-tube is followed by a condenser, a manometer, valve and vacuum line. The round bottom flask and the U-tube are filled with some liquid. First, the pressure is lowered to a sufficient extent so liquid boils for some seconds up to a minute, ensuring that the gas phase between vessel and U-tube is made up only by vapor of the liquid. Now the pressure is adjusted so the liquid in the U-tube achieves even level. At even level, the manometer shows the vapor pressure of the liquid at the given temperature. The static method allows to detect vapor pressures in the range of 10 to 100 000 Pa and is suitable for liquids and solids.

The gas saturation method can be used for vapor pressures in the range of 10^{-4} to 10 Pa. Here, a stream of inert gas is passed through the substance (either liquid or coated inert carrier for solids) so that finally the gas is saturated with the substance (Fig. 4). The analyte is collected in an efficient sampler (active charcoal; freeze trap) and quantified. According to the ideal gas law:

$$p = \frac{m \times RT}{V \times M}$$

M is the molar mass, m the collected mass and V the volume purged through/over the substance. Care has to be taken that saturation of the gas phase is achieved. Increasing the residence time by either increasing the pipe length or decreasing the flow will result in a higher saturation of the purge gas, and more material (m) will be collected per purge volume (V) (Fig. 5).

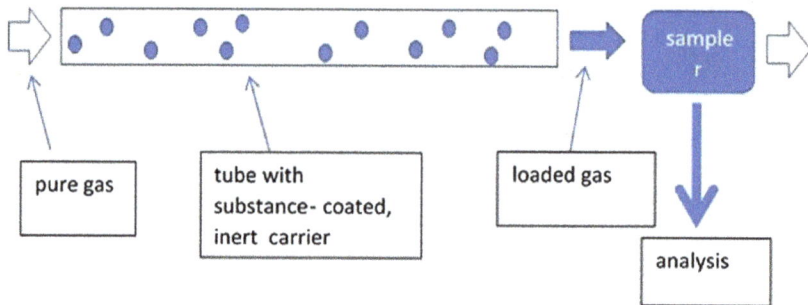

Fig. 4: Principle of the gas saturation method. The longer the tube, or the longer the residence time of the gas, the more likely saturation will be achieved.

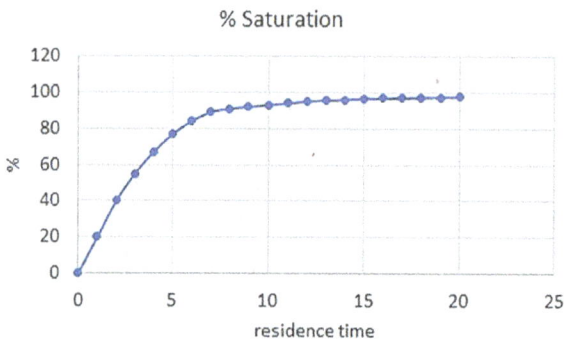

Fig. 5: At given temperature, gas saturation depends on the residence time; residence time depends on gas flow and tube length.

2.2.2 Water solubility

Like the vapor pressure, the water solubility is also dependent on the temperature. To check for water solubility, the substance under scrutiny is contacted with bidistilled water for a certain period of time at a set temperature. Care has to be taken with critical substances that may form micelles (surface active agents). For such substances, it is difficult if not impossible to measure water solubility. After a set contact time, the substance under question is analyzed in the water. Specific analytical methods are preferred. p.e. separation by GC or HPLC and identification by MS (GC-MS, HPLC-MS). More general detectors like light scattering detector or flame ionization detector are less suited but acceptable if via retention times or retention indices of the chromatographic peaks the substance identification, and via integration of the signals the quantification of a substance is possible.

More general analytical methods may be used if specific methods are not applicable. General methods or "sum-parametric methods" are the detection of the total organic carbon (TOC), dissolved organic carbon (DOC) or chemical oxygen demand (COD). In TOC or DOC tests, a sample of the water is either incinerated (TOC, DOC) and the formed CO_2 is analyzed by FT-IR, or organic compounds are oxidized with chromate in sulfuric acid at 140 °C, catalyzed by Ag_2SO_4 (in replacement of the previously used $HgSO_4$); residual chromate is detected photometrically or by titration with Fe^{2+} solution:

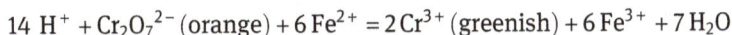

$$14\ H^+ + Cr_2O_7^{2-}\ (\text{orange}) + 6\ Fe^{2+} = 2\ Cr^{3+}\ (\text{greenish}) + 6\ Fe^{3+} + 7\ H_2O$$

The results delivered by these latter methods, delivering "sum-parameters," are error prone. For example, assume that you are investigating the water solubility of pentachlorphenyl-methylether, $C_7H_3\ Cl_5O$, $M = 280.5$ g/mol. About 10 g is equilibrated with 1 L water. The DOC measurement revels 0.11 g CO_2 per L water. What is the water solubility, if the ether was 100% pure? What is the water solubility, when the ether contained 1% methanol as impurity, which is known to be very well water soluble? Nevertheless, "sum-parameter" methods have their justification as they may be the only sensible access to water solubility for complex mixtures or substances of unknown composition, and to identify "unknowns" in environmental chemistry.

The water solubility is of importance with regards to ecotoxicological measurements as substances are either tested at a maximum concentration of 100 mg/L (see later chapters) or at its maximum aqueous concentration, if this is below 100 mg/L.

2.2.3 Surface tension

To detect whether a substance is to be rated as surface active, a solution in water at 90% solubility, but not more than 1 g/L is prepared. The force necessary to draw up a ring or a plate from the surface of the solution, or to expand a liquid film is measured (Fig. 6). Pure water at 20 °C has a surface tension of 72.75 mN/m (0.0723 N/m). Substances causing a reduction of the surface tension down to at or below 60 mN/m at 20 °C are rated as *surface active*. This may have an impact on further measurements. For example, liquid chromatographic methods which may be applied to identify the octanol–water partition coefficient (K_{OW}) or the soil adsorption constant (K_d) are not reliable for surface-active substances.

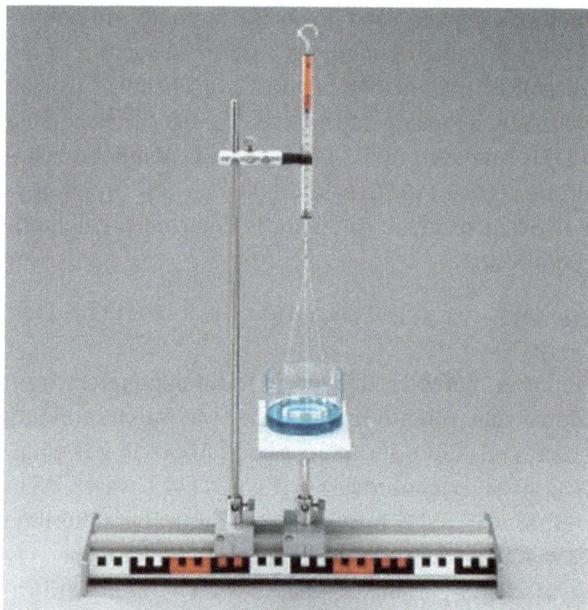

Fig. 6: Measuring the surface tension with the ring method.

For the ring method, a planar ground metal ring (iridium or simply aluminum) fixed to a spring balance is just dipped into the liquid (Fig. 6). The extra force necessary to pull the ring of radius R out of the liquid is a function of the surface tension, as

$$F = 2 \times (2\pi R) \times y$$

where F is the force, R is the radius of the ring and y is the surface tension.

Capillary action is another clue to the surface tension of a liquid. At a given temperature, the rise of a liquid in a capillary depends on the surface tension of the aqueous solution.

$$y = \frac{1}{2} \times \left(h + \frac{r}{3} \right) \times r \times \rho \times g$$

with y = surface tension (N/m), ρ = density (kg/m^3), r = radius of the capillary (m), h = height of the liquid in the capillary, g = earth acceleration, 9.81 m/s^2. The more exact calculation requires a factor cos Θ the surface tension has to be multiplied with, Θ being the contact angle between liquid and solid surfaces.

2.2.4 Octanol–water partitioning

Octanol–water partition coefficient, K_{OW}, is the equilibrium distribution of a substance (i) between water-saturated 1-octanol and 1-octanol-saturated water:

$$K_{OW} = \frac{C_{i,\,octanol}}{C_{i,\,water}}$$

This coefficient is temperature dependent and can be measured by checking the equilibrium distribution of the substance between water and octanol. A specific analytical method is to be applied to check the concentration of the substance in octanol and in water.

A quick, high-throughput screening method makes use of the retention factors in reversed-phase HPLC:

$$R_f = \frac{t_r - t_0}{t_0}$$

where t_r is the retention time of the substance and t_0 is the "dead-time" of the column, detected with nonretained substances like dimethyl formamide or thiourea. The $\log(R_f)$ is linear related to $\log(K_{OW})$ and a calibration line is generated with standard substances of known K_{OW}. This HPLC method is not applicable for surface-active substances, as these may creep along the interface between mobile and stationary phase instead of distributing into the phases.

2.2.5 Air–water partitioning: the Henry constant

Air–water partitioning coefficient is important to characterize the environmental behavior of a substance, and for the workplace it may be interesting for airborne exposure if a hazardous substance is used in form of aqueous solutions. Frequently the Henry coefficient, H, is approximated by water solubility, S, and vapor pressure, VP:

$$H' = \text{VP}/S \ \left[\text{Pa} \times \text{m}^3/\text{mol}\right]$$

The dimensionless Henry coefficient which is the equilibrium concentration of the substance between gas phase (C_{gas}) and aquatic phase (C_{water}), $H = C_{\text{gas}}/C_{\text{water}}$, is calculated via the ideal gas law, with $n/V = \text{VP}/(R \times T)$, therefore,

$$H = H'/(R \times T)$$

2.2.6 Adsorption/desorption on soil

Adsorption and desorption investigates the distribution behavior of a substance be-tween soil solid particles and soil water. As soil can be very locally different, eight different soil samples with varying contents of clay minerals, sand and organic car-bon have to be used in the test. This tedious approach can be circumvented by a HPLC method where the stationary phase is a cyanopropyl-modified silica. The polar cyano-group of the stationary phase can provide dipole-dipole interaction sites, whereas the alkyl chain presents hydrophobic interaction sites. However, the HPLC method is regarded as a screening method only, and not applicable for sur-face-active substances. The pH value is important when measuring compounds that can be protonated or deprotonated like carbonic acids, phenols and amines. The requirement for testing eight different soils can be explained by the different factors influencing the distribution of the substance between soil and water:
-   Hydrophilic-hydrophobic properties; soils can have different amounts of or-ganic fractions, and these are not necessarily uniform. A lipophilic substance is prone to prefer the organic fraction in soil.
-   Organic material in soil has aromatic structures for π-electron interactions and polarizability; alcohol, amide, acid and keto groups acting as hydrogen-bond do-nators and acceptors; phenolate and carboxylate groups for ion–ion interaction.
-   For certain organic compounds, namely primary and secondary amines, organic matter in soil provides covalent binding sites. These amines may undergo Michael addition reactions with quinones and α, β-unsaturated carbonyl compounds.
-   Ion-ion interaction sites are also provided by clay minerals, and some of them can intercalate other cations.

Under regulation [EU] No. 1272/2008, adsorption/desorption is allocated to "envi-ronmental behavior." For the physical exchange processes between water and soil, a distribution constant, K_d, is defined as the equilibrium concentration in soil di-vided by the concentration in water. For some persistent and bioaccumulative organic compounds that caused historical (and todays) soil contamination, the frac-tion of organic carbon in the soil has a great influence on the distribution constant. For these substances, K_d determined in one soil type can be estimated for other soils, if this value is corrected for the organic carbon content. This transformed, more universal applicable constant is the K_{OC}

$$K_d = \frac{C_{i,\text{soil}}}{C_{i,\text{water}}} \left[\frac{m^3}{kg}\right]$$

$$K_{OC} = \frac{K_d}{\% \text{ organic carbon}} \times 100 \left[\frac{m^3}{kg}\right]$$

Note that for environmental modeling, typically SI units are used, that is m^3 for the volume and kg for the mass.

2.2.7 Dissociation constants

Certain substances may release or attach a proton in the environment. Phenols, for example, are weak acids; carboxylic acids are somewhat stronger acids than phenols. Amines behave like bases in the environment and in organisms. Water solubility is increased, and vapor pressure is decreased when phenols or carboxylic acids are deprotonated and if amines are protonated. The octanol–water partition coefficient decreases when substances become charged. For the adsorption–desorption behavior, a prediction is impossible, so this property has to be measured not only with different kinds of soils, but also at different pH values. To mimic environmental conditions, pH values to be chosen are pH = 4, 7 and 9.

Melting point, boiling point, vapor pressure, water solubility and octanol–water partition coefficient are the minimum data points which are needed to model the distribution of a substance in the environment. For substances that can be protonated/deprotonated, the dissociation constants need to be measured as well.

HPLC-based screening methods for the octanol-water partition coefficient and for the adsorption/desorption cannot be used for substances which are surface active or which are present in a charged form (ions).

2.3 Physical chemical hazards

This chapter deals with physical–chemical hazards: explosive properties, fire (ignitability), oxidizing properties and corrosion of materials. In the globally harmonized system for the classification and labelling of hazardous substances (GHS), the hazard statements of the 200 series are reserved for physical hazards.

2.3.1 Explosion

Explosion hazard is defined as a rapid decomposition of material where a multitude of gas volume, respectively pressure is generated in a short time, typically associated with the release of heat and generation of fire (although the latter is not always the case and not a precondition to rate a material as explosive). Explosive substances are sensitive to temperature, mechanical shock and/or friction. The test methods are standardized under the scope of guidelines for dangerous goods. Depending on the outcome of the tests, the explosive substances are allocated to one of the total six classes. The EU guidance and Regulation (EU) No. 1272/2008 allows a theoretical assessment whether or not a substance may have explosive properties. There are several suspect groups that may raise concern and usually would trigger testing, listed in Tab. 1:

Tab. 1: Functional groups that may indicate explosive properties of the molecule; R means organic and – if applicable – inorganic residues.

Substance group	Critical structure
Hydrazines	R-NH-NH-R
Azides	R-N=N=N
Nitroso compounds	R-N=O
Nitro compounds	$R-NO_2$
Chloro amines	$R-NH-Cl, R-N(Cl)_2$
Cumulative double bonds	$R_2C=C=CR_2$
Alkines	$R-C\equiv C-R$
Peroxides	R-O-O-R
Oxychlorides	$R_3C-O-Cl$

If such groups are present, the oxygen balance, OB, should be calculated:

$$OB = -1,600 \times \frac{\{2x + \frac{y}{2} - z\}}{MW}$$

for a substance of formula $C_xH_yO_z$ and with MW being the molecular weight. If OB < −200, the substance is unlikely to be an explosive. For example, the known explosive trinitrotoluene (TNT), $C_7H_5N_3O_6$, has MW = 227 g/mol, $x = 7$, $y = 5$ and $z = 6$. The calculation for the OB gives

$$OB = -1,600 \times \frac{2\times7 + \frac{5}{2} - 6}{227} = -74$$

As −74 is larger than −200, by calculation this substance is possibly an explosive. Tests need to be run to check for the appropriate class of explosives.

Mixtures of flammable substances with inorganic oxidizers may be explosives (gunpowder!). In any case of doubt, the mixture should be tested. Inorganic oxidizers are allocated to classes 1 (very strong) to 3 (rather weak). If the content of an oxidizer class 1 or 2 is below 15%, or a class 3 oxidizer is below 30%, the mixture is expected not to be an explosive.

In principle, explosive properties need to be checked. Certain groups in organic (and inorganic) molecules are suspicious of providing explosive properties to the substance. Examples are peroxo, nitro, hydrazido and azido groups.

2.3.2 Flammability

Flammability is the property of a substance to catch fire. Gases are classified as extremely flammable, category 1 (H220) if a content of 13% or less in air can be ignited, or if the ignitability range stretches over at least 12% points. Other flammable gases are category 2, H221 (flammable gas).

Substances brought on the market as aerosol cans need to be tested concerning their flammability, irrespective of whether they release gas-dispersed liquids or solids or foams, pastes or powders. Based on the test results they are either non flammable, or they belong to the flammable aerosols category 1 (H222: extremely flammable aerosol) or category 2 (H223: flammable aerosol).

Flammable liquids have frequently been the reason for major accidents and, therefore, are "classics" in chemical substance regulations. The *flash point* is that temperature at which the liquid generates enough vapor so that a flame brought to the surface of the liquid spreads over the liquid (Fig. 7). The substance may be heated in an open cup (open cup test) or, typically preferred today, in a closed cup, were the lid of the cup is briefly removed while the flame is brought near the substance surface whenever a temperature is reached where the flammability shall be tested.

For mixtures containing flammable liquids, the flash point should be measured. However, a method for the calculation of flashpoints has been published [1]. The idea behind that method is that at the flash point, the vapor pressure of the substance is high enough that the vapor–air mixture above the liquid contains a high enough concentration of the substance, so its combustion delivers sufficient heat to make the flame spread. Let L_j be the vapor pressure of substance j at flash point temperature, P_j^0 the vapor pressure of the pure component, P_j the effective vapor pressure of the substance in the mixture, x_j its mole fraction and y_j its activity coefficient. P_j is dependent on the temperature. For a given temperature, the mixture is flammable if

$$\sum_j \frac{P_j}{L_j} = \sum_j \frac{y_j \times x_j \times P_j^0}{L_j} \geq 1$$

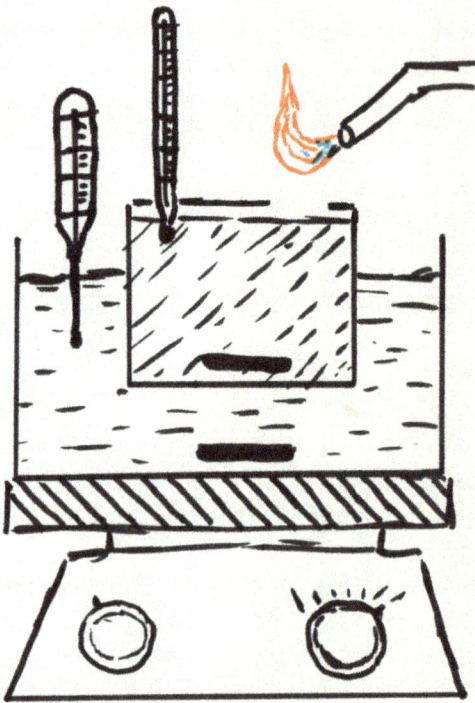

Fig. 7: Principle of flash point testing.

The activity coefficient of the components can be calculated with the UNIFAC method which is frequently imbedded in process simulation tools for chemical engineering.

For plant safety and safe handling, there are other properties connected to flammability which do not have necessarily a direct influence on classification and labelling. These are the *explosion limits* and the *self-ignition temperature*. Both data points have to be submitted to the European chemicals agency for substance registration.

The upper and lower explosion limits are the concentration of a gas or a vapor in air which will cause an explosion if ignited. The self-ignition temperature is that temperature of a surface which sets the substance under investigation under fire. To test this, an empty vessel is heated up in a heating block. At certain temperatures, a drop of the test-substance is added, releasing an audible "plop" when the self-ignition temperature is reached or exceeded (Fig. 8). Most flammable gases (besides hydrogen and methane) and vapors of flammable solvents have a higher density than air. Vapors may creep along the floor, entering gutters and the sewage system and might create ignitable atmospheres some distance away from the point of release (Fig. 9). Any open flames, sparks (electric switches!) and

Fig. 8: Principle of testing the self-ignition temperature.

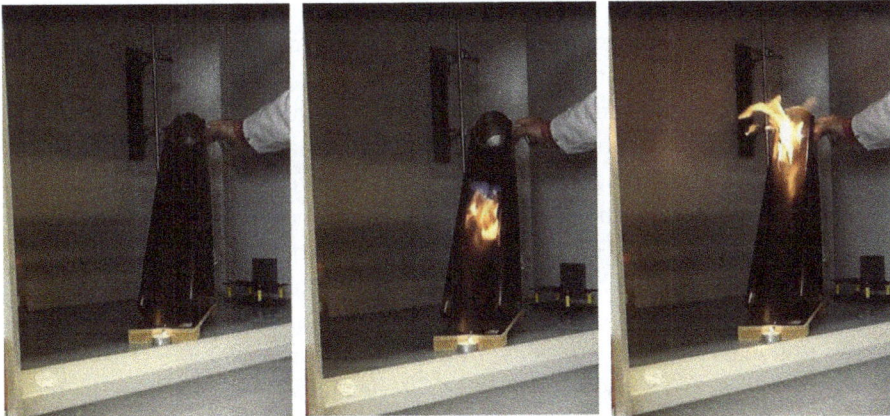

Fig. 9: Vapor of methyl-tertiary-butyl ether ((1,1-dimethyl-ethyl)-methyl ether) glides down the gutter and is ignited by a candle flame at the bottom. The flame propagates to the top.

static electricity discharge can ignite these vapors. This has to be born in mind when working with flammable gases and liquids.

Flammable solids are classified according to their burning behavior, that is, the spread velocity of a flame when set on fire and the capability to jump over a defined wetted zone. Both categories are allocated to the H228: Flammable Solid (phosphorus pentasulfide; red phosphorus).

The flash point of a liquid is the minimum temperature that ensures a sufficient formation of vapor which can spread a flame briefly hold over the liquid.

The self-ignition temperature for a substance is that temperature at which the substance ignites spontaneously when exposed to ambient air (20.5 vol% oxygen).

2.3.3 Pyrophoric and water-reactive substances

Pyrophoric substances are those that cause fire within 5 min when exposed to air. They are labeled as H250: Catches fire spontaneously if exposed to air (white phosphorus). Some substances heat up when exposed to air and are *self-heating substances* category 1 (H251: Self-heating; may catch fire. Examples: sodium dithionite; sodium methanolate) or category 2 (H252: Self-heating in large quantities; may catch fire. Example: magnesium powder).

Substances and mixtures reacting dangerously with water deserve extra attention (Fig. 10). They are allocated to category 1 (magnesium alkyls; calcium phosphide; phosphorus pentasulfide), category 2 (aluminum powder; calcium) or category 3 (slower reaction than category 2).

Oxidizers may easily release oxygen or oxygen equivalents, p.e. active halogens, which make flammable substances highly flammable. The categories are allocated by experimental comparison to defined standards.

For *oxidizing* liquids, the substance is a category 1 oxidizer if the 1:1 mix with cellulose shows an ignition or an increase in pressure equivalent to a 1:1 mix of cellulose/50% perchloric acid (H270: may cause fire or explosion; strong oxidizer. Chlorine dioxide; oxygen). Category 2 is a substance when mixed 1:1 with cellulose causes a pressure increase equivalent to 40% $NaClO_3$ (aq.) mix with cellulose, but the reaction does not justify category 1 (H271: May intensify fire; oxidizer. Hydrogen peroxide solution; sodium peroxide). For category 3, a 1:1 mix with cellulose is equivalent to a pressure increase caused by a 1:1 mix 65% HNO_3/cellulose (H272: May intensify fire; oxidizer. Examples are diammonium peroxydisulfate; calcium hypochlorite).

Fig. 10: Mixture of finely ground aluminum with iodine catches fire after adding a drop of water.

Oxidizing solids are classified in three categories as well, having the same H-phrases, but the categories 1 to 3 are compared to cellulose/$KBrO_3$ mixtures. *Category 1*: 4:1 to 1:1 mix of the oxidizer with cellulose has a faster burning rate than 3:2 $KBrO_3$/cellulose. *Category 2*: 4:1 to 1:1 mix of the oxidizer with cellulose has a faster burning rate than 2:3 $KBrO_3$/cellulose. *Category 3*: 4:1 to 1:1 mix of the oxidizer with cellulose has a faster burning rate than 3:7 $KBrO_3$/cellulose.

When oxidizers are mixed with flammable substances, the mixture itself may have explosive properties. Gunpowder is a mixture of milled charcoal and sulfur (both flammable) with potassium nitrate (an oxidizer).

2.3.4 Self-reacting substances and organic peroxides

Self-reacting substances do not fulfill definitions of explosives, but they may either release a critical amount of heat (more than 300 J/g) at decomposition or a 50 kg package (or more) may suffer self-accelerating decomposition at temperatures at 75 °C or below. For testing, a certain amount of substance can be heated up in a calorimeter, and the heat balance is recorded.

Organic peroxides – though showing a comparable behavior – are dealt with separately. A reason may be that organic peroxides may as well act like explosives, oxidants and irritant/corrosive materials. They tend to possess explosive decomposition, burn rapidly and/or are sensitive to shock-impact and friction. A common feature is the C-O-O-R element in the structure. Dibenzoyl peroxide and di-tertiary-butyl peroxide are radical initiators as they disproportionate at elevated temperatures (Fig. 11).

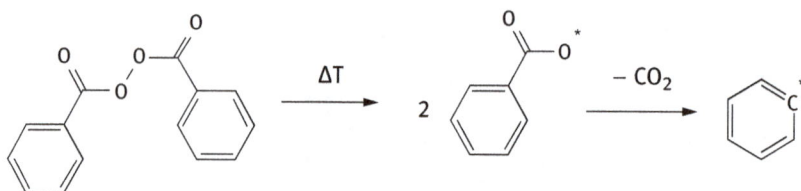

Fig. 11: Thermal decomposition of dibenzoyl peroxide: fragmentation of the molecule.

Hydroperoxides (besides use as intermediates for further rearrangement reactions or oxidizers) do not have a direct technical application. They play a role in the hardening of linseed oil. Some organic substances are sensitive to hydroperoxide formation, namely diethyl ether, tetrahydrofuran, dioxane and tetralin (bicyclo[4.4.0] decane). In the presence of air and sunlight, omnipresent OH* radicals initiate the activation of these molecules which then undergo a chain reaction leading to an enrichment of hydroperoxides (Fig. 12). When the original compound, having a higher vapor pressure and lower boiling point, evaporates, the concentration of the hydroperoxide increases. This makes the mixture more and more heat sensitive, and violent explosions may follow.

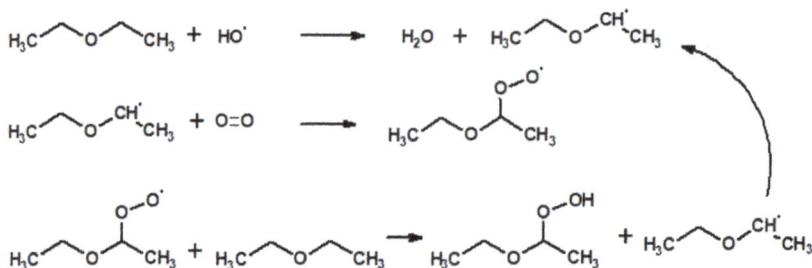

Fig. 12: Hydroperoxide formation in di-ethyl ether.

It is of utmost importance to check such solvents for hydroperoxides. This can be done with potassium-iodide paper which turns blue on the presence of hydroperoxides, or Fe^{2+} solutions which change color from greenish to pale yellow.

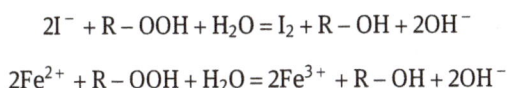

$$2I^- + R-OOH + H_2O = I_2 + R-OH + 2OH^-$$

$$2Fe^{2+} + R-OOH + H_2O = 2Fe^{3+} + R-OH + 2OH^-$$

Shaking with aqueous solutions of $FeSO_4$ or $Na_2S_2O_5$ destroys the hydroperoxides.

Ethers and branched hydrocarbons may form hydroperoxides when exposed to air and light. As these hydroperoxides have higher boiling points than the original solvent, they may concentrate up over time. Concentrated hydroperoxides tend to have violent decomposition.

2.3.5 Compressed gases

Containments with compressed gases pose a hazard in so far as that overheating or mechanical stress may end up in heavy rupture of the containment. The label and hazard statements shall ensure that the containments are prevented from overheating and are secured against mechanical shock.

It should be noted that refrigerated gas or expanding compressed gas may be very cold and can cause burns on direct skin contact. Violent evaporation of deep-freeze liquefied gas indoors may expel oxygen in air so people inside are in danger of suffocation.

2.3.6 Substances corrosive to metals

In previous EU regulations addressing hazardous substances, corrosivity was originally a hazardous property only in the area of toxicology. However, corrosion to materials can pave the way to major incidents by failure of piping and storage tanks. Further, corrosivity to metals may cause release of hydrogen. The build-up of pressure may cause rupture of the containment, and mixtures of hydrogen with air are explosive.

Author's experience: Following a spillage, a mixture containing 2.5% formic acid was cleaned up. The liquid was absorbed with the clay mineral bentonite. The mud was shoveled into standard steel drums which were closed tightly. A few hours later, the drums were bulged by hydrogen released by corrosion.

2.3.7 Powders, dusts and aerosols

Timber wood is not to be classified as something that is hazardous. However, by sawing, drilling and grinding, wood dust will be generated. Wood dust dispersed in air may be a flammable or even an explosive mixture. The same holds true for flour, hard coal, charcoal and many other products. Therefore, milling and grinding makes substances more reactive and can promote fire and explosion hazard. For example, Fig. 13 shows the thermite reaction. Finely ground iron oxide is thoroughly mixed with aluminum powder. Once ignited, the reaction generates enough heat to release molten, elemental iron. Flammable solids dispersed in air present an explosion hazard. The finer ground the material is, the less easily does it settle (the longer it stays in the air) and the more rapid is the combustion reaction.

Fig. 13: Thermite reaction.

Flammable liquids are classified due to their flash point. As long as a liquid that may burn at ambient air is handled at temperatures well below the flash point (say, 20 K below the flash point), the risk of fire is comparatively small. Processes may heat up the liquid, so it reaches its flash point. Furthermore, when by accident water is added to a hot, organic liquid that is not miscible with water, an azeotrope is formed with a much higher vapor pressure at the given temperature than the isolated components. As a result, instantaneously evaporation forms an aerosol cloud which is prone to ignition which may result in an explosion. Such an effect is known and feared in the private area as "fat explosion," when water is poured in a hot deep-fat fryer (Fig. 14).

Fig. 14: "Fat explosion:" On the left, you see a medium-sized pot with boiling cooking fat. A glass of water is added which causes an immediate boiling eruption with ignition of the hot fat aerosol.

2.4 Exercises

(1) By which quick methods can the purity of a solid and of a liquid substances/product be checked?
(2) What is the relevance of the molar mass, the density, the melting point and the boiling point in case of accidental release of a substance in the environment?
(3) To measure the vapor pressure, a substance with a molar mass of 520 g/mol was coated on glass beads. Air was passed at 293 K through this pipe at a flow rate of 1 L/min. In a cool trap, the substance is condensed and the amount analyzed. Depending on the filling height in the tube, the following data were generated for 3 h collection times (Tab. 2). Calculate the vapor pressure, assuming the substance in the gas phase shows ideal behavior; comment the data.

Tab. 2: Mass substance sampled in
dependence on the height of the column.

Test run	Filling height (cm)	µg substance
1	2	4
2	4	8
3	6	9
4	8	9.5
5	10	10

(4) Why is the measurement of the vapor pressure more accurate if the gas phase is analyzed specifically for the compound of interest compared to other methods like detection of the actual pressure?

(5) When testing water solubility of stearic acid ethyl ester ($C_{20}H_{40}O_2$), the equilibrated water had a content of DOC of 2 mg/L. In a separate test, the COD for the saturated aqueous solution was 4.45 mg O_2/L. A calculation program predicted a water solubility of 0.1 mg/L. Compare these results; where are areas for potential error, and how can the "real" water solubility be found out?

(6) What is a surface-active substance, and can surface activity be measured?

(7) What does the Henry constant of a substance tell us? Why can a substance have two different values for the Henry constant at a fixed temperature?

(8) At 298 K, a substance of molecular weight of 123 g/mole has a water solubility of 1.8 g/L and a vapor pressure of 0.01 Pa. Calculate the Henry constant, H', and the dimension-free Henry constant, H.

(9) For the distribution of a substance between soil and water, the K_{OC} is the constant mostly used. What does it stand for?

(10) Can you name at least five different functional groups that indicate that the substance may be an explosive?

(11) Calculate the oxygen balance of a) dinitrophenol ($C_6H_4O_5N_2$), b) 2-azo-acetic acid ethyl ester ($C_4H_7O_2N_2$) and c) oxalic acid ($C_2H_2O_4$) and decide, whether these substances are suspected to be explosives.

(12) What is meant by flammability, and how is it defined for gases, liquids and solids?

(13) If you are working with a flammable solvent, what do you need to be aware of to avoid accidents?

(14) Why are oxidizers hazardous?

(15) If in a lab you have the solvents dichloromethane, tetrahydrofuran, decalin (decahydro-naphthalin), dimethylformamide and toluene, which of these are susceptible for the formation of hydroperoxides and why?

(16) How can you check for hydroperoxides? How can you deactivate them?

(17) 10 g of a solvent were shaken with a surplus of KI solution. The Iodine formed was titrated with 0.1 N $Na_2S_2O_3$-solution, and 14 mL were consumed up to the equivalent point. How much mg hydroperoxide per g did the solvent contain, expressed as H_2O_2?

(18) What is the sense behind classification of materials as corrosive to metals in terms of product safety? Why not exclusively use plastic containers for wetted absorbent?

(19) The speed of wood combustion shall be linear proportional to the surface of the wood. Assume you have a ball of wood, density 1 g/cm^3, volume 1 cm^3. If you grind it down, and you would generate uniform spheres, how much would the surface increase (and, therefore, the speed of combustion) if you generate particles with a radius of 1,000, 100 or 10 µm?

3 Basics in toxicology

Toxicology is the discipline that looks into the interaction between an organism and an exogenous substance the organism is exposed to. Tests are run to discover the toxic mechanism of the substance (mode of action), the quality of the interaction (i.e., does the substance cause fainting, headache, kidney failure etc.) and quantitative relations to find out what minimum dose is required to trigger an observable effect, at which dose you cannot see an effect, and if there is a threshold or not. In the general public, "toxicity" is discussed as a yes/no issue, that is, a substance is either toxic or not. However, already in the early sixteenth century, the Physiologist Theophrastus Bombastus von Hohenheim (artist name: Paracelsus) wrote down the statement, that

> Everything is toxic. The dose alone determines whether something acts as a venom or not.

Compared to the other chapters in the book, this chapter may appear to be very long. The reason is that besides physical–chemical properties, it is the toxicity of substances that threatens health of chemical engineers at work. Further, some mechanisms discussed in this chapter are also applicable to ecotoxicity. Nevertheless, this chapter is far from being a complete introduction into toxicology. For more in-depth information, the interested reader is referred to toxicology textbooks, for example, *Toxicology and Risk Assessment: A Comprehensive Introduction* [2] or *A Practical Guide to Toxicology and Human Health Risk Assessment* [3].

Since ancient times, a focus was set on acute poisoning and murder by toxins. However, for everyday handling of substances, the prime interest is to know what amounts may be taken up without causing damage to health. These are the "No Observed Adverse Effect Levels," NOAEL. The terminus "adverse" is added because dosages being so low they do not affect health, or which even might be beneficial, do not need to be regulated.

NOAEL/NOAEC = No observed adverse effect level/concentration: highest dose or concentration that did not create an adverse response in the exposed species.
LOAEL/LOAEC = Lowest observed adverse effect level/concentration: the lowest dose or concentration in the test that caused an adverse response.

Now, what is the meaning of "adverse"? Adverse means that the effect is reducing the fitness of the affected organism. Due to the exposure, it may respond with reduced growth, lifetime, mating, reproduction success or simply reduced motility. For example, fat deposition in liver cells in Mammalia is rated as adverse, as over time this will reduce the fitness of the animal, and finally liver failure may result. Contrary to this, an increased liver weight without any other signs of intoxication is not necessarily adverse; if the general fitness is not changed in comparison with a

https://doi.org/10.1515/9783110618952-003

control group, the increased liver weight may be just an adaptation to a stimulus posed by the dosed substance. Sometimes, the discrimination between "adverse" and "not adverse" is anything else than obvious and needs to be left to specialists with expertise in pathology.

With respect to the site of toxic action of a substance, *local toxicity* can be discriminated from *systemic toxicity*. *Systemic toxicity* is the toxicity to a *target organ* and not necessarily linked to the first contact site of the organism to a substance. For example, phenol can be resorbed through the skin. In the skin, on the contact site, phenol can kill skin cells and causes skin corrosion, which is a *local effect*. However, the phenol will also enter the blood circulation and may destroy red blood cells which is a *systemic effect* (hematological system), and it may cause damage to the liver, which is a *systemic effect*, as the liver was not the prime site of contact.

Some substances may be critical at too high and also at too low dosages. As an example, the element cobalt (Co) is an important trace element for enzymes, and life will not last for long if you would exclude intake of any cobalt atom. Higher amounts, however, can cause cancer.

In toxicity testing diverse laboratory animal strains may be applied to investigate acute and chronic toxicities. The amount of substance applied is normalized to the body weight (b.w.) of the respective species after dermal or oral application. For inhalation exposure, the concentration in the inhaled air is the measure for the dose.

3.1 Some biological background

Biology is a science that most students of chemical engineering were no longer confronted with after having left school. However, some basic understanding of biology is necessary to better understand how and why substances can act as toxicants. The challenge for the contents of this chapter is to find a balance between "too little" and "too much."

The single cell is the simplest form of life. Biology defines life as having the capability to reproduce itself, to move and to react on external stimuli.

Between many others, the main features of the cell are the confinement by a cell membrane which is build up by a double layer of lipids. This membrane circumvents the cell like a shell, but it is mobile and floating. Phospholipids are the molecules building up the membrane, as the very polar head turns toward water and the very hydrophobic tail avoids water (Fig. 15).

In the membrane, small pores allow the free diffusion of small molecules like water (Fig. 16). Lipophilic, small molecules like diethyl ether or trichloromethane, can freely diffuse across the cell membrane.

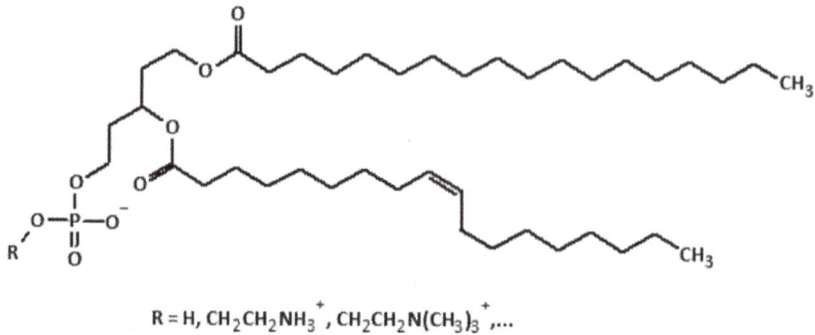

$$R = H, CH_2CH_2NH_3^+, CH_2CH_2N(CH_3)_3^+, ...$$

Fig. 15: Phospholipid, general formula.

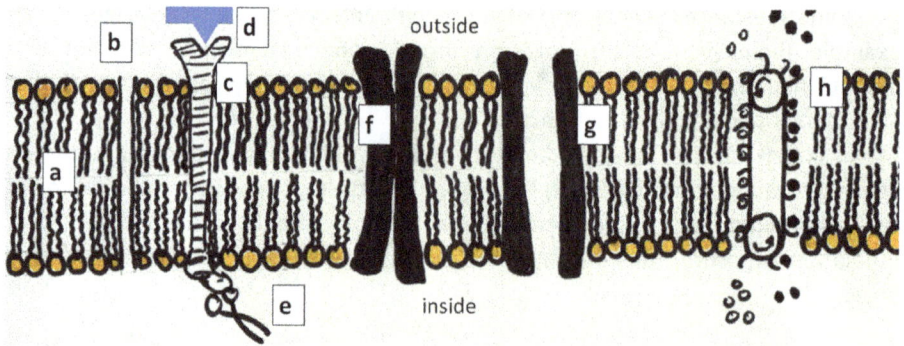

Fig. 16: Cell membrane, schematically; a: cell-membrane phospholipid bilayer; b: pore; c: receptor; d: ligand for the receptor; e: effect triggered by ligand binding to the receptor, symbolized by a pair of scissors; f: (ion-)channel, closed; g: (ion-) channel, opened; h: active transporter, symbolized by a conveyor-belt, transporting molecules against their gradient.

Large polar and/or charged molecules cannot enter the cell easily (if at all). They require special transport mechanisms.

Ions like potassium, sodium and chloride can hardly pass the pores. They need a passage called channels, which are specific for the ions. These channels open on specific stimuli which can be an electric pulse (nerve cells) or stimulating molecules that dock onto receptors on the channels. When a channel is opened, the specific molecules migrate according to their concentration gradient. Transporters can transport molecules against the concentration gradient; but doing so, they need to consume energy, which can be either released by degrading energy-rich molecules, or by cotransport, where the chemical gradient of another molecule is the driver (cotransport: one molecule is transported against its gradient while the other molecule moves according to its gradient). The monosaccharide glucose needs transporters to be able to pass the cell membrane. Glucose is the molecule every cell can make use of as energy source.

Some transporters are permanently active, and they are called "pumps" (Fig. 16). For example, potassium is permanently pumped inside the cell while sodium is permanently pumped to the outside. This process requires energy. As potassium ions "sneak" back to the outside more easily than sodium ions diffuse back into the cell, and because anions cannot follow to the same extend, the outside of the cell membrane is positively charged against the inside, and the electric potential across the membrane is about −60 mV. To keep ion concentrations inside the cell within a certain limit is very important for proper functioning of the cell and its organs (organelles); the cell is filled up with organelles, and there is nothing like an "empty chamber."

In multicellular organisms, cells have been specified for certain tasks. For communication between cells, signal molecules, called "transmitter" (small molecules, mainly for nervous pulses) and "hormones" (comparatively large molecules, released into the blood), are released when a certain service is required. These signal molecules may be able to pass the phospholipid bilayer and act at their targets inside the cell (hormones estradiol, testosterone), but in case of large hydrophilic molecules, they may bind on specific receptors outside the cell (hormone insulin, neurotransmitter acetylcholine) (Fig. 16). As the receptors range through the membrane, they convey the information to the inner side of the cell.

The schematic structure of a cell is shown in Fig. 17, and microphotographic pictures of human buccal cells are shown in Fig. 18. Membranes are not only the confinement of the cell, but also the confinement of the organs inside the cell (organelles). The nucleus contains the deoxyribonucleic acid (DNA) which is the carrier of the genetic information. Every cell contains the complete information on how to construct the whole organism! DNA replicates to DNA, which is one of the important parts when the cell grows and finally splits into two identical copies. For the synthesis of proteins, the DNA is transcribed into ribonucleic acid (RNA), which leaves the nucleus and migrates into the cytosol. In the cytosol, ribosomes add to the RNA, read it and translate it into polypeptides, the proteins. Proteins are important molecules in the cell if it comes down to functions like catalysis, synthesis and catabolism and structure, and ribosomes are the protein factories. The buildup of proteins in an aqueous environment requires energy input, and the universal energy carrier in cells is adenosine triphosphate (ATP). This molecule is the store for energy generated by the glucose degradation to water and carbon dioxide. The mitochondrion is the power plant inside the cell where an important part of this process takes place, and proper functioning of the mitochondria depends on membrane integrity of this organelle. The endoplasmic reticulum is a place where molecules are changed and transformed. The lysosome is the place where molecules are chopped into small pieces. Large molecules can be taken up and excreted by energy-consuming processes called endocytosis (uptake) and exocytosis (excretion).

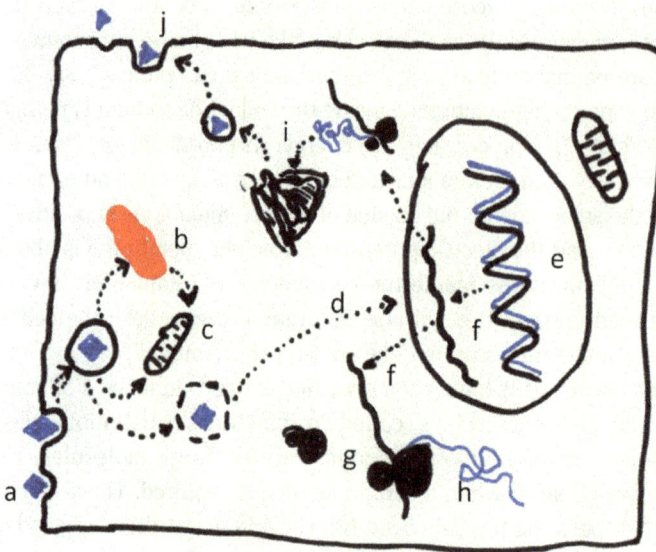

Fig. 17: The cell (simplified schematic sketch). Molecules too large for pores, channels and transporters may be taken up by endocytosis (a). In the lysosome (b), molecules may be digested and further transported to organelles like the mitochondrion (c, the "power-plant" of the cell) or to the nucleus (d). In the nucleus, the DNA (e) – as a result on stimulation – is transcribed into RNA (f) which leaves the nucleus and is read by ribosomes (g) which synthesize proteins (h). In the endoplasmic reticulum and the Golgi-apparatus, symbolized in (h), proteins are further progressed, and other molecules are synthesized. Products may be excreted by exocytosis (j), if they are too large for pores, channels.

Important structures of the cell:
The cell membrane, a lipid bilayer, separates the cell from the surrounding medium.
Pores, channels and transporters in the membrane ensure a defined homeostasis of molecules.
Nucleus, where the genetic information is stored and progressed as necessary.
Mitochondria are the main places of energy gain for the cell.
Ribosomes are the place of protein synthesis by reading the genetic code (which was transcribed from DNA to RNA in the nucleus).

The formation, transformation and decay of molecules inside a cell are taking place at moderate temperatures in an aquatic environment. This requires the recruiting of catalysts, and actually hundreds of different known catalysts – the enzymes – are active in the cell. These enzymes are in most case proteins. For proper function, they need to be in a certain three-dimensional state which requires a certain range of temperature, pH, redox-status and ion concentrations. Any disturbance in this respect can eliminate the enzyme function, sometimes irreversibly, and elimination

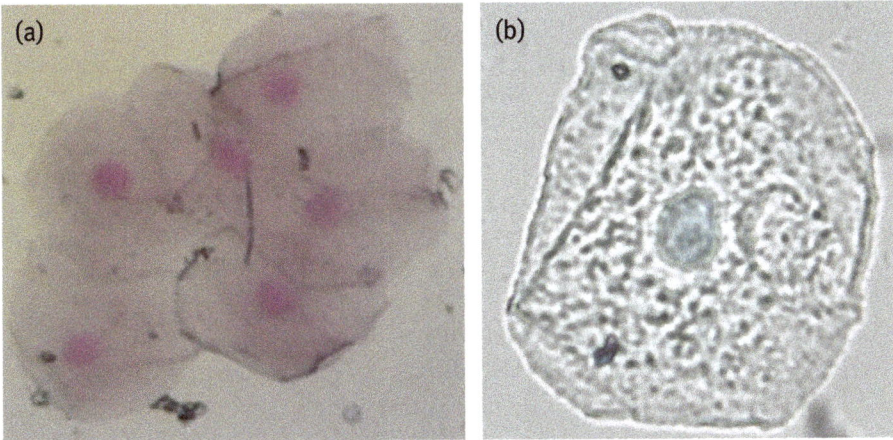

Fig. 18: Human buccal cells. a): magnification 100x, stained with eosin. b): magnification 400x, stained with methylene blue.

of the enzyme function may trigger cell death. Active centers inside the enzymes frequently comprise (and may be disturbed by) the following:

– The three-dimensional structure of the proteins, which is stabilized by hydrogen bridges and sulfur-sulfur bonds:
  – Splitting of these S-S bonds by reducing agents or strong oxidizers, and/or change of the hydrogen bonds, for example, by strong salts and acids disturbs the structure.
– Functional groups of the amino acid side chains may be involved not only in hydrogen-bonding, dipole-dipole interactions and London forces, but they may also participate in substrate reactions:
  – Oxidation or alkylation of OH, SH or NH groups causes denaturation of the proteins and/or inactivation of the active centers (Fig. 20).
– Basic or acidic substances catalyze the hydrolysis of peptide bonds and cause denaturation and dissolution (at least partly) of the protein (Fig. 19).
– Metal ions help to organize the orientation of reactants:
  – Strong complexation of the metals or their exchange by other metals can cause alteration of the enzyme activity.

Therefore, loss of function of ion channels, transporters or receptors on the cell, or rupture of the cell membrane causes loss/gain of ions, change in pH and ion concentration, this triggers loss of enzyme function and finally cell death. Also, loss of energy gain inside the cell, or loss of protein production, loss of molecule turnover and so on will impact cell function and may finally end in cell death.

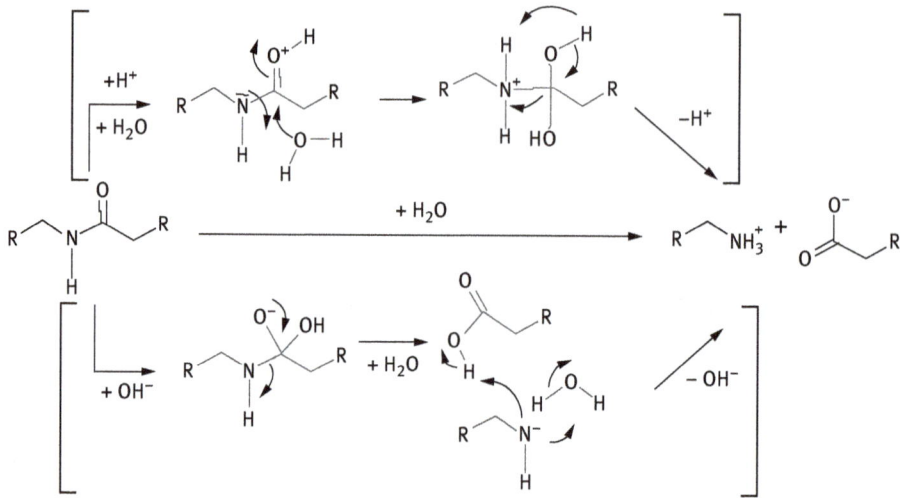

Fig. 19: Peptide hydrolysis, catalyzed by acid (top) or base (bottom).

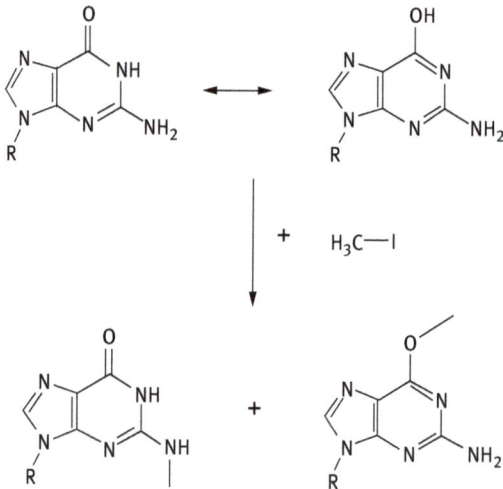

Fig. 20: Methylation of guanine in DNA or RNA: loss of hydrogen bridges.

Toxic mechanisms of deactivation of biomolecules (enzymes):
- Complexation of central metals.
- Block of R-SH, R-OH and/or R$_2$NH functions by oxidation or alkylation.
- Disturbance of structure by pH change, oxidation, alkylation, hydrolysis or surface-active agents.

Transport processes and synthesis in the cell require energy. Practically, every cell is capable to make use of the energy stored chemically in the monosaccharide (sugar) glucose ($C_6H_{12}O_6$) (Fig. 21). Whereas photosynthesis of green plants stores sunlight energy by synthesis of glucose from water and carbon dioxide, aerobic respiration is a kind of reverse process. The energy stored in glucose is converted into another universal "energy currency," the adenosine triphosphate (ATP), which itself can deliver energy by hydrolysis to adenosine diphosphate (ADP):

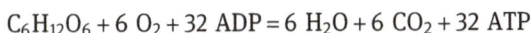

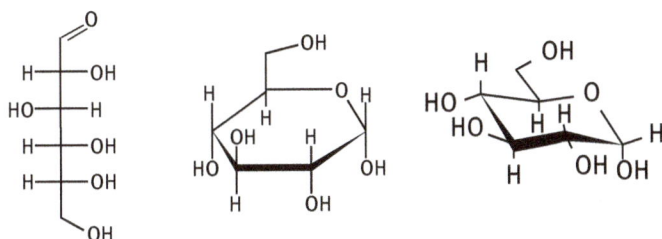

$$C_6H_{12}O_6 + 6\ O_2 + 32\ ADP = 6\ H_2O + 6\ CO_2 + 32\ ATP$$

$$ATP + H_2O = ADP + phosphate + {\sim}30\ kJ/mol$$

Fig. 21: Glucose ($C_6H_{12}O_6$) as aldehyde in Fischer projection, or as closed six-member ring in Haworth (middle) and chair presentation (right).

Supply with glucose and oxygen is important for every cell in the human body to fulfil its tasks and to survive. Organs with permanent very active cells, like the brain and the heart, followed by kidneys, are responding quickly to a lack of oxygen or glucose.

3.2 Absorption, distribution, metabolism and excretion of substances

This chapter describes the fate of a foreign substance from entering the body to leaving it, either as such or in a transformed state. In this chapter, the student of chemical engineering will be faced with pieces of knowledge he/she has gathered in courses of mass transfer, reaction kinetics, mass balance in chemical reactors, biocatalysis and organic chemistry. Therefore, this chapter goes more into detail than many other chapters of this book.

3.2.1 Absorption

Absorption plays a major role for systemic toxicity, and in toxicology, substance uptake is separated in oral ingestion, inhalation and skin contact. Oral ingestion should no longer play a role in workplace toxicology, but it is important for food toxicology and public health. As an example, if your company produces materials that have intended contact with food or drinking water, substances migrating out of the article are taken up via food and drinking water.

NOTE: For articles with intended contact to food or drinking water, special regulations and directives exist in the EU. This special legislation is not addressed in this book.

3.2.1.1 Oral uptake

Only if the substance is absorbed in the digestive tract and enters the blood circulation, it is systemically available. The digestive tract of an adult person has a contact area of about 100 m^2. The importance of absorption for toxicity becomes clear in two different forms of the metal mercury. If elemental mercury is swallowed, it is hardly absorbed into the blood, and the acute oral toxicity of it is comparatively low; centuries ago, it was used as a laxative. Dimethylmercury, on the opposite, easily passes cell membranes and is taken up very easily in the digestive tract; oral dosages as low as a few milligrams may be fatal.

Lipinski et al. had a look into properties a pharmaceutical should have if a high systemic availability after oral uptake shall be likely [4]. The trend they observed can be summarized as follows: if a pharmaceutical intended for oral medication has a $\log(K_{OW})$ greater than 5, and/or the molecular weight is above 500 g/mol, and/or the molecule has five or more H-bond donors and/or more than 10 H-bond acceptors and there is no specific transporter, there is a risk that oral absorption is low. This "Lipinski rule of 5" is not a quantitative structure–activity relationship, and it is more a representation of a trend [4]. Exemptions do exist. For example, from the authors own experience, it can be reported that polypropylene glycols with an average molecular weight below 500 or above 1,500 g/mol were not harmful to rats by ingestion (lethal dose for 50 % of the animals, $LD_{50} > 2,000$ mg/kg), but those with an average molecular weight between 500 and 1,500 g/mol, the oral LD_{50} in the rat was between 200 and 2,000 mg/kg, which is acute oral toxicity category 4: harmful by ingestion. It was hypothesized that the small molecules were too hydrophilic and the high molecular weight molecules were too large for uptake into the cell.

"Lipinski rule of five," as a rule of thumb to expect high systemic availability of a substance after oral exposure:

- The size: small molecules are taken up easier than larger ones. It is assumed that molecules below 500 g/mol are well resorbed in the digestive tract. However, this cut-off level is not very sharp.
- Molecules with a $\log(K_{OW})$ above 2 but below 5 are prone to good resorption.

– The molecule has no or only very few H-bond donors (OH and NH groups)
– The molecule has no or only very few H-bond acceptors (C-O, C = O, C-N, C = N and C ≡ N groups).
– There are specific transporter molecules that facilitate substance uptake.

Substances taken up via the digestive tract are first transported to the liver. The liver is the organ where substances are degraded and transformed. Because of this, substances taken up by ingestion may cause more severe liver damage when compared to other uptake routes, but they may as well undergo more rapid detoxification when compared to other uptake routes; this is called the "first-pass effect" after oral exposure.

3.2.1.2 Dermal resorption

With 1.8 m^2, the skin provides a much smaller contact surface to chemicals than the digestive tract. However, at workplaces and in the laboratory, skin contact is more likely to happen than oral ingestion.

The "Lipinski rule of 5" provides also an indication when a substance is prone for good skin resorption: small molecular weight, $\log(K_{OW}) < 5$, molecular weight <500 g/mol [4]; in addition, a low vapor pressure increases the contact time by reduced evaporation rate. Substances known to be taken up in critical amounts via the skin are aniline, phenol, dimethyl formamide and N-methyl pyrrolidone (Fig. 22).

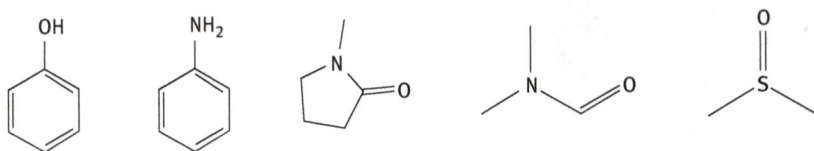

Fig. 22: Examples of organic compounds that penetrate the skin easily: phenol, aniline, N-methyl-p2-pyrrolidone (NMP), N,N-dimethylformamide (DMF), dimethyl sulfoxide (DMSO).

Dimethyl sulfoxide itself is not classified as toxic, but it can facilitate the dermal uptake of other compounds. This is also the case for other organic solvents. For that reason, organic solvents shall never be used to clean contaminated skin (unless they are explicitly tested for skin decontamination).

NEVER use organic solvents to clean contaminated skin!

The skin has a layered structure (Fig. 23, 24). The outer sphere, the epidermis ("skin toward outside"), is formed by (from inside to outside) the basal membrane below which there are the blood capillaries, the growth zone (permanent cell renewal), the stratum granulosum build by cells having left the basal membrane and slowly moving to the outside and the stratum corneum build by dead, keratinized cells.

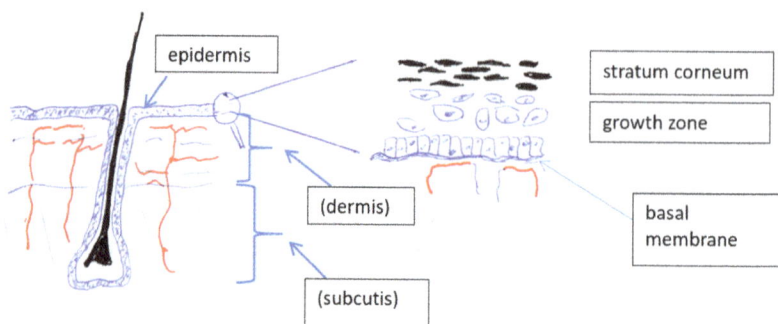

Fig. 23: Structure of the skin.

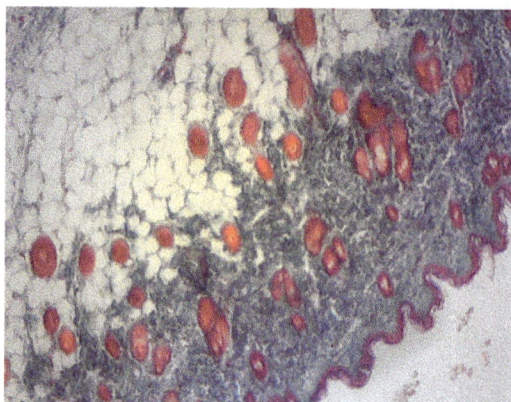

Fig. 24: Human skin (magnification 40×); basal membrane (right, bottom), followed by germinative layer and horn layer. Multiple red-stained ducts.

Hair follicles range deep into the skin. Whether or not a substance is resorbed well depends on the thickness of the stratum corneum (hand palms against inner side of the fore arms) and the number of hair follicles per area.

In wet skin, the stratum corneum is less tight, which is the downside of wearing protective rubber gloves permanently. Freeware programs exist to calculate the dermal uptake of a chemical. These programs do allow at least for a rough estimate of dermal uptake. The algorithm used in the US Environmental Protection Agency (US EPA) "Dermwin" program incorporates the $\log(K_{OW})$ and the molecular weight of the compound. This program is part of the open-source software EPISUITE® of the US EPA [5]. The amount of substance dermally resorbed per unit time (dn/dt) is proportional to the concentration difference between substance on the skin and in the blood (Δc), the thickness of the boundary layer (Δz) and the contact surface area (A). With a constant of proportionality, K' in comparison with the mass transfer as modelled in chemical engineering, the equation is:

$$\frac{dn}{dt} = -K' \times A \times \frac{dc}{dz} \approx -K \times A \times \Delta c,$$

where K values are given in dimension distance/time; the EPISUITE program, for example, delivers for 2-pentanone a value of $K = 0.0025$ cm/h. If, for example, both hands ($A \sim 420$ cm^2) are covered with an aqueous solution containing 20 mg/cm^3 2-pentanone for 4 h, and if it is assumed that inside the body the 2-pentanone concentration is zero, then the uptake is calculated as

$$\Delta n = -0.0025 \frac{\text{cm}}{\text{h}} \times 420 \text{ cm}^2 \times \left(-20 \frac{\text{mg}}{\text{cm}^3}\right) \times 4\text{h} = 84\text{mg}$$

3.2.1.3 Inhalation

Gases, vapors (gas above liquid), mist (liquid dispersed in air) and dust (solids dispersed in air) can be inhaled; vapor literally is a substance in the gas phase below its critical point, and it can be brought to condensation by increased pressure. Aerosols are either mists or dusts.

Gases and aerosols are inhaled via nose and/or mouth, pass the larynx and follow the trachea that splits up into bronchi, smaller bronchioles and finally alveoli. In the alveoli, the gas phase is in close contact with the blood, and a two-cell layer on a basal membrane is the only boundary between blood and air. Exchange of oxygen against carbon dioxide happens in the alveoli only. A surfactant produced by the lung cells spreads over the alveoli and ensures they do not collapse due to surface tension.

Where the substance acts in the respiratory tract, depends on the water solubility of the gas/vapor or the size of the aerosol (Fig. 25). Very water-soluble gases like ammonia or hydrogen chloride typically irritate the nose and the eyes, whereas sulfur dioxide acts more in the area of the larynx (creating a feeling as if the throat congests), and chlorine, nitrogen oxides, ozone and phosgene can be inhaled down to the far end bronchioles and alveoli, where these gases can cause the feared and potentially mortal lung edema.

Aerosols above 10 μm aerodynamic diameter are deposited mainly in the nose, throat and upper trachea, whereas particles smaller than 10 μm reach the bronchioles and the alveoli. For aerosols, it is frequently observed that the toxicity is a function of the size, smaller aerosols being more critical than larger ones.

The lung contact surface to the external air is between 60 and 100 m^2, depending on the depth of inhalation.

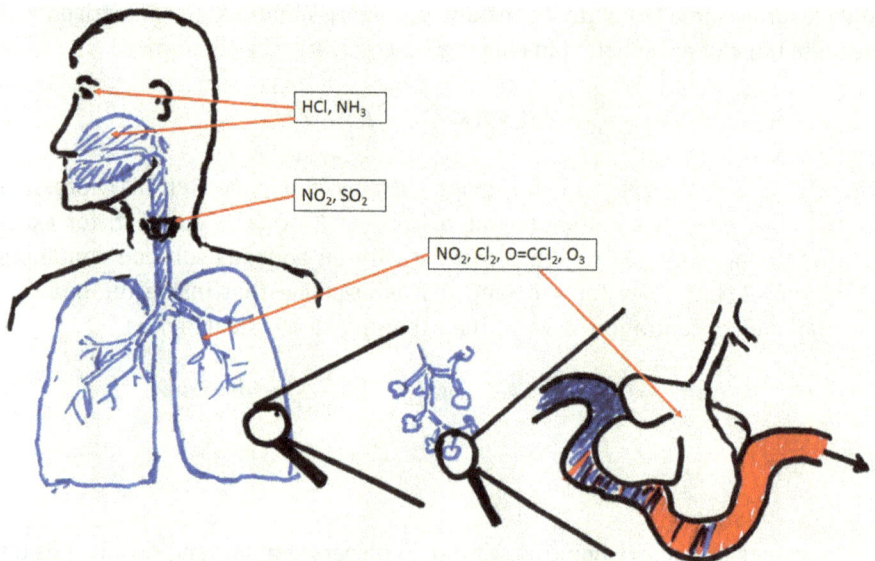

Fig. 25: Respiratory tract; arrows indicate the major target sites for inhaled gases. Magnification: alveoli, circumvented by blood capillaries.

Uptake of substances into the body occurs via
- oral uptake (swallowing of substances and subsequent resorption in the gut),
- dermal contact,
- inhalation.

3.2.2 Distribution

Irrespective to the port of entry, the substance is not earlier systemically available than having entered the circulation. In the blood, the substance is distributed through the body, and equilibria with different organs and tissues will be established. The more lipophilic and the less hydrophilic a substance is, the more it will be distributed from blood toward the organ and vice versa. By this way, a toxin may reach its target organ. The distribution is a function of the molecular size, the octanol–water partitioning and the water solubility.

In the blood, the substance may be dissolved in the liquid (plasma), bound on proteins or red blood cells or absorbed into blood cells (mostly red blood cells, the erythrocytes, due to their high abundance). Only the free substance dissolved in the blood plasma is available for uptake by the target organ.

3.2.3 Metabolism

Metabolism is the transformation of the substance by biochemical reactions in the cells. The phase I metabolism introduces or changes functional groups, phase II metabolism is a follow-up reaction of the phase I metabolites: they may become hydrolyzed or connected to endogenous, highly water-soluble molecules like glucuronic acid or glutathione (GSH). The main organ of metabolism is the liver; smaller – but occasionally important – metabolic activities are present in other organs as well, for example the kidneys or the lung. Due to the prominent role of the liver in metabolism, toxicity may be different for oral uptake on the one hand – where the substance is transported from the intestines to the liver and then is spread over the organism (first-pass effect) – and dermal or inhalation uptake on the other hand.

Metabolic turn over in general can be described by the Michaelis–Menten kinetics:

$$V = \frac{V_{max} \times [S]}{K_m + [S]}$$

with V being the reaction rate [mol/(L × s)], [S] being the substrate concentration, V_{max} the maximum reaction velocity and K_m as Michaelis–Menten constant.

3.2.3.1 Phase I metabolism

The most important reaction of the phase I metabolism is the introduction of oxygen into organic molecules. A large enzyme group, catalyzing this reaction, are the cytochrome P450 mono-oxygenases (CYP450) (Fig. 26). The reactive center is an iron complex, and these enzymes were originally quantified by photometry ("P"), as after exposure to carbon monoxide, they show a strong absorption at 450 nm ("450"). "Mono-oxygenases" means these enzymes transfer one oxygen atom of dioxygen onto the substrate molecule, whereas the other oxygen atom builds water. The cofactor nicotin-adenin-dinucleotide-phosphate is required to complete the reaction.

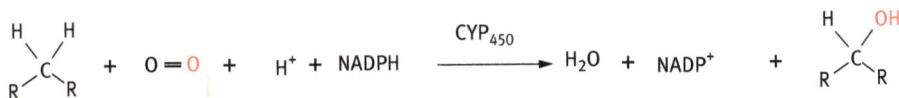

Fig. 26: General reaction catalyzed by CYP450 mono-oxygenase.

The CYP450 enzymes are categorized into classes and subclasses, having different substrate specificities. Other oxidizing enzymes like peroxidases or amino-oxidases may be important for specific target molecules. Examples of reactions catalyzed by CYP450 are given in Fig. 27.

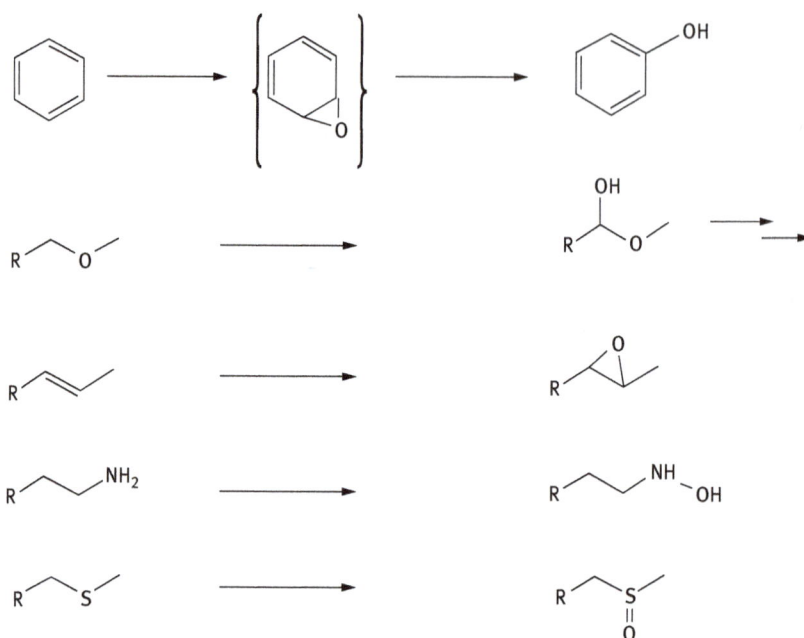

Fig. 27: Transformations catalyzed by CYP450 mono-oxygenase.

CYP450 may create molecules that are more reactive than the parent molecule, p.e., epoxides generated from alkenes. Such epoxides are electrophilic and can comparatively easily attach on NH groups of biomolecules.

Another class of enzymes for phase I metabolism that shall be mentioned are the alcohol and aldehyde dehydrogenases. Alcohol dehydrogenase oxidizes primary and secondary alcohols to aldehydes and ketones, respectively; aldehyde dehydrogenase oxidizes aldehydes to carboxylic acids (Fig. 28).

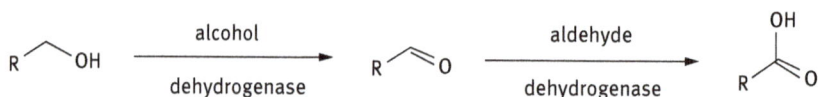

Fig. 28: Reactions of alcohol and aldehyde dehydrogenases.

3.2.3.2 Phase II metabolism

Of the several enzymes existing for phase II metabolism, the glucuronosyl-, glutathionyl (GSH)- and acetyltransferase and the epoxide hydrolase shall be mentioned.

The glucuronosyl transferase adds the polar glucuronic acid to the molecule, so a previously hydrophobic molecule builds a hydrophilic product that is important for the excretion (Fig. 29). Glucuronides are built in the liver and excreted via bile into the intestine. Bacteria in the large intestine may cleave the glycoside bond, so that a certain proportion of the molecule is re-absorbed. Bilirubin, a degradation product of blood hemoglobin, needs to be glucuronized before it can be excreted. In newborns, the glucuronation capacity of the liver needs some days to achieve its final activity; as a result, some of the bilirubin (a breakdown product of heme from blood) piles up in the liver and finally enters the blood causing "newborn jaundice." Jaundice can be an indication of liver intoxication; the affected liver is no longer capable of performing glucuronidation at the required level.

Glutathion (GSH) is a tripeptide containing the amino acid cysteine, whose SH-group may undergo nucleophilic substitution (Fig. 29). By this reaction, not only reactive electrophilic molecules are deactivated, but they become more water soluble. (For some molecules, the GSH adducts are degraded to reactive metabolites in the kidney.)

Epoxide hydrolases cleave epoxides generating the less reactive glycols (Fig. 29).

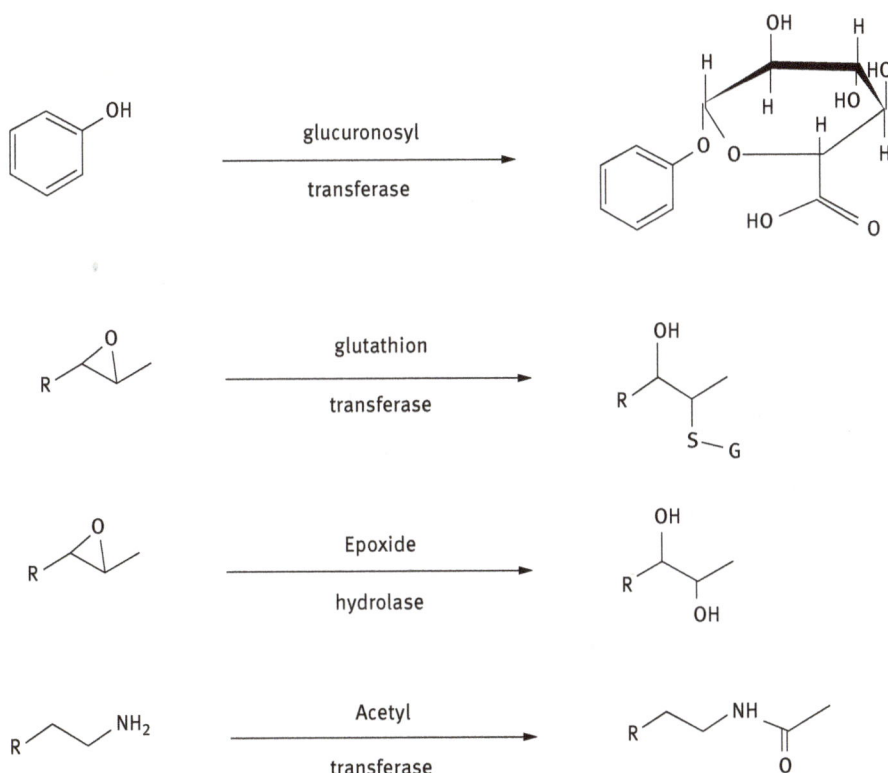

Fig. 29: Examples for phase II metabolism.

Acetylation is a phase II reaction which results in a less polar and less water-soluble product. Acetyl groups can be transferred to NH, OH and N–OH groups (Fig. 29).

Differences in the metabolic capacity can sometimes explain differences in the reaction toward an exposure. A classic in this respect is methemoglobinemia caused by aniline. In the twentieth century, it was observed that workers with equal exposure to aniline showed different levels of methemoglobin. "Fast" acetylators were less sensitive, and aniline was rather rapidly acetylated in their body, whereas in "slow" acetylators, methemoglobin levels were comparatively high and aniline was much less rapidly acetylated.

Distribution and metabolism
Substances having entered the blood are distributed by circulation. The more lipophilic a molecule is, the more it will tend to enter specific tissues. Metabolism introduces functional groups into molecules (phase I) and links them to polar, endogenous molecules (phase II) so they become more water soluble. The more water soluble a substance is, the easier it can be excreted.

3.2.4 Excretion

Molecules with a comparatively high vapor pressure can be excreted via the lung (exhalation). For other molecules, it is important that they are transformed to water soluble substances. Molecules up to a molecular weight of 500 g/mol are excreted via the kidneys. The smallest functional unit in the kidney is the nephron (Fig. 30). Small

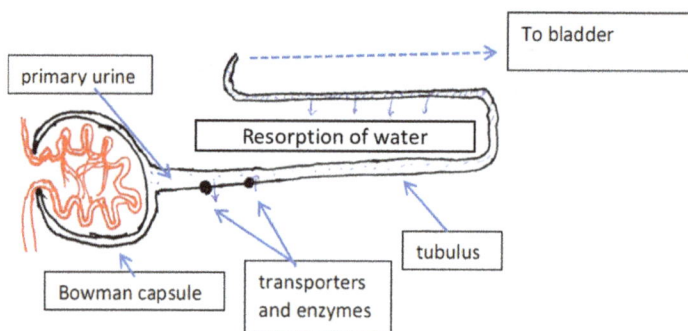

Fig. 30: Sketch of a nephron, the basic functional unit in the kidney. In the Bowman capsule, blood plasma is squeezed into the tubulus, large molecules are hold back. Valuable substances in this primary urine are resorbed, and some substances may actively be excreted into the urine.

arteria (arterioles) transport blood into the Bowman capsule, and blood plasma is squeezed out together with small molecules. Many of these small molecules are valuable substances that are reabsorbed from the primary urine, like the mono-saccharide glucose or sodium ions. Small proteins are cleaved by enzymes in the tubules wall, and the amino acids are re-absorbed. Elevated levels of glucose (sugar) and/or proteins in the urine are an indication of disease or intoxication of the kidney. Further down the tubulus, water is re-absorbed, and the primary urine is concentrated 100 fold. That means, substances in the blood plasma may achieve hundred times higher concentrations in the urine. This up-concentration is one reason, why the kidney is a typical target organ for intoxications. Larger molecules than 500 g/mol may still be excreted via the kidneys, at least partly, depending on the molecular charge, but a molecular mass higher than about 70,000 g/mol is the ultimate cut-off. Molecules above 500 g/mol may be subject to excretion via the bile, and biliary excretion becomes more important with increasing molecular weight. Frequently, such molecules are added to glucuronic acid and transported with the bile into the small intestine. In the caecum, bacterial activity can cleave the glucuronic acid adducts which facilitates re-uptake of substances (enterohepatic circulation).

As urinary excretion is an important aspect also in pharmacology, it was given special attention, and the "Clearance," CL, was defined.:

$$\text{concentration in urine} \left[\frac{mg}{L}\right] \times \text{urine flow} \left[\frac{L}{h}\right] = \text{concentration in plasma} \left[\frac{mg}{L}\right] \times CL \left[\frac{L}{h}\right]$$

Clearance defines the volume of blood plasma that is purged from the substance per unit time. In this respect, it should be noted that the available blood volume is taken as "apparent," or distribution volume: If 1,000 mg substance is injected into the blood stream (intravenously = i.v.) and after a few minutes the concentration in the blood is 125 mg, the apparent distribution volume is

$$\text{Distribution volume } V_d = 1,000mg/125mg/L = 8L$$

although the average adult person has not more than 5–6 L blood. But this theoretical entity, the distribution volume, allows to run calculations concerning the residence time of a substance in the body.

The clearance increases with increasing water solubility of the substance. By medical treatment, urine can be made more alkaline so it better can dissolve acidic molecules and vice versa. This is important in case of intoxications for phasing out the toxin. Molecules that bind to plasma proteins are less well excreted, and lipophilic small molecules are prone to reuptake from the primary urine in the tubulus.

Excretion of a substance may be controlled by measuring its concentration (or the concentration of one of its metabolites) in the blood or in urine. If the excretion is determined by saturated enzymatic processes or by saturated transporters, it follows a kinetic of zero order. Once the enzymes or transporters are no longer saturated, the kinetic of substance decay is usually first order in concentration. An example is the decay of the blood alcohol concentration; if the level is at $1^0/_{00}$ and higher, it is linear due to the high alcohol (substrate) concentration. This can be deduced from the Michaelis–Menten equation for enzyme kinetics:

$$\frac{d[S]}{dt} = v = \frac{v_{max} \times [S]}{K_M + [S]}$$

with $[S]$ being the substrate concentration, v_{max} the maximum reaction rate and K_M the Michaelis–Menten constant. If $[S]$ is very large, the equation becomes $v = v_{max}$; for very small $[S]$, it resembles a first-order rate equation, $v \sim k \times [S]$ (Fig. 31). Ethanol elimination from blood is a suitable example as ethanol is a small molecule that is mixable with water at any ratio.

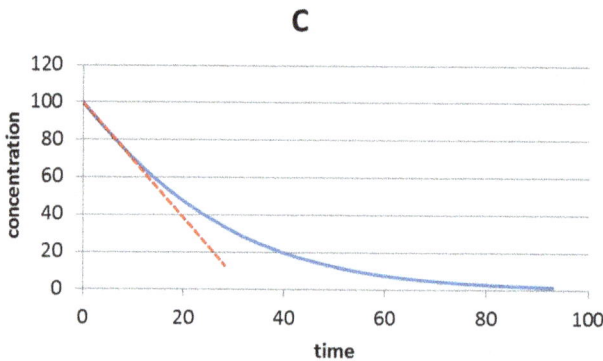

Fig. 31: Elimination curve by metabolism, starting with a saturated enzyme.

Excretion/elimination
Small molecules below a molecular weight of 500 g/mol are excreted via urine (kidneys). With increasing mass, excretion via bile and feces becomes more important, and molecules with a molecular weight higher than 70,000 g/mol are excreted almost exclusively via the bile and feces. In the kidney, the primary urine has a similar composition as the blood plasma but lacks the large molecules. The primary urine is concentrated 100-fold.

3.2.5 The time-curve of tissue concentrations and the question of sensitivity

Uptake, absorption, distribution ("disposition" in pharmacology), metabolism and excretion frame the boundaries for the concentration of a substance and its metabolites in the different tissues of an exposed organism. Modeling the distribution and fate of the substance in experimental animals helps to understand why the substance or its metabolites cause effects, and how these can be transferred in predicting toxicity to human beings. The process is typically broken down to distinctive steps where changes in concentrations are modeled by mathematical equations. As the chemical engineer is familiar with mass-transfer mathematics and reaction kinetics, this section is somewhat more exhaustive than the others in this chapter.

How knowledge gained in toxicological experiments can be translated into protection of human beings is always a challenge, and translating toxic or safe level data gained with selected populations (workforce) to the general population is also a critical task of finding the right balance between mandatory protection, precautionary principle and panic paralysis.

The interspecies differences can be split up into two factors: a *toxicokinetic factor* and a *toxicodynamic factor*. The toxicodynamic factor describes the different sensitivity organisms may have; for example, a substance may severely block an esterase in human beings but may hardly block the equivalent esterase in the rat, and vice versa. The toxicokinetics describe the time-dependent concentration of the parent molecule or of relevant metabolites at the site of action. Toxicokinetics is dependent on the rate of uptake, available metabolic pathways, the reaction rates of these pathways and the rate of excretion of the parent compound or its metabolite(s).

Toxicokinetics addresses the time-dependent concentration of a substance in different tissues. Toxicodynamics addresses the interaction between the substance and the target tissue.

For toxicokinetics, the question arises how a dose for an experimental animal can be extrapolated to human beings. The smaller an animal is, the higher is its relative metabolism rate and the clearance. Certain parameters are proportional to the body surface, not to the : breathing rate, food demand, drinking water demand, metabolic turnover and renal clearance. *Interspecies* differences address the extrapolation from experimental animal results to human beings. The *intraspecies* extrapolation addresses the variability in sensitivity of the human population.

If dosages are given as ppm substance in drinking water or food, or as mg/m^3 in air, a correction of the toxicokinetics for human exposure is not necessary, as the higher metabolism and clearance rate in smaller animals are leveled off by their higher intake rate.

If dosages are given as mg per kg body weight per day (mg/kg b.w./d), an allometric correction is required, which is calculated as

$$\left\{ Dose\ man \left[\frac{mg}{kgb.w.} \right] \right\} = \left\{ Dose\ animal \left[\frac{mg}{kgb.w.} \right] \right\} \times \left\{ \frac{\{kgb.w.\}_{animal}^{0.25}}{\{kgb.w.\}_{man}^{0.25}} \right\}$$

$$= \left\{ Dose\ animal \left[\frac{mg}{kgb.w.} \right] \right\} \times AF$$

For rat and mouse, these allometric factors (AF) are 1/4 and 1/7, respectively. That means, for example, if a certain substance has an acute oral toxicity of $LD_{50} = 4$ mg/

(kg b.w.) in the rat, the LD_{50} in man is expected to be 1 mg/(kg b.w.), if the difference in metabolic degradation and excretion was the only difference between rat and man.

An established standard used in regulatory toxicology is the use of a factor of 10 each for inter-species and intra-species extrapolation. As the rat is the standard species in toxicity testing, a factor of 4 would be used to correct the dose for toxicokinetics (i.e., 40 mg/kg b.w. in the rat are expected to achieve the same tissue concentration as 10 mg/kg b.w. in human beings); a further factor of 2.5 is used as a multiplier to cover remaining species differences based on toxicodynamis [6]. Exposure limits for human beings start with the so-called NOAEL or the NOAEC from an animal experiment. This animal experiment should have covered chronic exposure (if not, further factors for exposure-time adjustment need to be applied). As an example, for a NOAEL of 200 mg/kg b.w./d in an animal experiment in a chronic study, the simple equation reads:

$NOAEL_{animal, chronic}$	200 mg/kg b.w./d
animal to human beings (interspecies) :	10
to protect sensitive individuals (intraspecies):	10
exposure limit for the population	= 2 mg/kg b.w./d.

That means, a person of 70 kg b.w. should not ingest more than 140 mg of the substance every day. Instead of NOAELs or NOAECs, benchmark doses may be chosen (see Chapter 5).

For animal to man and man to general population extrapolation, an inter-species factor and an intra-species factor is applied. Default values are 10 for each of them.

In the following, the toxicokinetics after inhalation, dermal or oral uptake of a substance shall be modelled. If a certain amount of substance, N, is applied externally (i.e., outside the blood circulation), its disappearance at the application site (skin, intestines or lung) toward the circulation is modeled as a first-order process, with k_1 as uptake rate constant:

$$\frac{dN}{dt} = -k_1 \times N$$

and the solution by integration is.

$$N = N_0 \times \exp[-k_1 t]$$

N_0 is the dose, D, that was applied. What has entered the circulation is the difference between the amount originally applied (N_0, which is D) and the amount that is left over at the application site, N. That is, the substance that has entered the circulation after a time t has passed is:

$$N_{in} = N_0 - N = N_0 - N_0 \times \exp[-k_1 t] = N_0 \times (1 - \exp[-k_1 t]) \leftrightarrow N_{in} = D \times (1 - \exp[-k_1 t])$$

N_{in} divided by the distribution volume V is the concentration that would result by the uptake of the substance at time t:

$$C_{in} = \frac{D}{V} \times (1 - \exp[-k_1 t])$$

The change of the concentration due to uptake is the derivation after time:

$$\frac{dC_{in}}{dt} = k_1 \times \frac{D}{V} \times \exp[-k_1 t]$$

As soon as substance is present in the blood, elimination processes start, which are modeled as first-order rate processes, with k_2 as elimination rate constant,

$$\frac{dC_{ex}}{dt} = -k_2 t$$

The concentration in the blood compartment is the result of both processes, $dC/dt = dC_{in}/dt + dC_{ex}/dt$:

$$\frac{dC}{dt} = k_1 \times \frac{D}{V} \times e^{-k_1 t} - k_2 t = k_1 \times \frac{D}{V} \times \exp[-k_1 t] - k_2 t$$

This differential equation is solved by the method "variation of the integration-constant," where first the homogeneous part is solved, introducing the integration constant A:

$$\frac{dC}{dt} = -k_2 C => LN(C) = -k_2 t + A \leq > C = C_0 \times \exp[-k_2 t] \times \exp[A]$$

As exp[A] is a constant again, exp[A] = B, the homogenous solution is $C = C_0 \times \exp[-k_1 t] \times B$. Now, B is regarded as a function dependent on t, the derivative dC/dt is formed, and dC/dt and the homogeneous solution are introduced in the original equation. Bearing in mind that $C = 0$ when $t = 0$ yields the final solution for the time-dependent concentration in blood, C_t:

$$C_t = \frac{k_1 \times D}{V \times (k_2 - k_1)} \times (\exp(-k_1 t) - \exp(-k_2 t))$$

In pharmacology, this equation is named Bateman equation, and the concentration-time curve is shown in Fig. 32. The integral over time delivers the "area under the curve" (AUC), which is a measure for the biological availability of an applied substance:

$$AUC = \frac{k_1 \times D}{V \times (k_2 - k_1)} \times \left(\frac{1}{k_1} - \frac{1}{k_2}\right)$$

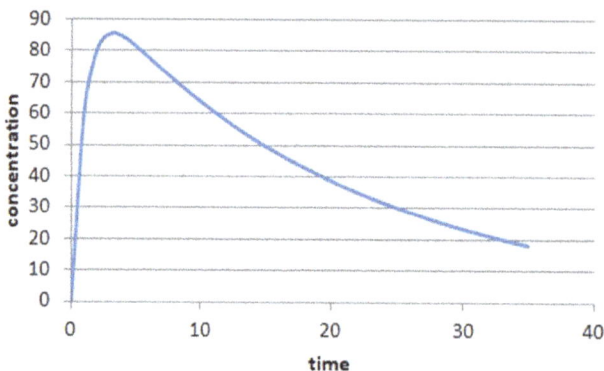

Fig. 32: (Hypothetical) blood concentration of a substance over time following singular oral intake.

with D = dose [mg], V = distribution volume [L], k_1 uptake rate constant and k_2 = elimination rate constant (both for the blood).

Note: More precisely, it is the concentration of the free compound in blood plasma that is of importance, because these are the molecules that may interact with receptors on target organs. Molecules that are tightly bound to plasma proteins or that are taken up by blood cells hardly interact with the target tissue unless there is a rapid exchange between free and bound substance molecules.

The area under the time-concentration curve (AUC) is a measure for the availability of the substance. This AUC is also an important measure for comparisons between species. Besides the AUC, the peak concentration can be very important especially for those substances, where a certain minimum concentration needs to be achieved before an effect can be triggered. The peak concentration can be calculated by setting dC/dt = 0 which delivers for the time point of maximum concentration

$$t_{max} = LN\left(\frac{k_1}{k_2}\right) \times \frac{1}{(k_2 - k_1)}$$

The value for t_{max} is used in the Bateman equation, which then delivers the concentration peak, C_{max}. For pharmaceuticals, typically a certain minimum blood concentration is required to induce the desired therapeutic effect. However, C_{max} shall not arrive at or exceed a level where toxic effects are triggered.

The area under the blood concentration – time curve, the "area under the curve" (AUC) is a measure for the biological availability of a substance.

If a repeated daily oral dose is taken up, every time when substance is taken in, the plasma level is refreshed, ending finally in a curve given in Fig. 33. This curve was created by using the Bateman function in a spread-sheet program (Microsoft Excel®).

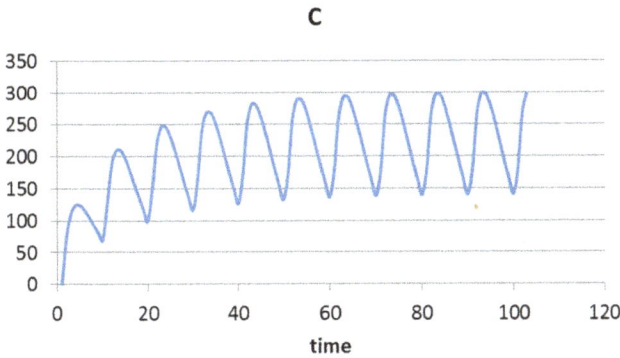

Fig. 33: Blood concentration following a repeated oral dose.

Every day, a certain dose D is applied. Over the day, the concentration of the substance in blood follows a distinct curve as described by the Bateman function. The following day, the next dose is applied while the blood concentration caused by the dose of the previous day has not yet dropped to zero. For the average dose, a rate equation can be set up:,

$$\frac{dC}{dt} = \frac{D}{V} - k_2 \times C$$

and the solution is

$$C_t = \frac{D}{k_2 \times V} \times (1 - \exp[-k_2 t]) = \frac{Y}{k_2} \times (1 - \exp[-k_2 t])$$

with D = dose, V = distribution volume, C = blood concentration, k_2 = first-order elimination rate constant. D and V may be combined in a new constant, $Y = D/V$.

Figure 33 shows that the excursions in the concentration may be considerable; that is, while the average concentration may be fine, the peak concentration could be too high to avoid a damaging effect. To derive an equation for the minimum and maximum concentrations over long term, consider that at the end of the first, second and third day, the concentration is

$$C_1 = Y \times \exp[-1 \times k]; \ C_2 = Y \times \exp[-2 \times k] + Y \times \exp[-1 \times k]$$
$$C_3 = Y \times \exp[-3 \times k] + Y \times \exp[-2 \times k] + Y \times \exp[-1 \times k]$$

with $Y = D/V$; for N days, we can write

$$C_{N,\,\text{min}} = Y \times \exp[-k] \times \sum_{n=0}^{N-1} \exp[-n \times k]$$

$C_{N,\,\text{min}}$ indicates the minimum concentration at just before the next dosage is applied. From textbooks of mathematics, for infinite series, it is known that

$$\sum_{n=0}^{N-1} x^n = \frac{1-x^n}{1-x}$$

With $x = \exp[-k]$, the equation changes to:

$$C_{N,\min} = Y \times \exp[-k] \times \sum_{n=0}^{N-1} \exp[-n \times k] = Y \times \exp[-k] \times \frac{1-\exp[-n \times k]}{1-\exp[-k]}$$

For large n, $\exp[-n \times k] \sim 0$, and the equation simplifies to:

$$C_{N,\min} = \frac{Y \times \exp[-k]}{1-\exp[-k]} \quad <=> \quad C_{N,\min} = \frac{Y}{\exp[k]-1}$$

For the maximum concentration, the next dose was just added, and therefore:.

$$C_{N,\max} = Y + \frac{Y}{\exp[k]-1} = Y \times \left(1 + \frac{1}{\exp[k]-1}\right)$$

The average concentration is calculated as

$$C_{N,av} = \frac{Y \times \exp[k] + Y}{\exp[k]-1}$$

Medication can be prescribed that, for example, sometimes a pill is taken only once a day; in this case, the AUC is probably more important for the substance activity than the maximum concentration. If pills shall be taken, for example, three times over the day, it is likely that a certain maximum concentration shall not be exceeded. Depending on the rate constants for uptake (k_1) and elimination (k_2), the results for the plasma peak concentrations and the AUC can be very different (Fig. 34).

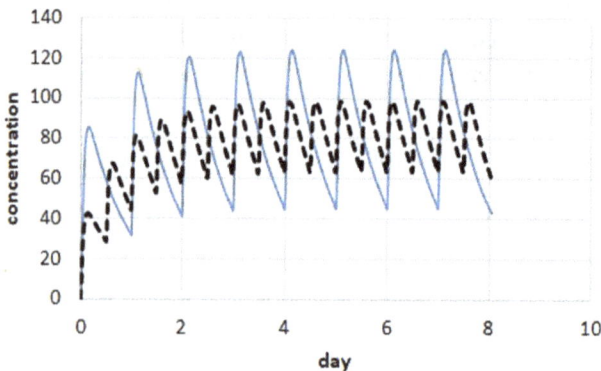

Fig. 34: Daily dose of 100 units given once a day (blue, solid line) or in two portions á 50 units, twice a day (black, dotted line). Here, $k_1 = 1/h$ and $k_2 = 0.05/h$.

Keeping the daily dose constant, shorter time intervals for dosing result in less differ-ence between C_{max} and C_{min}. This has not only implications for the dosing of pharma-ceuticals. In toxicity testing, differences in toxicokinetic due to the dosing regime might be an explanation if with the same substance in the same animal species, differ-ent results are observed. Further, Fig. 34 shows that after 2–3 days, a steady state is achieved. This time frame is about 5 half-lives ($k_2 = 1.2$ / days, $=> t_{0.5} = 0.58$ days). Depending on k_2, longer time periods may be required before a steady state is achieved. This is one of the reasons why a single dose study may perhaps not reveal the full toxic potency of a substance, and repeated dose studies are required.

For inhalation exposure, it is generally assumed that substances are more rap-idly and completely taken up by inhalation than via the digestive tract. A typical situation is at the workplace where a person is exposed for 8 h and then enters a clean area. Over the first 8 h, the concentration buildup can be described as

$$C = \frac{I}{k} * (1 - \exp[-kt])$$

with I = intake (concentration in air × breaths per minute × volume per breath/volume of body or compartment), k = first-order elimination rate constant (1/min) and t being the time after onset of exposure (min). After 8 h, being back in the clean air, the con-centration in blood decays according to a first-order kinetic: $C = C_{8\,h} \times \exp[-k \times (t - 8)]$ (Fig. 35):

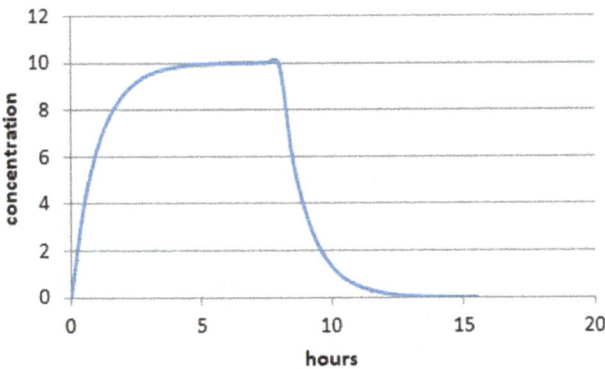

Fig. 35: Blood concentration–time curve after entering an area with contaminated air (at 0 h) and leaving it (8 h).

From the examples discussed so far, theoretical and practical data match quite well if a so-called one-compartment model describes the substance's behavior. This is usually the case for small, hydrophilic substances. Less hydrophilic and more lipophilic molecules may show a different behavior, and their elimination can be described by a two-compartment model. Compartment 1 is the blood (aqueous compartment), compartment 2 is the so-called deep compartment, which might be, for example, the body fat (Fig. 36).

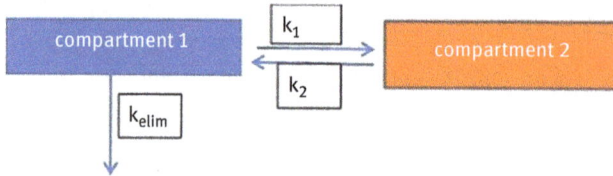

Fig. 36: Two-compartment model.

A mathematical description of the concentration-time curve in compartment 1 (usually called the "central compartment") can be derived for an i.v. injection (i.v.) with follow-up of the blood concentration over time. For the two-compartment model, the kinetic rate equation is

$$\frac{dC_1}{dt} = k_2 \times C_2 - k_1 \times C_1 - k_e \times C_1$$

where C_1 is the substance concentration in compartment 1 (blood) and C_2 is the concentration in compartment 2. An analytical solution of this equation is published by Mayersohn and Gibaldi [7]. After some mathematical transformations, the analytical solution is:

$$C_1 = A \times \exp[-k_1 \times t] + B \times \exp[-k_2 \times t]$$

At $t = 0$, the concentration obviously is $C_1 = A + B$, given by the amount injected divided by the apparent volume of distribution. Interpretation of data at very early time points would allow to derive values for A and k_1, whereas interpretation of data at comparatively late time points allow to derive values for B and k_2. If the concentration–time curve is followed experimentally, a picture like that shown in Fig. 37 results if a two-compartment model describes the substance behavior best. Half-lives as quotients of LN(2) divided by the rate constants are different in the beginning (primary half-life) and the end of the experiment (secondary half-life). The AUC for a two-compartment model is.

$$AUC_{two-compartment} = \frac{A}{k_1} + \frac{B}{k_2}$$

As a result, if the excretion of the substance directly after application, and then again for a later time interval, can be described by first-order kinetics, but if the rate constants of the initial period and the later period differ from each other, a two-compartment model is better suited to describe the time course of the substance in the body than the one-compartment model. Even three-compartment models for

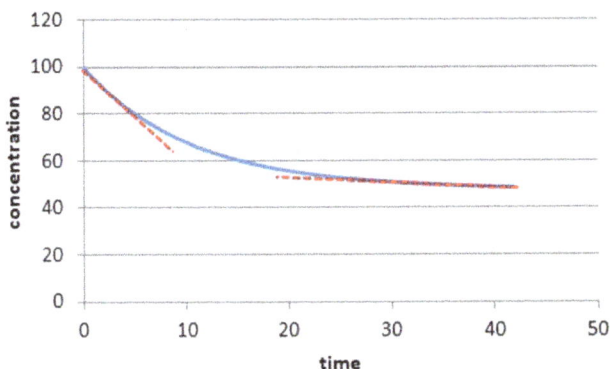

Fig. 37: Hypothetical, experimental concentration–time curve; a two-compartment model may describe the substance behavior correctly.

substances are not uncommon, and first-, second- and third-half-lives with their corresponding rate constants are reported.

With the ready availability of sufficiently effective calculators, it is possible to model the time-dependent distribution in almost all organs of the organism. Data required are physical–chemical constants like molecular weight, water solubility, vapor pressure and octanol–water partition coefficient; this, again, highlights the importance of a sound and reliable measurement of these data. With these data, the equilibrium distribution of the substance between blood and organs as well as blood and air can be estimated. Every organ can be represented as a kind of ideal, continuosly stirring reaction vessel: the typical blood flow for the organ Org, Q_{Org} (L/min), carries the amount $Q_{Org} \times C_A$ of the substance to that organ, C_A being the concentration in the arterial blood. In that organ, the substance will redistribute between blood and organ tissue, and the concentration C_V in the venous blood leaving the organ results from this tissue-blood-distribution coefficient P_{Org}:

$$P_{Org} = C_{Org}/C_{V,Org}$$

In this way, every organ that is deemed relevant is modeled. Within the different organs, the substance may be metabolized which is calculated by the Michaelis–Menten kinetics. Uptake of the substance may happen via inhalation, skin contact or via oral uptake. Following oral uptake, the substance will first be transported to the liver, where significant metabolism may happen. After inhalation or dermal uptake, the substance will be more evenly distributed to the richly perfused organs like kidney, brain and liver.

Fig. 38 illustrates the process of substance balance in an organ without any metabolism. In this simple model, the organ has just the function of a storage. The change in mass balance per unit time is

$$\text{Mass Balance}\left[\frac{\text{kg}}{\text{h}}\right] = Q' \times (C_A - C_V)$$

As the liver is the most important metabolizing organ, outcomes of toxicity studies may be different in terms of target organ and severity of effects depending on oral, dermal or inhalation exposure; this is to be expected if metabolism is comparatively

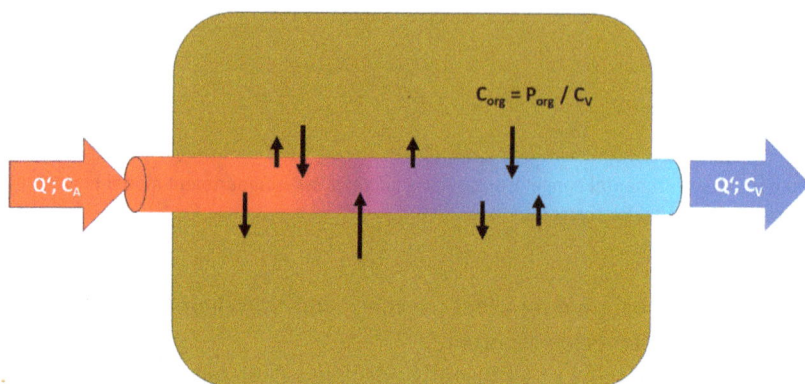

Fig. 38: Arterial blood flow (Q_{art}'), coming from the heart, carries the concentration C_A of the substance. While passing the organ, the substance equilibrates between blood and organ. Venous blood flow (Q_{ven}') leaves the organ having the concentration C_V of the substance. $Q_{art}' = Q_{ven}'$.

fast and contributes to either the detoxification or the toxification of a substance. The differential equation for the substance in a metabolizing organ like the liver, as an example, is as follows:

$$\frac{d[S]}{dt} = Q_L \times (C_{a,t} - C_{vL,t-1}) - \frac{V_{max} \times C_{vL,t-1}}{K_M + C_{vL,t-1}}$$

with Q_L: liver blood flow; $C_{a,t}$: arterial blood concentration at time interval t; $C_{vL,\,t-1}$: venous blood concentration at time interval $t-1$. The liver is the organ which typically shows the highest enzymatic activity. However, metabolism in other organs, namely the lung and the kidney, may play an important role as well.

The substance or its metabolites can in principle be excreted via exhalation, urine or bile. In the kidney, active excretion may increase the clearance, while reabsorption in the tubule decreases it. Bile adducts of the substance are excreted with the feces. In the large intestine, however, bacteria may cleave bile adducts, and the liberated substance may be reabsorbed (enterohepatic circulation). These situations are sketched in Fig. 39.

Simple spreadsheet models are available in the public domain, for example, the Canadian Centre for Environmental Modelling and Chemistry [9], or the program IndusChemFate provided by the European Chemical Industry Council (Conseil Européen des Fédérations de l'Industrie Chimique, CEFIC) [10, 11].

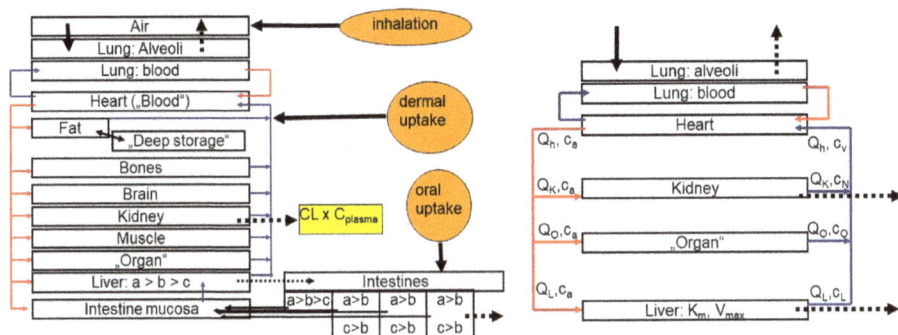

Fig. 39: Physiology-based toxicokinetic (PBTK) model. Black, solid arrows: uptake; black, dotted arrows: excretion; red arrows: arterial blood flow; blue arrows: venous blood flow (between lung and heart: blue = arteria, red = veins). The letters a, b and c stand for the parent substance (a) and the first (b) and second (c) metabolite. This figure was drawn according to [8].

As illustrative example, the IndusChemFate program v2.0 was used to calculate the time-concentration curve of aniline and its metabolites in an adult person at rest that inhales 150 mg/m³ aniline for 4 h. Chemicals modeled are as follows (Fig. 40): The model outputs of the physiology-based toxicokinetic (PBTK) modeling are shown in Fig. 41–43. During the exposure period, aniline (0) builds up in the venous blood, but its concentration drops rapidly once exposure stops. N-Acetyl-aniline (1) and the N-acetyl-amino-benzene sulfate (3) are deemed to be the prominent

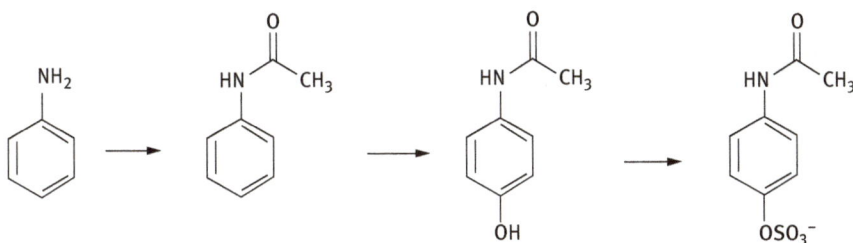

Fig. 40: Aniline (0) and its metabolites N-acetyl-aniline (1), 4-N-acetyl-aminophenol (2) and 4-N-acetylamino-benzene sulfate (3).

metabolites. In excreted urine, 4-N-acetylamino-benzene sulfate is the only relevant compound. This is obviously attributable to its presumably high water solubility (Fig. 42). In the liver, the compounds with highest concentrations are N-acetylaniline (1), followed by the 4-N-acetylaminophenol (Fig. 43).

PBTK modeling not only allows to estimate the concentrations of compounds and their metabolites in different tissues, it is also helpful to compare concentrations in different species, or in the same species but with diverging metabolic capacity. For these reasons, PBTK modeling can be an important tool for inter- and intraspecies

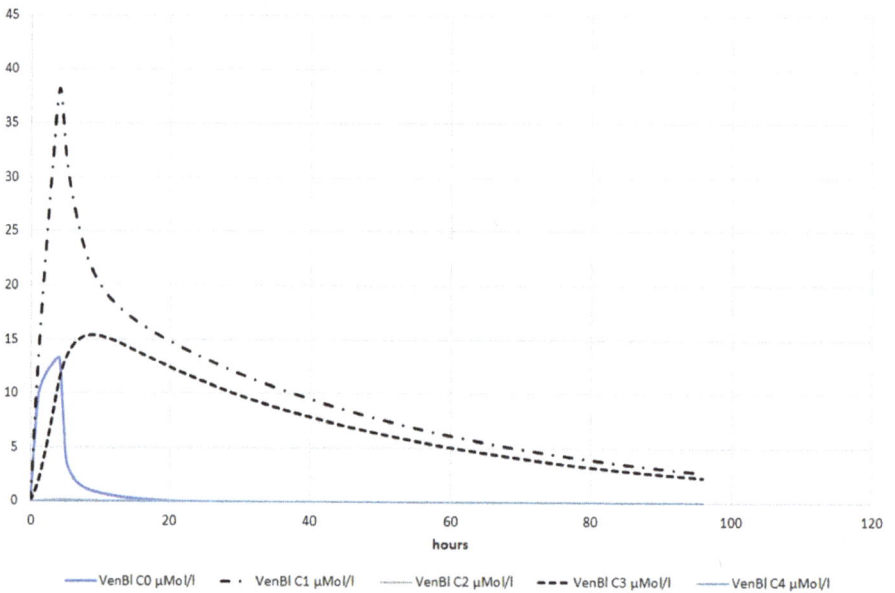

Fig. 41: Venous blood (VenBl C) concentration of aniline (0) and its metabolites *N*-acetylaniline (1), 4-*N*-acetylaminophenol (2) and 4-*N*-acetylamino-benzenesulfate (3) during 4 h inhalation exposure and thereafter.

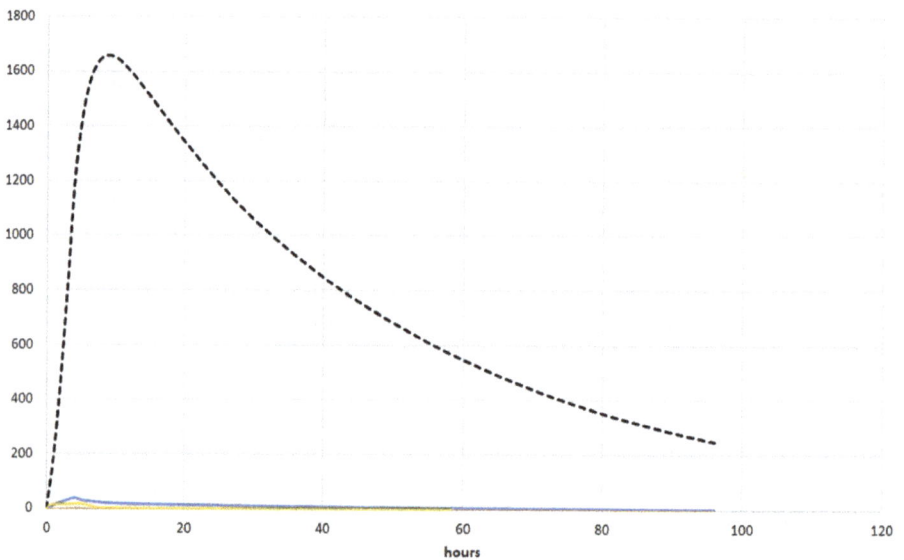

Fig. 42: Urine concentration (UrinConc) of aniline (0) and its metabolites *N*-acetylaniline (1), 4-*N*-acetylaminophenol (2) and 4-*N*-acetylamino-benzenesulfate (3) during 4 h inhalation exposure and thereafter.

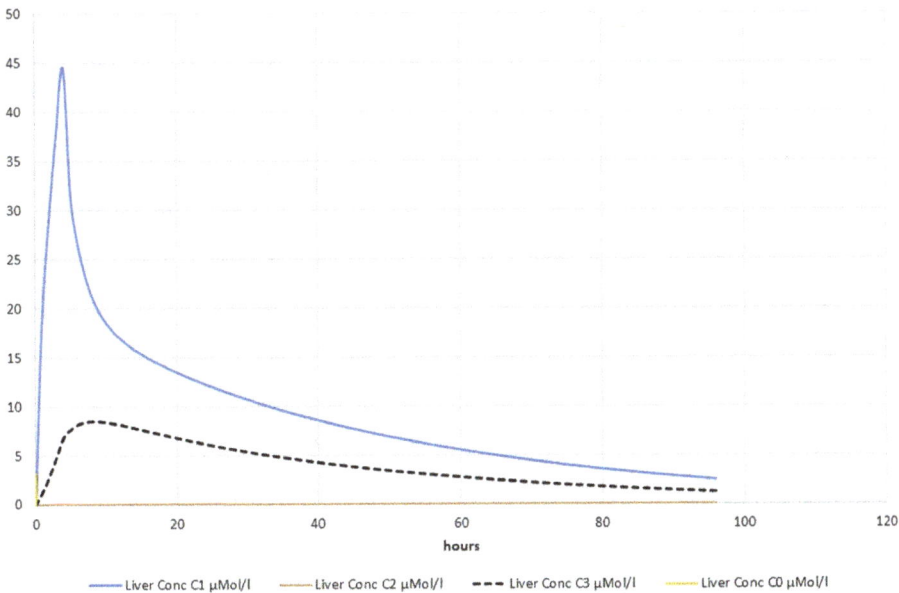

Fig. 43: Liver concentration of aniline (0) and its metabolites *N*-acetylaniline (1), 4-*N*-acetylaminophenol (2) and 4-*N*-acetylamino-benzenesulfate (3) during 4 h inhalation exposure and thereafter.

extrapolation factors. However, modeling on its own can be misleading, especially if only estimated values instead of measured values were taken for physical properties and metabolic turnover. For example, open access programs like EPISUITE® [5] allow to calculate water solubility and the log K_{OW}; however, calculated values may diverge from measured values by a factor of 2–3 for water solubility, or by 0.5–1.0 units for log K_{OW}. Metabolic turnover in organs is difficult to estimate, and enzymes may react very specific. Therefore, the more robust data points are available, the more likely will the modelled data match experimental data. For critical questions to be solved, modeled data should be substantiated by measured in vivo data.

3.3 Brief overview of organ toxicity

As is the case with biology, the subjects of this chapter are usually not very familiar to chemical engineers. Therefore, the content is limited to an extent that allows some basic understanding of organ toxicity. Where molecular mechanisms come into play, the text may be somewhat more elaborate.

3.3.1 Liver toxicity

As mentioned earlier, the liver is the main organ for metabolism. Functions of the liver are, to name a few, as follows:
- transformation of molecules to make them ready for excretion: excretion of ammonia as urea, bilirubin as glucuronide, bile acid formation and excretion;
- synthesis of functional molecules: enzymes, precursors (p.e., those for the blood clotting system), amino acid synthesis and turnover, gluconeogenesis, fat synthesis;
- synthesis of transport molecules: proteins in blood serum, fat transport vesicles;
- storage and release of glucose: glucose homeostasis.

Toxicity to liver can result in the loss of such functions. For example, bilirubin, a heme degradation product, is excreted via bile into the gut lumen in healthy individuals. If the excretion or formation of bile is hampered, bilirubin is deposited in the skin, causing a yellow discoloration, known as jaundice. The endogenous degradation of peptides would release ammonia, which is very toxic; in the liver, ammonia is transformed to and bound as urea, which can easily be excreted with urine (see textbooks of biochemistry, urea cycle). If the urea formation is disturbed, ammonia concentration in blood increases and the brain is affected, leading finally to unconsciousness ("liver-coma"). If synthesis of clotting factors declines in the liver, extended internal and external bleeding may result. Fat deposition in the liver is a result of disturbed fat turnover, uptake and excretion. Permanent or repeated inflammation may end up in the decay of functional liver cells and their replacement by nonfunctional, connective tissue: liver cirrhosis.

As a classical example of toxic action in the liver, acute and chronic intoxication with lead (Pb) can disturb heme synthesis. Incomplete heme production can cause anemia (lack of red blood cells); certain precursors in the heme production chain may pile up in the liver; finally, they are released into the blood and affect the autonomic nerve system. This leads to symptoms known as porphyria (obstipation, heavy gut pain, numbness in limbs and finally paralysis).

3.3.2 Kidney toxicity

One main role of the kidney is the formation of urine and recovery of useful molecules from the primary urine. The primary urine is concentrated 100-fold; consequently, the concentration of molecules in the remaining urine increases accordingly. This is one of the several reasons why the kidney is a common target organ for toxic action. Metallothionein in the tubulus epithelial cells can bind metals which form stable complexes with sulfur groups in the metallothionein; if the binding capacity is exhausted, tubulus cells die away. Compounds active in this field are cadmium, lead and mercury.

Some chlorinated solvents known to be typical liver toxins, for example trichloroethylene, are affecting the kidney via their degradation products.

Toxicity to the kidney may appear as changes in urine volume: polyuria (very much urine), oliguria (little urine) and anuria (no urine). Blood in urine indicates damage in the kidney-uric tract, proteins and glucose in urine indicate loss of resorption functions in the tubulus. If there is no more formation of urine, compulsory urine components level up in the blood and will be fatal in short time. External dialysis has to be run.

3.3.3 Blood

Main tasks of the blood and the circulation are oxygen transport, transport of nutrients, distribution of hormones, immune function and temperature regulation. Blood contains about 45% liquid phase (serum with proteins) and about 55% blood cells (Fig. 44). The red blood cells (erythrocytes, about 5 million per μL) are responsible for oxygen transport. White blood cells (leukocytes, about 5,000 per μL) are important for the immune defense, and the platelets (thrombocytes, about 5,00,000 per μL) are important for blood clotting. Some substances may cause specific blood cell losses; others cause a general anemia ("lack of blood"). This can happen by affecting the bone marrow, where blood cells grow and differentiate from precursor stem cells, but also by damage to the matured blood cells. Anemia may present itself by pale skin, tiredness and susceptibility to infections.

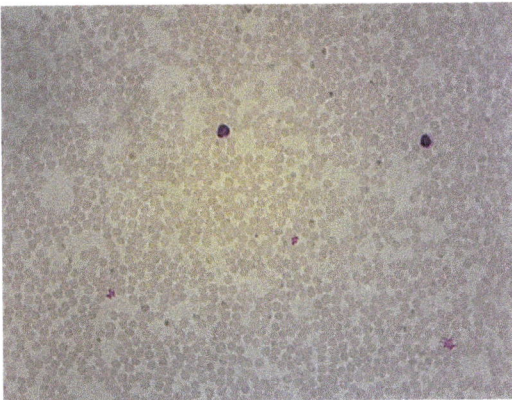

Fig. 44: Blood smear; pale red blood cells (erythrocytes) and a few pink stained white blood cells (leucocytes): here lymphocytes and granulocytes.

3.3.3.1 Oxygen transport

The main task of the red blood cells (erythrocytes) is oxygen transport. Molecular oxygen is loosely bound end-on to the Fe^{2+} in heme (Fig. 45). In tissue poor in oxygen and with lowered pH (CO_2 high), the oxygen is released.

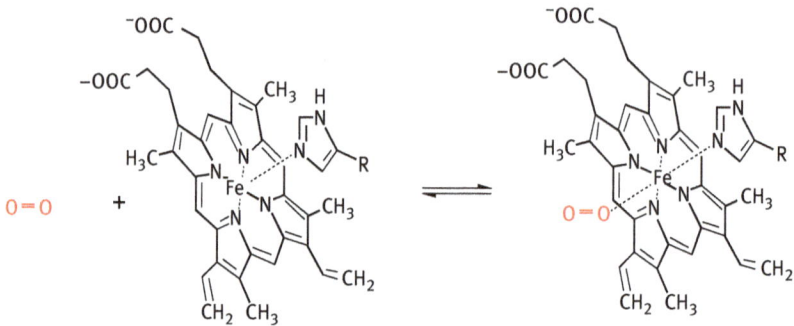

Fig. 45: Molecular oxygen is loosely bound to the Fe^{2+} in heme. The heme molecule is imbedded in the globin protein, which delivers an imidazole from a histidine side chain (R) as additional ligand for the iron.

The equilibrium constant of the reaction of oxygenated heme with carbon monoxide (CO) is 250; that is, if the concentration of oxygen to carbon monoxide in air is 250:1, 50% of the heme is occupied by CO. The carboxy-heme can no longer transport oxygen. Aromatic amines and aromatic nitro compounds are oxidized in such a way that the iron in heme is co-oxidized to Fe^{3+} which can no longer bind oxygen. Oxidizers like chlorates, Cu^{2+}, chromates and nitrites oxidize the iron in heme directly.

Acute intoxication with phenol, quinones, Cu^{2+} and arsine (AsH_3) can cause a hemolysis: a decay of erythrocytes, with dark discoloration of the urine.

Benzene depresses the genesis and maturation of blood cells in bone marrow (erythropoiesis, lymphocytopoiesis and thrombocytopoiesis), Pb intoxication hampers the heme synthesis. Symptoms are – between others – a lack of erythrocytes: anemia.

Cyanide ions (CN^-) bind tightly to Fe^{3+}. They block cytochrome-a inside the cell. Cytochrome-a has structural similarities with heme; Fe^{3+} in that enzyme is complexed and blocked by cyanide, so the cell can no longer transfer electrons to oxygen; this situation is sometimes named "internal asphyxiation."

3.3.3.2 Blood clotting

In the circulatory system, there are always micro-injuries. However, they are kept under control by the detailed orchestrated blood clotting system. Besides clotting factors synthesized by the liver (see above), the thrombocytes play an important role in the formation of plugs. A lack of thrombocytes results in impaired blood clotting which may become life threatening. Benzene as a general depressant of the bone marrow may cause a thrombocytopenia (lack of thrombocytes).

3.3.3.3 Reduced immune function

Although immune function is not confined to the blood but also comprises other parts of the immune system like lymph nodes, spleen and thymus, a first indication of depressed immune function may be indicated by reduced leukocyte counts in blood, detected in the scope of repeated dose toxicity tests. More specific investigations may follow, for example checking the activity of killer cells, or response to injected, foreign erythrocytes or infection models. Substances known to suppress the immune system are those suppressing the bone marrow (cytostatica for cancer treatment, benzene and chloramphenicol). Some dialkyltin compounds like dibutyltin and dioctyltin salts depress the thymus gland and the maturation T-lymphocytes.

3.3.4 Nervous system

The brain and the spine form the central nervous system (CNS), nerves going to and coming from the different tissues are the peripheral nerve system (PNS). The CNS is preserved by the blood–brain barrier: epithelial cells of the blood vessels from the one side and astrocytes from the nervous side build up a tight barrier that may be passed only be specific molecules. Small lipophilic molecules, however, can pass the barrier easily. A further distinction of the nervous system is that of the autonomous system which controls elemental life processes (breathing, digestion, reflexes) and the CNS, which is subject to our willing.

The nervous system collects information from the body and environment (afferent system), analyses the information and sends action potentials to receiving organs (efferent system), p.e., for a muscle to move. The impulse is conveyed electronically along the nerve (Fig. 46). At the synapsis, the electric pulse releases

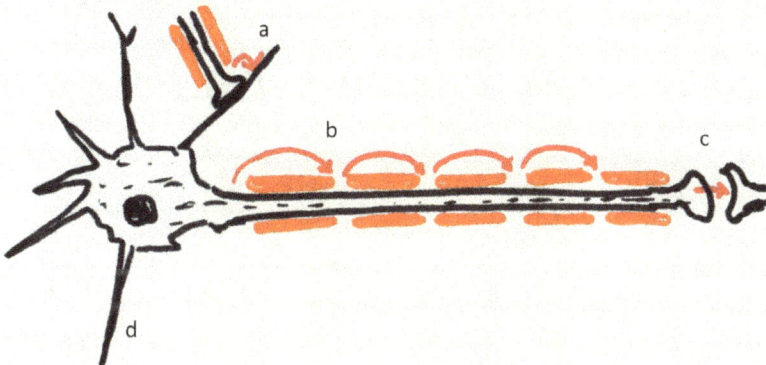

Fig. 46: Schematic sketch of a neuron, receiving, propagating and delivering an electric pulse. Incoming pulse (a) at one of the dendrites (d) is propagated along the axon (b) in a saltating way over fatty sheets wrapped around the axon and delivered via the synaptic gap to a receiving cell.

small molecules – the neurotransmitters – that evoke an electric pulse on the other side of the synaptic gap, and this pulse is conveyed along the next neuronal cell, the neuron (Fig. 47). Neurons accept pulses via their dendrites and forward them along their extended body, the axon. The synapses are like valves guaranteeing a one-way transport of the signal. Fatty sheets wrapped around the axon accelerate the transport of the pulse. The postsynaptic neuron carrying the receptors for the neurotransmitters are as long activated as a certain concentration of neurotransmitters is present. Reuptake of transmitters by the pre-synaptic neuron, or decay of transmitters by certain specified enzymes, stops the pulse.

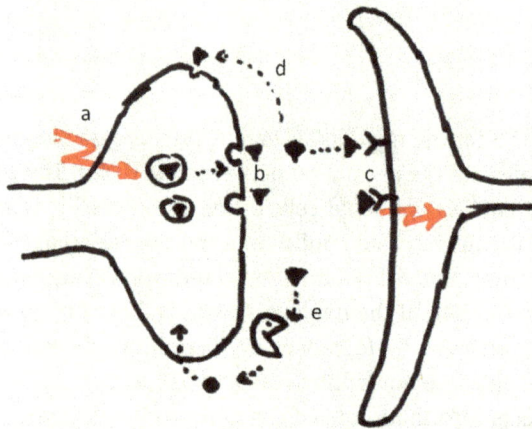

Fig. 47: Synaptic cleft between delivering and receiving neuron, or between neuron and effector organ. The propagated pulse (a) results in a release of vesicular stored transmitter molecules (b) into the synaptic cleft (c), where they bind to their receptor, triggering a pulse in the receiving cell. The pulse ends by reuptake of the transmitter (d) or enzymatic decay of the transmitter (e), followed by reuptake of the transmitter decay products.

Maintaining the ion gradient along the membrane, and permanent neurotransmitter turnover causes a permanent energy consumption which makes neurons very sensitive against lack of oxygen and glucose.

The conductance of the nerve pulse along the cell and its axon is managed by a brief opening of the sodium and potassium channels in the cell membrane. This inverses the charge on the surface of the cell membrane and sensitizes the neighborhood region, and it causes a release of neurotransmitters into the synaptic gap. Short opening of chlorine channels restores the polarization and makes the cell ready for the next pulse. Sodium and potassium pumps restore the original sodium and potassium gradient along the membrane.

Neurotoxic substances can – between others – act on the following levels:
-   killing the neuron,
-   destroying the axon,
-   destroying the fat sheets (myelin sheets),
-   over-release of neurotransmitters,
-   inhibition of neurotransmitter release,
-   inhibition of neurotransmitter reuptake (cocaine),
-   inhibition of neurotransmitter decay (nerve gases),
-   disturbing the ion gradient along the membrane by disturbing ion channels, disturbing ion pumps, disturbing the energy budget of the cell, as the ion pumps consume much energy.

Symptoms of neurotoxicity can be headache, dizziness, drowsiness, sleepiness, coma (all at CNS level), impaired vision, impaired hearing, gait disturbance, uncontrolled movement, trembling, convulsions and weakness in limbs.

3.4 Testing for toxicity

This section is important for the understanding of the toxicological background for the classification and labeling of hazardous substances. In addition, it provides some insight in the understanding of toxicological test results and their meaning for human health risk assessment. For these reasons, some sections in this chapter are more elaborate than other sections in this book.

Originally, whether a substance was more or less toxic was discovered by experience: workers either got ill or stayed being healthy. Of course, discovering toxic properties by workers exposure is absolutely unacceptable, and in the late nineteenth century, scientists started to test substances – in many instance candidates for pharmaceuticals – in experimental animals. The aim was (and is) to find out which dose should not be exceeded to avoid an adverse effect in human beings. The issue of extrapolation from animal to mankind and from a standard human being to workers or the general population was briefly addressed in Section 3.3.

Originally, people were mainly interested in acute toxicity: does a single dose or a few doses result in death or not. However, with the developing industry in the nineteenth century, occupational medicine and toxicology became more and more important as certain industry branches became identifiable by typical diseases observable in their work forces. However, such observations were recorded already a few centuries before, when the German Engineer Georg Bauer (in Latin: Georgius Agricola) around year 1540 recommended to stand upwind while fumigating mercury from gold-amalgams in open pans, because standing downwind can cause you loss of teeth [12]. It became apparent that daily small dosages are not necessarily ending in early death but severe illness by organ failure, cancer etc. Therefore, testing regimes had to be established that allow to evaluate the toxicological hazards of a substance and its NOAEL/NOAEC, as appropriate.

REMINDER:
NOAEL/NOAEC: No observed adverse effect level/concentration: highest dose or concentration that did not create an adverse response in the exposed species.
LOAEL/LOAEC: Lowest observed adverse effect level/concentration: the lowest dose or concentration in the test that caused an adverse response.

Toxicological tests are working with different kinds of animals, and biological systems tend to have a considerable standard deviation concerning the results. This can be controlled to a certain extent by standardization. In the EU and in many other regions worldwide, agencies accept animal test data only if these comply with international standards, for example the OECD guidelines for testing of chemicals [13]. Further, in the EU and many other regions, the number of animal tests has to be reduced to an unavoidable minimum for animal welfare reasons. Concerning vertebrate animals, but also for some invertebrates, only certain animal species can be used. For several species, regulations request to purchase them from certified breeders. The rat is the standard mammal for toxicology tests; for some tests, mice, rabbits, guinea pigs and dogs might be used, mostly to clarify findings in the rat, or to increase confidence that the substance is not harming. Most agencies all over the world require running tests under good laboratory practice (GLP). GLP is a quality assurance system to guarantee that the complete process of data generation is retrievable, and test institutions, their equipment and their personnel must fulfil certain quality standards.

For testing toxicity, the substance shall be applied in a way that is representative for the substance use. For industrial chemicals, dermal administration and inhalation are the most relevant pathways of exposure. However, frequently testing starts with oral exposure. The substance is either mixed into the drinking water or the feed for the test animals, or it is instilled via a dull steel needle directly into the stomach (gavage). This oral exposure has no direct relevance for workplace exposure (or better: should not have relevance). As is described in later chapters, eating, drinking and smoking at workplaces while handling dangerous substances is generally forbidden for industrial hygiene reasons. Background for this regulation are intoxications at workplaces in older times, where people touched their food with contaminated hands and doing so ingested toxic compounds (p.e., Pb exposure), or vessels for food were mixed up with containments of chemicals. These poor practices are hopefully over, now. So, why testing chemicals by oral exposure of animals? There are two main reasons. The most important one: you can be sure, that a certain, defined amount of substance has reached the intestines and is subject to high uptake into the blood. If the substance poses a toxicological hazard, you should be able to detect it. The second reason is that oral dosing is typically easier than dermal dosing, and definitely easier than inhalation exposure with its demands for air analysis. In addition to that, the tested substance may end up in food or drinking water due to migration of traces out of consumer goods; in this case, oral exposure has most relevance for assessing the risk of consumers.

Skin exposure is relevant for workplaces, and there is always the question how much of the substance can actually penetrate the skin. For better understanding dermal exposure hazards it might be desired to perform *in-vitro* or *in-vivo* skin penetration tests. In many instances, test animals show a higher dermal uptake rate than human beings due to their higher density of hair follicles, but there is no general rule to quantify the difference. In dermal exposure experiments, the substance is applied on clipped skin and covered by a badge to prevent animals from swallowing the substance by grooming.

Inhalation exposure is workplace relevant and, due to the high lung surface and the thin barrier between air and blood, it will ensure a high substance uptake rate. However, inhalation exposure experiments are a technical challenge: you need to generate and analyze an appropriate test atmosphere, need to hold this atmosphere constant for the testing time, and you may wish to make sure the test animals do only inhale the substance; licking stained fur (grooming) would provide an oral exposure on top. For these reasons, inhalation exposure experiments are much more a terchnical challenge and expensive than those for oral or dermal exposure.

Other exposure regimes are much less relevant for the workplace, but they may be applied for certain products like pharmaceuticals. In that case, the application shall mirror later use of the pharmaceutical and may be applied directly into the blood stream (i.v.), into the tissue of the skin (intradermal or subcutaneously, s.c.), into a muscle (intramuscularly, i.m.) or into the subcutaneous tissue of the belly (intraperitoneal, i.p.). Some of these application regimes may be used for industrial chemicals if the investigator wants to make sure that the test substance arrives at the target tissue.

If the substance causes damage at the site of contact, this is called a local effect. If the substance is taken up and causes damage at a certain target organ, this is called a systemic effect.

For animal welfare reasons and for reliability in test results, agencies all over the world accept test results only if these were generated in standardized, guideline tests. Tests have to be performed according to GLP, at least if tests with vertebrate animals are involved.

In toxicological tests with mammals, substances are either applied via the mouth into the digestive tract (oral exposure), via skin painting (dermal exposure) or via inhalation (inhalation exposure). Damage may occur on the site of contact (local effect) or at a target tissue (systemic effect).

3.4.1 Acute toxicity

Testing for acute toxicity is one of the eldest test regimes established. The prime target is to find out whether or not a single exposure can cause fatality. For example, a single dose is instilled into the stomach (oral exposure; rats cannot vomit!), or the substance is painted on the clipped skin and covered for 24 h, or the animals are exposed to an atmosphere containing the substance (typically for 4 h). After this exposure, following today guidelines the animals are observed for 14 days, and the number of deceased animals in that period is counted. The need for this observation period is dictated by substances which caused a damage which mortality turned out over time; for example, inhalation exposure to a critical concentration of 1,3-bis((1-isocyanato-1-methyl)ethyl) benzene caused fatalities not earlier than about 7 days after exposure.

That dose or concentration that causes death in 50% of the exposed animals is called the LD_{50} (lethal dose with 50% mortality) or the LC_{50} (lethal concentration with 50% mortality). Different species may react with different sensitivity to certain substances. Rabbits, for example, are quite insensitive against atropine [14]. Theobromine, an ingredient in cocoa and chocolate, is harmless for human beings, but a small dog will probably die after having swallowed 40 g of chocolate [15]. For that reason, the LD_{50} and LC_{50} values are listed together with the species and the exposure regime. For oral and dermal exposure, the dose is referenced to the body weight (b.w.); for inhalation exposure, the concentration is given as mg/m^3 or, alternatively, in ppm, and the exposure time is mentioned. For example, for aniline, you can find the values [16]:

- LD_{50}/oral/rat: 250 mg/kg b.w.
- LD_{50}/oral/rabbit   1,000 mg/kg b.w.
- LD_{50}/oral/dog: 195 mg/kg b.w.
- LD_{50}/dermal/rat: 1,400 mg/kg b.w.
- LD_{50}/dermal/cat: 254 mg/kg b.w.
- LC_{50}/4 h/rat: 839 ppm (=3,270 mg/m^3, calculated by ideal gas law).

According to these data, when rats are exposed to an atmosphere containing 839 ppm aniline for 4 h, 50% of the rats will probably die. If a group of dogs is dosed orally with 195 mg aniline per kg b.w., 50% of the dogs will probably die. These are estimates based on linear correlations between a parameter representing the dose and the number of dead animals within a certain period of time. However, although inbred animals are used, it is not unusual that at repetition of the test in the same lab, the same aniline sample and the same animal strain, the new result may diverge by a factor of 2–3. The point estimates LD_{50} and LC_{50} are not that precise as a number like 254 mg/kg b.w. may imply. For dose–response and data interpretation, see Chapter 5.

Decades ago, there was a great emphasis laid on a precise identification of the LD_{50}. The testing regimes required a high number of test animals. With increasing knowledge in toxicology, it became obvious that ranking and evaluation of substances due to their acute toxicity, approximate LD_{50} and LC_{50} values would be sufficient. The OECD testing guidelines were revised accordingly, now starting with comparatively few animals (typically rats) and first evaluating the reaction against a certain selected dose, and then going to next lower or higher dosages. This procedure spares experimental animals and delivers estimates for the LD_{50} and LC_{50} values that are sufficiently precise to allocate the substance to an acute toxicity category.

The target of acute toxicity testing is mortality. However, other effects of obviously unhealthy state will be recorded and may end up in classification and labeling.

3.4.2 Irritation/corrosion to skin and eyes

Skin irritation and corrosion are local effects where the substance causes tissue damage at the side of contact. Classical animal tests were developed in the 1940s, typically using the rabbit as sensitive species. The test substances are applied on the shaved skin for 24 h (during which the substance is occluded so it cannot be wiped off), or as liquid or finely ground powder in the eye. After the prescribed contact time, the reaction of the skin or eye is recorded. Due to shifts in the pH value, or because of cytotoxicity, the substance may induce an inflammatory reaction. Inflammation is a sequence of reactions the body triggers to counter the insult, which might have been induced by germs (or at least germs may be involved): pain raises attention and turns the behavior toward general protection and care of the affected area; arteries and veins relax and become "leaky" so circulation is increased in the affected area and white blood cells can more easily sneak into the affected region (to look for, and battle invaders), and doing so causes swelling and reddening. However, if the insult is too strong, cells may die away and disintegrate causing wounds.

Animal tests are performed to clarify whether or not the substance is irritating or even corrosive. Irritating and corrosive substances require special attention in terms of personal protective equipment like gloves and goggles and with respect to first aid. International test guidelines have standardized testing regimes. So, for example, the OECD guideline No. 404 standardizes tests for skin effects, and guideline No. 405 describes testing regimes to check for effects on the eyes. For *skin effects*, the endpoints are as follows:
-   erythema (reddening), grade and size (from 0 {nothing to see} to 4 {beef red});
-   edema (swelling), grade and size (from 0 {nothing to observe} to 4 {raises more than 1 mm and beyond of contact area});
-   eschar formation (pealing of dead tissue);
-   reversible (= irritation); or
-   not reversible (= corrosion): scar formation, or contact side shows 14 days later still ulcers, bleeding and so on.

For the *effects on eyes*, the endpoints are as follows:
-   cornea: opacity, grade 0 (= no effect) to 4 (completely opaque);
-   iris damage: grade 0 to 2;
-   conjunctivae: redness 0 to 3;
-   swelling grade 0 to 4.

Based on the results and their duration, the testing lab will draw a conclusion on classification and labeling. As positive substances in these tests are obviously very painful for the animals, irritation/corrosion testing was one of the first areas where scientific and regulatory institutions looked for animal-free alternatives. Meanwhile, many nonanimal test systems have been evaluated and standardized. Some physical–chemical properties like extreme pH-values may already tell the substance must be corrosive. Test systems have been developed using isolated eyes (chicken, cows), udders as left over from the butchering, or reconstituted skin systems in petri-dishes. The OECD test guidelines 404 and 405 now request that first a substance undergoes alternative tests. Only if these are passed with the result "no classification," the test in a mammal is required to ensure the substance is really neither corrosive nor irritating.

3.4.3 Sensitization

A substance is rated as sensitizer if it can trigger an over-reaction of the immune system, commonly described as allergy. Allergies in occupational settings are subclassified as skin allergies (where typically, but not necessarily exclusively the skin reacts) and respiratory allergies. Common examples are as follows:

- Nickel allergy: contact to metallic articles containing nickel causes a delayed swelling and itching of the contact area. The delay may range from a few hours up to 2 days.
- Nut allergy: eating raw hazelnuts causes gingiva swelling and an itching sensation in the mouth.
- Hay fever: exposure to grass pollen may cause red and itchy eyes and/or running nose and sneeze attacks.
- Bee-sting allergy.

The symptoms caused by a sensitizer resemble to a large extent an inflammation but may become much more severe. How can people acquire a sensitization? Immune cells (white blood cells) patrol through the body, searching for invading and parasite germs with the target to eliminate them. These cells are trained to separate "self" from "strange." Some of these immune cells – the B cells – can produce antibodies: these are proteins that are released and that can bind on the so-called antigen, a structure on the surface of the strange organism. They tag the organism and make it more visible for the germ-eating macrophages or killer cells, or they can precipitate the strange germs, triggering further attack by other immune system cells which then bombard the tagged germ with aggressive chemicals like peroxide and hydroxide radicals. These aggressive chemicals, however, can also cause damage to own tissue. The T cells may either support the identification of strange structures and help the appropriately fitting B cells to mature or they can be present as cytotoxic T cells. Cytotoxic T cells recognize strange structures and then attack the strange area with reactive molecules. The strange structures are presented on the surface of infected body cells; by destroying infected body cells, the cytotoxic T cells eliminate the breeder for strange germs.

As described so far, the immune system is important for the body to defend against invading, parasitic microorganisms. However, sensitizing chemicals are not organisms. How can they trigger an immune response? Typically, the "triggers" on the surface of B and T cells – the antigen binding structures – need to "see" a structure of at least six amino acids in a chain which have to be different against own tissue, before an immune response is triggered. The assumption is that almost all small, sensitizing chemicals can change protein structures in the body so they "look strange." For example, the skin sensitizer 2,4-dinitro-chloro-benzene (DNCB) binds covalently to amino groups in proteins. The covalent binding of this molecule will change the structure and the binding capabilities of peptide area in the protein.

The immune system attacks strange protein structures to avoid germs.
Small molecules may attach to proteins of the body, making them look strange to the immune system.
Allergy is an overshooting immune reaction.

The problem with allergy is that the reaction of the body against nonexisting germs is overshooting. Having antibodies is not the problem; if a medical practitioner detects antibodies in your blood against, for example, mold fungus, it is not necessarily saying that you will always overreact against mold. It simply says that your body has had contact to these molds previously, and your immune system has recognized the strange tissue and – therefore – synthesized antibodies. This is called the induction (the immune system is induced). In the induction phase, the allergy inducing substance may probably need to be present in an at least minimal irritating concentration to trigger the immune system. If you are allergic, then a second contact will either cause a mass release of antibodies (the so-called humoral response) or an induction and mass release of cytotoxic T cells (cellular response). On the second or further repeated contacts, a sub-irritating concentration of the substance is sufficient to trigger the immune response (elicitation). The immune system generates and preserves memory cells which can provide a much quicker and more profound immune response on the next contact to the identical antigen (Fig. 48). If you compare the whole process to vaccination (take tetanus vaccination as an example) you recognize many parallels.

In medicine, there are four major types of allergic reactions described. For our purposes, it is sufficient to differentiate the immediate response from the delayed response.

Induction: first contact to the sensitizing molecule that induces the immune system and results in amplification of antibodies or cytotoxic T cells; it is assumed that an at least minimal irritating amount of substance is required for the induction.
Elicitation: a repeated contact releases the immune response with typical symptoms of an allergy. A smaller dose / concentration than in the induction process is sufficient to cause the response.

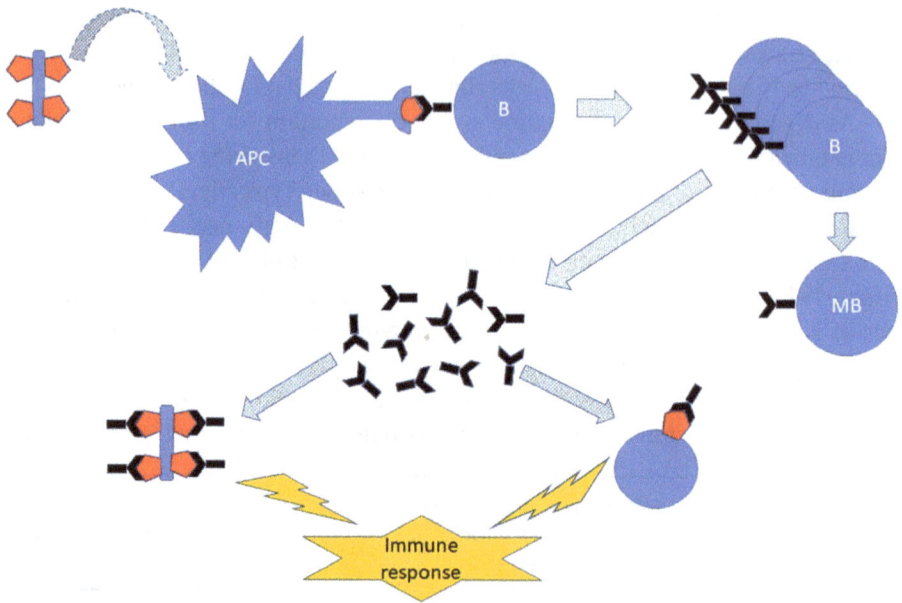

Fig. 48: Simplified sketch of the humoral immune response. A germ is taken up and processed by an antigen-presenting cell (APC). Parts of the germ (antigens) are presented at the special surfaces of the APC. As soon as a B cell having a fitting antibody structure to the antigen binds, it becomes activated and multiplies. Part of these fitting B cells becomes memory cells. Others release a plethora of specific antibodies. When these free antibodies bind to their antigens – either on the surface of the germ or on infected body cells – they inhibit them and they trigger an immune response, releasing inflammation signals and attracting other immune cells to combat the intruders.

In the immediate response, antibodies bind to the antigen of the allergen. Directly after binding, they trigger the release of inflammatory signals, so blood vessels are opened widely and become leaky (to ensure a quick supply of macrophages and white blood cells); this can cause a drop of blood pressure (up to fainting) and swelling of mucous membranes. Typical examples are bee stings, hazelnut allergy, hay fever or asthma attack. People having an asthmatic attack sense breathing difficulties. They have the impression their chest cannot wide open enough to ensure the required ventilation. The allergic reaction causes a constriction of the bronchi and an overproduction of mucous; this would normally help to get germs out of the airways, but as overshooting reaction it causes shortness of breath and makes affected people scary, the latter for a good reason: an asthma attack may become life-threatening.

In the delayed response, cytotoxic T cells identify their antibody (Fig. 49). After that contact, they multiply, which takes some time. Within 48 h, the multiplied T cells attack all cells showing the fitting antigen and release reactive chemicals to kill these cells.

The responses to irritation and sensitization may seem to be identical on a first glance, as the symptoms have many in parallel: rash, swelling, itching and possibly

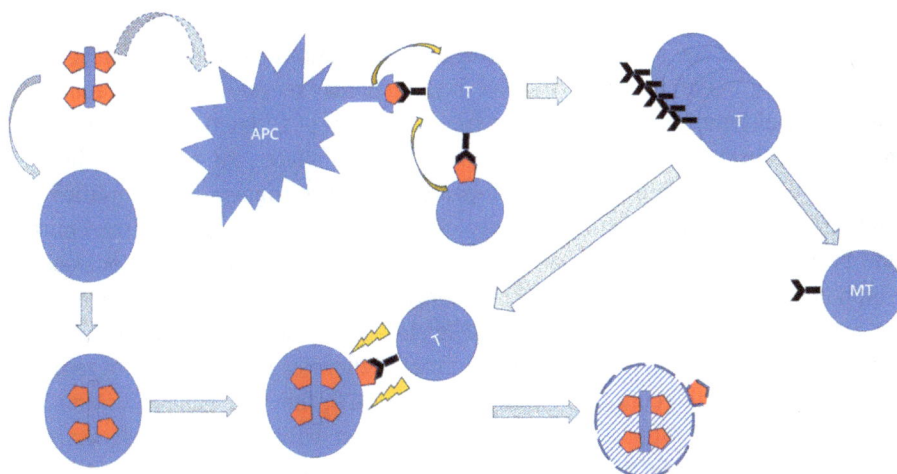

Fig. 49: Simplified sketch of the cellular immune response. Germs may highjack body cells making them to slaves for the reproduction of the germ. On the surface of infected cells, but also on specialized surfaces of antigen-presenting cells (APC), structures of the germ will appear as antigens. T cells having the fitting antibody-structure bind to these antigens and become activated. Cytotoxic T cells will be multiplied and venture through the body. Whenever they bind to their antigen, they release reactive chemicals which destroy the structure that presents the antigen. Some T cells are stored as memory cells.

blisters. However, for irritation, these symptoms are confined to the area of contact, whereas for sensitization the symptoms may spread beyond the contact area. *Irritation* has a threshold, and irritation responses on repeated contact are triggered only if the threshold is exceeded. *Sensitization* has a moving threshold, and repeated contact to the sensitizer can trigger adverse responses at declining dosages, the more frequent a contact has happened; in addition, the symptoms may increase in strength after repeated contact. However, the decision whether a worker has acquired a sensitization has to be left to the company physician or external medical experts.

Experiences at workplaces show that handling of sensitizers can induce a sensitization in exposed workers. However, who becomes sensitized at which dose, and after how many repeated contacts a sensitization will be induced, is unpredictable.

3.4.3.1 Skin sensitization

Test systems for skin sensitization are available since decades. The preferred animal species of the past for testing skin sensitization was the guinea pig. The guinea pig test is still a valid method today, but due to need for a comparatively high number of guinea pigs, and because it provides rather a qualitative than a quantitative output, today the mouse local lymph node assay is the preferred in vivo method to identify skin sensitizers.

The test runs as follows (Fig. 50): four groups of four mice each receive different concentrations of the substance (control; low dose; mid dose; and high dose) in a vehicle on day 1–3 on the ears. After six days, they may receive a substance making the activity of lymph nodes detectable (p.e., injection of tritiated thymine which is incorporated in the new synthesized DNA). A few hours later, the mice are killed. In case of a contact to an allergen, the repeated exposure triggers and amplifies the immune response, of which the activity of the lymph node(s) next to the contact side is elevated (incorporation of radioactivity from tritiated thymine; increase in lymph node weight). In case of a sensitization, after two days with no further contact, the activity of the local lymph node still is elevated compared to controls (control group). If the stimulation is at least three times higher (for incorporation of radioactivity) than that of the control (stimulation index, SI3), the substance is a skin allergen. Depending on what concentration of the test substance was necessary to achieve that benchmark, a categorization as moderate, strong or extreme skin allergen is possible. For the protocol with the radiolabeled thymine, the LC3 is used as a categorization benchmark. It is the concentration that can elevate the activity of the lymph node three times above background. To find the LC3, draw a linear line between the highest dose below a stimulation index of 3 and the lowest dose with a stimulation index above 3 and read the LC3 from that line.

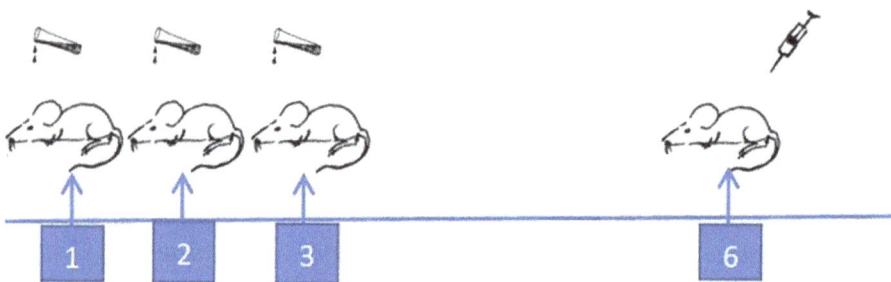

Fig. 50: Test protocol of the local lymph node assay (LLNA) in mice. Mice are dosed on three consecutive days with the test substance at different concentrations. On day 6, a marker for the activity of lymph nodes is injected and the mice sacrificed shortly thereafter.

To reduce animal testing, in silico, in chemico and in vitro test methods were developed. QSAR programs can search for alarming molecular structures known to prone the substance for triggering immune response. The OECD 442 C described the reaction of an oligopeptide with lysine ($R-NH_2$) and cysteine ($R-SH$) side chains which may react with the substance, causing decay in the HPLC peak of the oligopeptide. Such reactivity can result in altering proteins in the body and making them looking "strange" to the immune system. In vitro tests look after the activation of cells involved in the immune response.

3.4.3.2 Respiratory sensitization

Different to skin sensitization, there is currently no generally accepted standard test method for respiratory sensitization classification and labeling. There are in vivo test methods available, p.e., making use of guinea pigs, mice or rats. As for skin tests, there is an induction phase and an elicitation phase. Responses can be measured as labored breathing, narrowing of airways or infiltration of inflammatory cells into the airways and/or airway remodeling, the latter being typical for chronic damage and/or asthma. The debate in how far these parameters are relevant for human beings and in how far these test results allow to conclude on respiratory sensitization and asthma induction in human beings is ongoing. Today, substances are classified as respiratory sensitizers due to experience at workplaces. Typical symptoms (together or isolated) can be coughing; sneeze attacks; labored breathing; running nose; chest tightness; and fever.

3.4.4 Repeated dose toxicity and carcinogenicity

When handling chemicals, employers and employees are interested in whether or not the substance can cause damage to health at all, what kind of damage can be caused (i.e., what are the target organs) and which dose or concentration shall not be exceeded to make sure that the health is not affected. Acute tests look at the effects of a single, comparatively high dose. However, a dose that does not cause an observable adverse effect in an acute test might well trigger undesired responses when given every day. As an example, imagine a substance that reversibly stops the synthesis of blood clotting factors in the liver. If this substance is easily excreted, an isolated, singular dose might not be a problem as the body has a reserve of clotting factors. However, if the dose of this substance is given every day, the pool of existing clotting factors becomes exhausted and after several days of exposure, the organism runs into trouble as bleeding would not be stopped any longer by endogenic control mechanisms.

In repeated dose toxicity, tests animals receive a daily dose of the substance for several days in series. The mostly used animal for these studies is the rat, but mice, rabbits and dogs might be used as well.

Repeated dose toxicity tests follow up the acute tests. The latter gives an indication which dose is the maximum dose that can be tested. For the repeated dose toxicity tests, all animals shall survive until the end of the test period, but the highest dose shall create some adverse effect. In addition to the top dose, exposure groups for mid dose, low dose and control are required. The repeated dose toxicity test shall reveal whether the substance can cause disturbances of organ or system function. As the tests are run with juvenile animals, a sensitive but generic adverse effect is reduced b.w. gain. A reduction in b.w. gain of more than 10% is regarded as adverse. Other effects are changes in the blood functions (number of cells, blood clotting, etc.), nervous system (gait disturbance, paralysis, behavior, etc.), organ dysfunction (liver, kidney, etc.), together with morphological changes of the organ tissues, to name a few.

For oral exposure, the substance can be mixed into feed or drinking water, but the stability of the substance in these matrices over the testing period has to be demonstrated by appropriate analytical measurements. Direct injection via the mouth into the stomach with a blunt steel tube is an alternative. This gavage administration has the advantage that the dose administered is known. However, the handling of the animals is more laborious. The test substance may need to be diluted in an inert carrier, which may be water, olive oil, corn oil or others. Again, stability and concentration in the carrier has to be verified by analytical measurements. Oral dosing via gavage represents a bolus of the substance, and peak concentrations in the blood and target tissue may be higher compared to uptake via drinking water or food.

For inhalation exposure, it has to be considered that whole-body exposure may cause deposits on the fur. Licking the fur (grooming) creates an additional oral exposure on top of the inhalation. In nose-only exposure, the rats are in body-sized glass tubes directed to the exposure chamber where a certain air concentration of the test material is maintained; the air concentration has to be verified analytically. Respiratory exposure experiments are technically challenging; for that reason, and because of animal stress in nose-only exposures, exposure is typically not longer than 6 h per day and 5 days per week.

The subacute study with rats is run for 28 days, and five animals per gender and dose are exposed. The subchronic study runs for 90 days, and 10 rats per gender and dose are exposed. The subchronic study typically is the required minimum data point for the derivation of safe dosages for human beings, because all relevant organs are investigated, and the number of exposed animals allows for a sufficient statistical power: to find out whether an increase of an adverse effect against the natural background is statistically significant or not. The chronic study runs for at least 52 weeks, and in case of rats with at least 20 animals per gender and dose. The cancer study in rats lasts 24 months (with mice 18 months) with 50 animals per dose and gender.

Tab. 3: Repeated dose toxicity studies with rats.

Study type	Rats per dose and gender	Total number of rats	Study duration
Subacute	5	≥40	28 days
Subchronic	10	≥80	90 days
Chronic	20	≥160	≥12 months
Cancer	50	≥400	24 months

At the end of the tests, rats of the high exposure group are compared to rats of the control group. When adverse effects are detected, rats of the low and medium exposure group are controlled as well. In general, an adverse effect should become worse with increasing dose (dose–response; see Chapter 5). Because the identification of the NOAEL/LOAEL is of great importance for the derivation of exposure values like the Derived No Effect Level (DNEL; see under "Dose–Response"), guidelines limit factors between the dose groups. For example, if the acute oral toxicity test did not indicate relevant toxicity of the test substance up to 2,000 mg/kg b.w., a typical dosing for the oral subacute study would be 0 (control group), 100, 300 and 1,000 mg /kg b.w.

If the interest is in identifying the NOAEL for chronic exposure, why are subacute and subchronic studies run at all? The reason is that in spite of having acute toxicity data, the substances tested may disclose surprises in repeated dose toxicity testing. If you go directly to the expensive chronic test, premature losses of animals may render the test results invalid. Therefore, the NOAEL and LOAEL identified in the subacute test give an idea for the dosing regime in the subchronic test, and the NOAEL and LOAEL delivered in the subchronic test provide an idea what would be recommended doses for the chronic test. Results from the subacute test may be that severe that interest in following a marketing of a substance vanishes; as unfavorable as this finding is, at least the expensive and animal consuming tests were not initiated. The increase numbers of animals in the longer duration tests make it less likely that an adverse effect is missed out due to limited statistical power.

The investigation of repeated dose toxicity may end with the subchronic test. If in comparison with the subacute test no more effects were observed, the NOAEL /LOAEL did not change, and if the substance is neither accumulating in the body nor does it have mutagenic potential, a chronic test may not be necessary.

3.4.4.1 Carcinogenic substances and cancer risk

In cancer tests, at least 50 animals per gender and dose, and typically three dosages beneath the control group are used. The high number of experimental animals can be explained by the fact that even with these in total more than 400 animals, a substance may need to induce about 5% additional cancer cases in the exposure group compared to the control group before the result achieves statistical significance. To avoid false-negative results, the high-dose group should induce clear signs of toxicity, and the animals in the group should "just survive" the exposure period.

Cancer is a process where some genetic programs in the cell are switched on which should stay on "off" and vice versa. In a multicellular organism, all cells have their place and job to do: liver cells perform anabolism and catabolism of molecules, skin cells form the boundary to the outer world, muscle cells deliver mechanical work, and so on. Most cells are organized in tissues, imbedded in a three-dimensional structure. In case of tissue damage, lost cells are replaced by growth and mitosis of neighbor cells until the gap is closed. That is, the cell in the tissue goes into mitosis only if there are positive signals for growth and mitosis, and if in addtion there are no stop signals. In case of close contact to other cells in all relevant dimensions, this cell–cell contact creates a cell-contact inhibition. A cancer cell has undergone the following changes:
- it is overreactive to mitotic signals: mitosis goes beyond just tissue repair;
- it does no longer react to contact inhibition;

- it may start to produce its own positive mitotic signals;
- it may become immortal (like stem cells);
- it reduces and finally closes down the altruistic services for the body;
- it becomes insensitive to suicide signals of the immune system.

These changes result finally in a *benign tumor*. The benign tumor may be oversized, but it is not penetrating into other tissue. Further acquirements change the benign tumor into a malign tumor:

- Secretion of proteases and other tissue loosening enzymes: the tumor is no longer confined to a certain region but penetrates into other tissue and damages it.
- Splitting off cancer cells that then seed into other tissues, that is, formation of metastases.

All these properties are available as genetic programs in every cell, and for certain purposes and time frames, the cell needs these programs. For example, permanent growth is needed in germ cells. Macrophages, which digest infectious germs, need to venture through the body and tissues. Over all it can be summarized that in a cancer cell, "good" genes are switched off (or are lost) and "bad" genes are switched on.

As damage to the DNA is a natural process, cancer is a matter of time. The processes that cancer cells undergo can be split into at least three stages (Fig. 51):

Initiation: This is the first act on the genes. The cell becomes over-reactive to growth-stimuli and inherits this property to the daughter cells.

Promotion: This is a process not necessarily linked to DNA reactions. Cell death in the tissue by irritation, to name an example, leads to the release of mitotic signals, and the initiated cells grow preferentially. During mitosis daughter cells may be formed, which gain additional properties which are "good" for cancer development.

Progression: Due to further genetic changes, the cell enters a stadium where it no longer needs external stimuli for mitosis, becomes insensitive to contact inhibition and penetrates into other tissues.

Before a cell becomes malign, several hits have to be set in the DNA of this cell. Repair processes may reverse these hits. Only if the repair is too slow to reverse the malign changes, the cell may finally go through the process of progression. Therefore, a singular high and short exposure to a carcinogen is unlikely causing cancer in human beings; repeated exposures to low levels with a high exposure frequency are much more critical.

The cancer stages and the transformations cells undergo can be demonstrated on a cellular in vitro level. In the cell-transformation assay, cells are exposed to initiating and promoting agents, and transformed cells will become visible as nodes (Fig. 52).

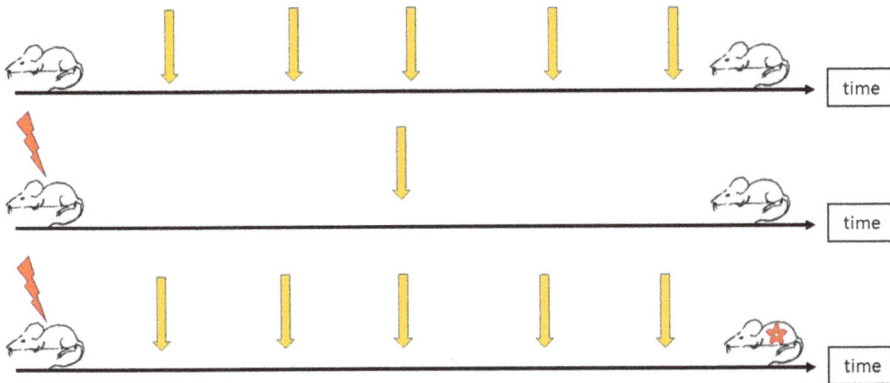

Fig. 51: Cancer model: a substance that is active as promotor only (yellow arrow) is unlikely causing an increased tumor incidence (top line). An initiator (red arrow) can induce DNA damage, but if additional insults are too rare, an increase in tumor incidence is unlikely as well (median). However, if the DNA attack by an initiator is followed by rather high frequent booster from a promotor, a detectable increased cancer incidence will result (bottom).

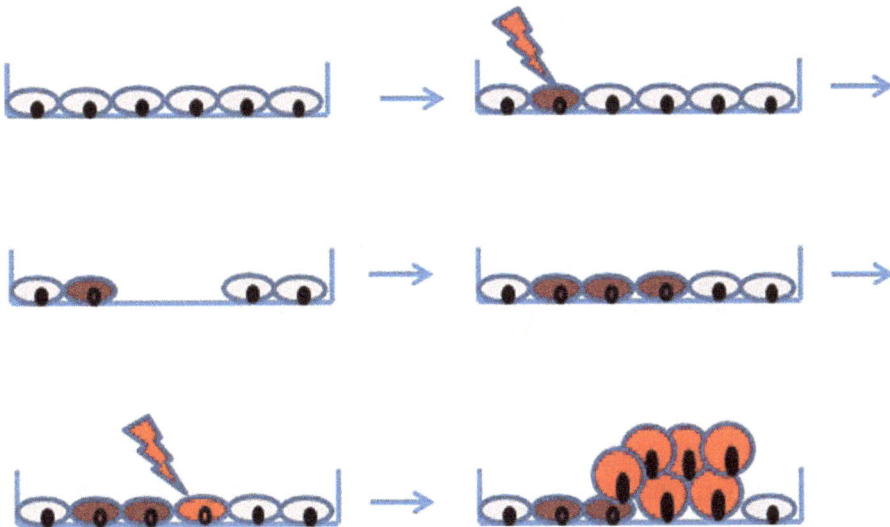

Fig. 52: Sketch of the cell transformation assay, "cancer on a petri-dish": cells form a closed structure, here a mono-layer. A mutagenic impact initiates one cell which becomes hyper-sensitive for growth signals. When gaps in tissue have to be filled up, this cell splits predominantly, whereby it acquires additional features of a cancer cell (promotion). A further hit in one of these cells of the clone transforms one of the cells into a cancer cell (progression) which growths and splits unlimited.

The likelihood P $(0 < P < 1)$ of getting cancer can be generally described by the equation

$$P = 1 - \exp[-\exp(A + B \times \log(D) + K \times \log(T)]$$

where D is the dose of the carcinogen, T is the time passed after beginning of dosing and K are the cancer stages a cell has to pass (minimum $K = 3$); the variable A stands for a background probability for cancer. For a given cancer probability, say 50% $(P = 0.5)$, this relation finally means:

$$D^B \times T^K = \text{constant}$$

For a fixed dose this means that the logarithm of the probability of acquiring cancer matched against ln(time) is a straight line with slope K, and for a fixed time, the sketch against dose is a straight line with slope B. For $D = $ constant, the above equation can be simplified to.

$$P = 1 - \exp\left[-A \times T^K\right]$$

Now assume that the exposure to the permitted limit of aflatoxins in peanuts increases the liver cancer risk by 1 in 1 million. The "background" for liver cancer is assumed to be 0.8% for 75 years lifetime $(P = 0.008, T = 75$ years). For $K = 3$ stages to cancer, the value for A is

$$A = 1.90164 \times 10^{-8}$$

Due to peanut consume, the risk increases to $P = 0.008001$. With this P value, leaving all other parameters constant, A changes to

$$A = 1.90405 \times 10^{-8}$$

At which time will the original background cancer risk be arrived at? With $P = 0.008$, $A = 1.90405 \times 10^{-8}$, now, and $K = 3$, the result is $T = 74.998$ years. That means the liver tumor or death by liver cancer appears one day earlier in the exposure group than in the control.

Another way of reading "increased cancer risk of 1 in 1 million" is as follows: assume two cities, 1 million inhabitants each. In city A, there is only the background cancer incidence (0.8 % for 75 years), city B has an additional cancer risk of 1 in 1 million. After 75 years, in city A, 8,000 persons have died from liver cancer, in city B 8,001 persons. However, in cities C, D and E without exposure, due to statistical scattering, the number of liver cancer deaths may be 8,020, 7,705 and 8,012. Because cancer is associated with extreme physical and psychological suffering for the affected persons and their families, for exposure to carcinogens, the principle "ALARA" (as low as reasonably achievable) is justified. What is reasonably achievable is a matter of debate. When a certain low level of risk is not exceeded, it might be prudent to concentrate resources on other life-threatening issues.

3.4.5 Testing for mutagenic and genotoxic effects

In the nucleus of every cell, the plan on "how to construct the whole organism" is laid down and preserved in the structure of the DNA. This molecule is a linear polyester of phosphoric acid and the sugar deoxyribose, where the OH groups of the carbon atoms C-3 and C-5 are involved in the polymer chain. The anomeric C-1 is an β-aminoglycoside, and there are four different nitrogen-containing heterocyclic compounds – called bases – that are linked to that anomeric C-1. These bases are adenine, thymine, guanine and cytosine. When these bases are fixed on the sugar-phosphoric acid polymer, hydrogen bridges can be formed between the bases: two hydrogen bonds between adenine and thymine, three hydrogen bonds between cytosine and guanine. For that reason, cytosine always pairs with guanine (vice versa), and adenine always pairs with thymine (Fig. 53).

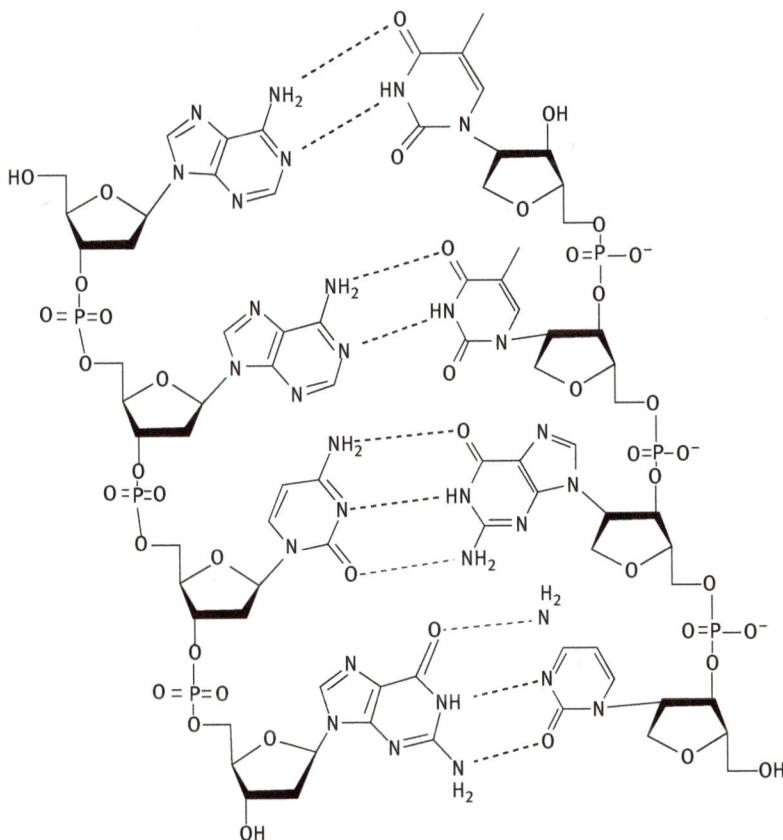

Fig. 53: The DNA network. Only four monomers are shown from a typically very long, double-strand sugar-phosphate backbone.

One major feature of the DNA is the sequence of bases; this is a code for the sequence of amino acids in a protein. Three bases in line stand for a certain amino acid (triplets). The sequence of amino acids in a protein determines the structure and function of that protein. The section of the DNA that codes one protein is called a gene. In simple organisms like bacteria, every part of the DNA is coding for at least one protein. In higher organisms, there are DNA parts that do obviously not code for proteins.

Changes in the DNA sequence are called mutation. For example, DNA bases may be exchanged against other bases. Such an event may result in an exchange of an amino acid in the protein the gene coded, and this exchange may change the protein function which then will, in principle, become detectable. If the protein function is not changed because the new amino acid does not change the protein function, the mutation is "silent." Some amino acids are coded by more than just one base triplet in the DNA, and this is another way how a mutation can be silent: the new triplet codes, the same amino acid as the old triplet.

Because the functioning genome – the summary of all genes in the organism – is very valuable, and because changes in the DNA are in most cases incompatible with survival (if they are not silent), the preservation of the sequence in the DNA is very important. For that reason, when the cell grows and splits in two daughter cells, the preservation of the DNA has to be guaranteed and two identical copies have to be synthesized (Fig. 54). This is managed by unwinding the DNA double-helix, and every single strand is a template for the new synthesized strands.

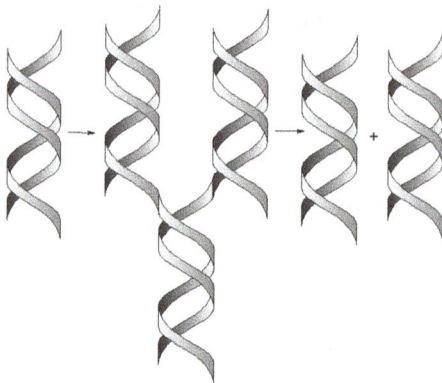

Fig. 54: Semi-conservative replication of DNA. The parent strand unwinds, and both single strands – the so-called coding and the template strand – are used as template to synthesize the other pairing strand.

In bacteria, the DNA double strand is a circular molecule. In higher organisms, the much larger amount of DNA is organized in chromosomes. The fruit fly has four chromosomes. Human beings have 46 chromosomes as a duplicate set: 23 from the mother and 23 from the father. Only in germ cells (eggs and sperm), there is a single set of 23 chromosomes.

When a mammalian cell prepares to split, all the chromosomes are duplicated. In this process – called mitosis – the chromosomes are condensed and can pick up special dyes (that is where their name comes from: chromosomes = colored). Genotoxic agents can interfere with whole chromosomes and their distribution in a way that
– chromosomes are broken (chromosomal aberration);
– chromosomes are unevenly distributed between daughter cells (aneuploidy);
– parts are exchanged between chromosomes (translocation).

"*Genotoxicity*" is the overarching terminus for all adverse effects on the genetic material. It may become apparent as cytotoxicity (daughter cell dies because certain genes are switched off), as cancer (parts of the genetic program are switched on and off, so the cell becomes immortal and insensitive for mutual action and starts unrestricted growth and multiplication) or as mutation which changes the phenotype of the organism (phenotype = how the organism appears and behaves).

Mutations fixed in the germ cells and then forwarded to the progeny, causing a change in the phenotype is the property which identifies the mutagenic substance. For example, one of the first test systems for mutagenicity was the fruit fly (*Drosophila melanogaster*). The wild-type animals have red pigmented eyes. Due to the action of a mutagen, the pigmentation may be lost (Fig. 55). This test is no longer used for regulatory purposes, but it is easy to understand and suitable to demonstrate what is meant by "mutation."

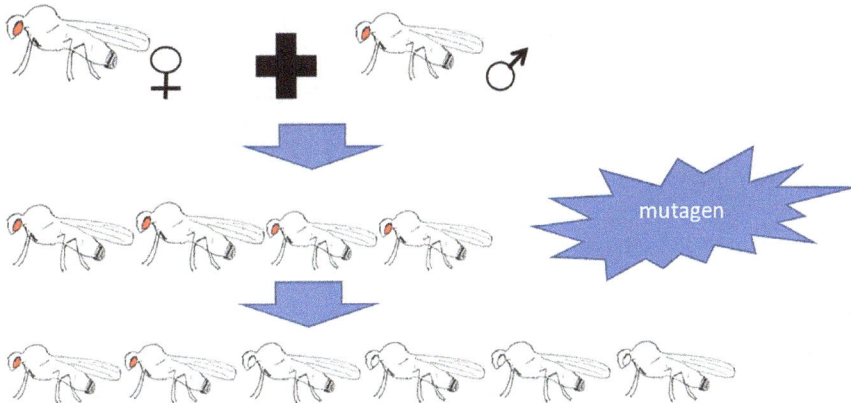

Fig. 55: Mutation test in the fruit fly *Drosophila melanogaster*. A mutagenic agent destroys the gene for the eye pigment. As a result, part of the progeny hatches having white eyes.

Mutations may happen if DNA bases are exchanged (for example, by alkylation, see Fig. 56), if parts of chromosomes are lost, or if parts of chromosomes are placed at other locations in the genome, so certain genes become hyper- or hypo-active

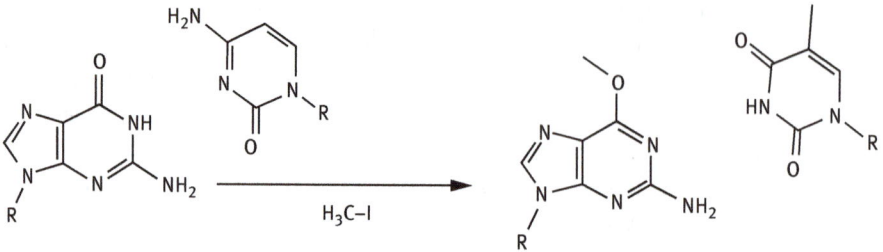

Fig. 56: O-6-Methylated guanine no longer pairs with cytosine (left, top) but with thymine (right, top).

compared to control. They may also happen if the DNA bases are modified (which may result in an exchange in daughter cells). Alkylating agents are known carcinogens. For example, dimethyl sulfate or methyl iodide can methylate the O-6 in guanine (Fig. 56). When the DNA is duplicated before the damage is repaired, adenine will take over the place of guanine. All alkylating agents, and all those substances which can be transformed to an alkylating metabolite, are potentially DNA-reactive and may induce mutations. However, diverse repair enzymes patrol the DNA and correct mistakes. Only if an alteration is "not seen" by the repair system, or if damages are so numerous that the repair systems cannot keep up, critical changes in the DNA may survive and be forwarded to the next generation. Repair systems for DNA are a necessity as endogenous processes alone cause damage to the DNA. It is assumed that every cell in the human body suffers about 3,000 DNA damages per day by endogenous, "natural" processes [17]. Theoretically, a single hit on the DNA may change the genetic program so an adverse outcome like cancer could result. From there, it is in principle not possible to define a safe level for the exposure against a direct or indirect DNA reactive agent. However, due to DNA repair, dose–response curves for genotoxic agents may show a "practical" threshold below which an adverse outcome is unlikely to be observable.

3.4.5.1 Test systems for mutagenicity

Mutations can lead to harsh health effects. Therefore, a proper identification of a mutagenic potential of a chemical is required. Many test systems do exist, and only a very small selection will be presented here. The fruit fly as shown in Fig. 54 was used in the past, but for regulatory purposes it is no longer a common test system. Nevertheless, it presents a test system where mutations become visible in the phenotype (appearance) of then organism.

Many in vitro and in vivo test systems for the identification of mutagens are established. Only two of them, the bacterial reverse mutation assay and the micronucleus test, shall be explained briefly as important, frequently used test systems. Interested readers are referred to textbooks in toxicology.

3.4.5.2 Bacterial reverse mutation assay

The test system is frequently named Ames test after its inventor, Bruce Ames at the University of California.

In the Ames test, modified bacteria of the strain *Salmonella typhimurium* are used. Due to mutations, these bacteria can no longer synthesize the amino acid histidine, so it has to be added in the nutrient solution. Other mutations facilitate the uptake of substances from the medium and reduce the DNA-repair capacity of the bacteria. Bacteria are cast on agar plates with a "minimal" medium: the medium contains a small amount of histidine (other nutrients are present in abundance), so the cells can split a few times (Fig. 57). Then, the histidine is consumed, and bacterial growth stops. However, if a mutagenic agent acts on the DNA, some bacteria may regain the capability to synthesize histidine. These bacteria can grow very well and form visible plaques on the agar within 48 h. In the test, a negative control (no addition of mutagenic substances), a positive control (addition of a known mutagen) and a dosage-series of the substance under investigation are applied. On the negative control plate, a few plaques will become visible due to spontaneous mutations. In the positive control, many plaques will be visible. The test substance is rated positive, if the number of plaques is twofold higher than in the negative control. A dose–response is expected but may not always be detectable. Increasing cytotoxicity is a factor than can produce bell-shaped dose–response curves (Fig. 58).

Using bacteria, it may be questioned how relevant the Ames test is for human beings. Generally, the test is designed to be sensitive to make sure that a substance with mutagenic potential is identified as such. Human beings and bacteria have the same DNA bases; the bacteria are much less active in metabolic processes which can

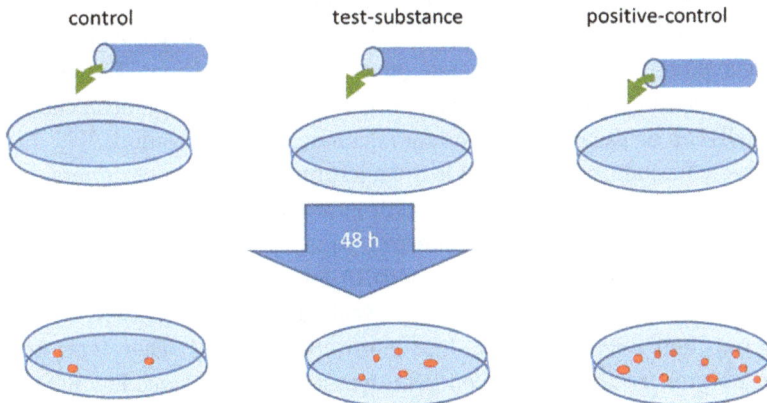

Fig. 57: Principle of the bacterial reverse mutation assay (Ames-Test).

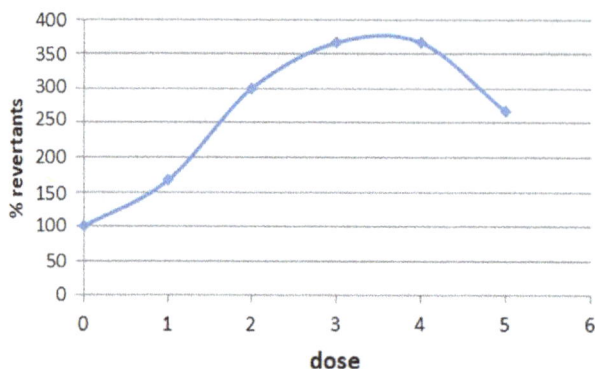

Fig. 58: Bell-shaped dose–response curve in the Ames test. % revertants = (number of plaques in dose group)/(number of plaques in control) × 100. Cytotoxicity at higher dosages reduces the number of revertants.

activate substances (p.e., buta-1,3-diene becomes the electrophilic 1,2–3,4-di-epoxy-butane). To support metabolism, the test system can be amended with homogenates of rat liver which contain phase I metabolism enzymes and the required cofactors.

3.4.5.3 Micronucleus test

The micronucleus test can be run in vitro with cell cultures or *in vivo*. Cells or animals are exposed to the test substance. If the substance creates breaks in chromosomes, fragments of the chromosomes are left over in daughter cells and form little entities besides the nucleus which can be stained with DNA dyes. Besides the nucleus, the cell shows one or more micronuclei. In case of malsegregation of chromosomes, micronuclei are formed as well (Fig. 59). In principle, the micronucleus test can be run in every tissue that can proliferate (growing by cell division). Tissue cells may either be seeded in Petri dishes and then exposed or the whole organism can be exposed, and cells can be isolated afterward for micronucleus investigations.

Comparatively frequently performed is the micronucleus test in peripheral erythrocytes in the mouse. This is an in vivo testing; other than in in vitro test systems, the whole organism may act on the test substance, either activating or deactivating it. Repair processes for damaged DNA are active. The simple reason why the erythrocytes are a popular target for investigation of micronuclei is that they squeeze out their nucleus during their maturation process. Therefore, any dot in the erythrocytes reacting positive on DNA dyes is a micronucleus. Frequently, this test is used to check the relevance of positive in vitro tests (which are more or less designed to avoid false-negatives) for in vivo systems. However, care has to be taken. Toxicologists need to answer the question whether the test substance (or its active metabolite) could reach the target tissue (bone marrow), and whether the

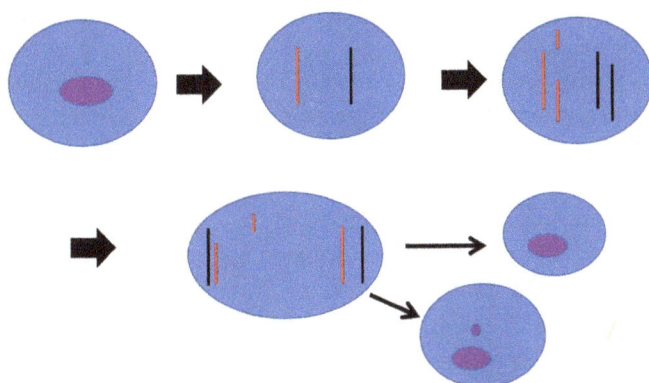

Fig. 59: Scheme of a micronucleus test. A hypothetical cell having two chromosomes proceeds to mitosis. The red chromosome is damaged when duplicated. The chromosomes move to antipodes in the cell, with exemption of the short fragment. This fragment is left in one of the daughter cells and forms a micronucleus.

system is sensitive in detecting the relevant kind of DNA interaction. For point mutations, that is, change of a singular DNA base, the micronucleus test is less sensitive than for chromosomal aberration.

3.4.6 Toxicity to reproduction

Interest in toxicity to reproduction has gained much impetus with the Thalidomide catastrophe in the 1960s, when an analgesic given to pregnant women caused crippled legs and arms in their children. Toxicity to reproduction has two main branches:
– Developmental toxicity
– Disturbed fertility

Concerning developmental toxicity, agents can affect the development of the young organism. The embryo may die in the uterus, an abortion may occur. The young organism may have lacking or delayed development on a physical basis (body mass; underdeveloped or missing bones or tissues; body size) and/or on a mental/ neurological basis. Retarded development may ameliorate over time, or it may be persistent. The malformation or absence of limbs, bones or tissues, or lack of limb formation is called *teratogenicity*.

Agents can disturb male and/or female fertility. This may happen by interference with sexual hormones (so the maturation of the germ cells is affected), or by damage to the relevant tissue or germ cells all over their different maturation stages. To detect effects on fertility, male and female animals receive the substance before mating, so a whole estrous cycle or the time span for complete sperm maturation is covered.

The preferred animal for reproduction toxicity testing is the rat. Mice or rabbits are the typical "second" species. Tests in a second species may be required by the agency if results from the first test are not clearly adverse, or if the exposure situation for the general population calls for a high reliability that substances toxic to reproduction are identified.

A screening study (OECD test guideline 421 or 422) is run with 10 male and female rats which receive the test substance prior to mating. After mating, the females are dosed until day 4 after birth. All animals are investigated histopathological, and litters are evaluated according to gender distribution, number of pups, birth weight and any visible damages. If the screening study shows clear adverse effects, the substance can be classified accordingly and further studies concerning reproduction toxicity are not required. However, if the screening study does not reveal adverse effects, the one- or two-generation study may be requested due to their higher statistical power based on the higher number of animals exposed per dose group.

A teratogenicity study starts with 20 pregnant dams which are exposed over the time-span starting shortly after implantation until just before birth. Pups are excised by caesarian section and investigated for viability, skeletal and visceral malformation. If there is a statistically significant increase in these endpoints in absence of maternal toxicity, the substance is toxic for the development. If developmental effects appear only in combination with maternal toxicity, its influence on the developmental effects has to be interpreted by experts.

In a one-generation study, 20 male and 20 female animals are exposed prior to mating, during gestation and after birth until the pups are juvenile. In a two-generation study, the progeny from the parent generation is mated and allowed to litter, and animals of all three generations are evaluated for effects, primarily concerning fertility, but also for any developmental effects.

Besides these special tests, results from a subchronic study may indicate whether the substance could affect male fertility. It turned out that reduced weights of the male sexual organs together with sperm motility and morphology in rodents are indicators for potentially reduced male fertility.

Epidemiological studies can investigate the sperm quality or the reproduction success of exposed workers.

3.4.7 Special tests

There are some other toxicological tests standardized, looking after special effects of chemicals. For example, structural alerts of substances, or indications from the repeated dose toxicity studies, may call for special investigations in the field of, for example, neurotoxicity, developmental neurotoxicity, immune suppression and endocrine activity.

For neurotoxicity, hens turned out to be a sensitive species for delayed peripheral paralysis. The developmental neurotoxicity test exposes animals (typically rats) in utero, during lactation and the juvenile status and checks for neuronal performance (e.g., gait, balance), learning behavior and social behavior.

Certain organotin compounds like dibutyltin or dioctyltin salts turned out to be toxic to the thymus, so a reduced T-cell count results. Special follow-up tests check the function of the immune system.

Endocrine modulators interfere with the endogenous turnover of hormones, or they are mimicking or inhibit endogenous hormones. Tests have been developed to check for interference with male or female sexual hormones and with thyroid hormones. Concerning endocrine modulation, NOAELs may turn out to be much lower than those derived in repeated dose toxicity tests.

3.5 A brief view on in silico tools

When a new substance is synthesized, initially there are in general no data concerning toxicity available. However, data mining might be helpful in cases where the substance was "re-invented," and in silico tools allow a first estimate of toxicological potency.

The OECD toolbox is a free online tool that has imbedded some in silico tools for toxicity and ecotoxicity as well as physical–chemical properties [18]. Via the link provided, a webinar is available as well as written instructions how to use the toolbox. With this toolbox, the user can define important structures of his target molecule and search for other molecules which share certain properties with his target molecule and where data are available, so a read-across might be possible. Predictions are done concerning the likelihood for certain toxicological properties, for example sensitization, DNA reactivity, genotoxicity and others. Although this tool is available for everybody, conclusions concerning product safety and the need for further testing or nontesting should be left to experts in toxicology. An example of how to incorporate the OECD QSAR Toolbox in risk evaluation is given in [19].

The OECD QSAR Toolbox sometimes requires lengthy time periods for the calculations (QSAR stand for "qualitative structure activity relationship"). Subprograms are available as freeware and might be preferred if a focused, cursory estimation of the molecule is required. One such program is "toxtree®" [20, 21]. Once the window is open, under the tag "Chemical Compound" you can draw your molecule of interest. This can be done, for example, for 2-(N,N-dimethylamino)phenol. Under the tag "Method" you choose "Select a decision tree," and you click one of several estimation targets, for example "In vitro mutagenicity (Ames test) alerts by ISS," "Skin irritation/skin corrosion," "Skin sensitization reactivity domains," "DNA binding alerts," "START biodegradability" and others. You select one of these, confirm with "ok" (literature references show up), and under the tag "Toxic Hazard" you press "calculate" and confirm with "ok" if asked to do so. For our example molecule, the program recognizes an alert for Ames test positive results (aromatic amine without sulfonic group in the same ring). For skin irritation/corrosion, you are asked to fill in some physical chemical data; write these data in the free box, or select the box "silent," if these are not available. Ticking always "silent," our target molecule is expected to be a skin irritant. After selecting the skin sensitization decision tree under "method," clicking "Estimate" under "Toxic hazard," our molecule is expected to be a skin sensitizer. Going on this way, the DNA-binding alert is positive, and the DART biodegradation decision tree predicts persistence of the substance. The decision tree "Cytochrome P450-Mediated Drug Metabolism" delivers after clicking "estimate" (tag "Toxic Hazard") results in the right hand box, and if you click on – for example – Q1.SMARTCyp primary sites of metabolism, a window opens showing the estimated metabolites (Fig. 60).

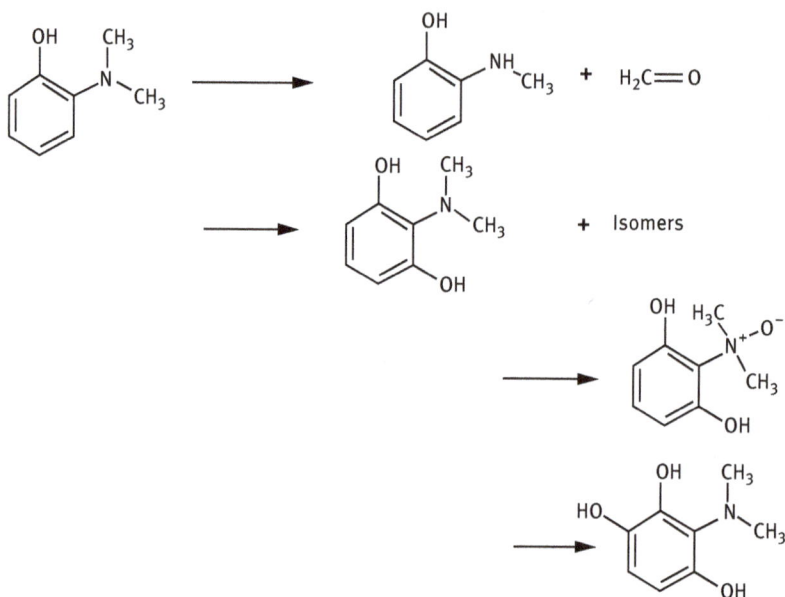

Fig. 60: 2-(Dimethylamino)phenol and its metabolites predicted by toxtree®.

In how far in silico predictions of toxicity are accepted as replacement for test in the laboratory is typically defined in the regional chemical regulations. Warning: never make use of such in silico tools instead of consulting an expert in toxicology! Instead, the appropriateness of these in silico tools requires expert judgment. The European Chemicals Agency has published a guidance document addressing the use of QSAR tools [22]. In any case, these QSAR tools allow to estimate properties of substances for endpoints where the generation of data is not (yet) required by the regulatory body because certain market volumes are not yet met, or because the substance is an intermediate which is further processed either on-site or within the chemical industry, only.

QSAR predictions can fill data gaps where the regulatory body does not yet request testing and where testing is overdone due to other considerations.
Whether and when QSAR can replace testing is typically defined by competent authorities.
QSAR tools must not be used as a replacement of consultation of experts in toxicology.

3.6 Questions

(1) How can molecules enter the cell? What is the effect of molecular size (molecular weight), net charge and lipophilicity?

(2) Certain hormones cannot enter the cell. Nevertheless, they can trigger a reaction in the cell. How is this possible?

(3) How can molecules exert a toxic interaction with the cell on a molecular level?

(4) How can the action of an enzyme be blocked?

(5) What is meant with systemic toxicity, what is local toxicity?

(6) What factors/situations do cause an increased inhalation uptake of a substance in air?

(7) What factors/situations cause an increased dermal uptake of a substance?

(8) At the workplace, a colleague got his skin stained with contaminated grease. As the grease is insoluble in water, he wants to remove it with gasoline. Provide arguments, why this is (not) a good idea.

(9) After inhalation, where does HCl and where does Cl_2 exert its toxic action?

(10) Of the molecules listed, which of the following belong to one group of parent compound, phase I metabolite(s) and phase-II-metabolite(s): 1,2-Ethanediol, o-xylene, propanal, ethene, 2-(glutathionyl)-ethanol, o-methyl-benzyl alcohol, 2,3-dimethylphenol, oxirane (ethyleneoxide), 1-propanol, 2,3-dimethyl-phenylsulfate, propanoic acid.

(11) What phase I metabolism products would you expect from (a) 1-butanol and (b) styrene?

(12) 1-Naphthol is a rather lipophilic molecule of poor water solubility. Which are potential metabolic processes that make it much more water soluble?

(13) What phase II metabolism products may you expect from 1,2-epoxybutane?

(14) An adult person excretes 2 L urine over 24 h. During that time, a substance has had an average blood concentration of 1.5 mg/L, and the concentration in the 24 h urine sample is 30 µg/L. What is the clearance of the substance?

(15) After intravenous injection of 30 mg of a substance, the highest blood concentration was 0.2 mg/L. What is the apparent distribution volume?

(16) Explain the expressions allometric scaling, interspecies-extrapolation and intraspecies extrapolation.

(17) If a chronic daily dose of 10 mg is the maximum tolerable dose for a rat (body-weight 0.5 kg), what is the maximum tolerable dose for a man (70 kg), extrapolated by allometric scaling? Assume that the tissues of man and rat react equally to a given local concentration of the substance.

(18) Assume uptake and elimination of 1-propanol can be described by a one-compartment model. Two persons, A (6 L blood volume, breathing rate 1.25 m^3/h) and B (5 L blood volume, breathing rate 2.5 m^3/h) are entering a room with an air concentration of 10 mg propanol per m^3. What is their blood level after 4 h, if they both have an elimination rate constant of (A) 0.5 h^{-1} or (B) 0.2 h^{-1}?

(19) Which organs would react comparatively rapidly to a lack of oxygen?

(20) In one acute toxicity study, propoxylated toluene-2,3-diamine (reaction product of tolene-2,3-diamine with 4 molecules propylene oxide) was dissolved in water and given to rats by oral gavage; the LD_{50} was about 1,000 mg/kg. In another oral gavage study with rats, the substance was dissolved in corn oil, and the LD_{50} was greater than 2,000 mg/kg. What could be a toxicokinetic based explanation for that finding?

(21) When a fixed dose (as mg/kg b.w.) was given to rats by oral administration, the peak concentration of the parent compound in blood plasma was 2 mmol/L, and the peak appears 1 h after dosing. In case of dermal administration, the peak appears after 4 h and achieves 2.5 mmol/L in blood plasma. How can these differences be explained? If the brain would be the target organ, and the parent compound is critical, is oral exposure or dermal exposure more critical?

(22) The plasma concentration of substance A and substance B can be described by a one-compartment model. Both substances have the same target organ and are acting via the same mechanism. Equal doses given orally to rats reveal that substance A achieves a two times higher peak concentration than substance B, but substance B causes a higher proportion of toxic effects than substance A. How can this finding be explained?

(23) Why can liver toxicity enter in problems of blood clotting?

(24) What kind of adverse effects can be introduced by a substance that is toxic to the kidneys?

(25) What are specific toxic mechanisms for nerve cells on a molecular level?

(26) Explain what is meant with LD_{50}, LC_{50}, NOAEL, NOAEC, LOAEL and LOAEC.

(27) How are genotoxicity and mutagenicity linked to each other? What is mutation?

(28) Explain the principle of the bacterial reverse mutagenicity test (Ames test).

(29) What mechanism of genotoxicity is checked with the micronucleus test?

(30) How can a substance interfere with reproduction?

4 Basics in ecotoxicology and environmental behavior

Substances entering the environment will distribute due to their physical and chemical properties. A highly water-soluble compound is most likely found in aquatic environments. A poorly water-soluble substance is expected to be found in the soil if the vapor pressure is low; with increasing vapor pressure, the substance is more likely to be found in the air. The compartments focused on in environmental chemistry are the air, water, sediment and soils. Substances may suffer a decay due to biodegradation or abiotic degradation. In biodegradation, bacteria and mold fungi digest an organic compound so that finely any carbon ends up either as carbon dioxide or as part of the body tissue. Examples for abiotic degradation are hydrolysis and photolysis. In case of continuous emissions, sooner or later an equilibrium will be established between substance input and substance output in a compartment (air, water, sediment and soil). In these compartments, the substances themselves, but also their metabolites, may interact with organisms, causing a toxic response: this is the field of ecotoxicology.

4.1 Environmental behavior

The environmental behavior of a substance is determined by some crucial physical–chemical properties which are addressed in Chapter 2: melting point, boiling point, vapor pressure, water solubility, octanol–water distribution constant, K_{OW} and the soil–water distribution constant, K_d. The K_{OW} can be used as a surrogate for accumulation in organisms. The Henry coefficient is frequently calculated via vapor pressure and water solubility and describes the water–air distribution. With these constants, an equilibrium distribution of the substance between the different compartments can be calculated. If the substance is not susceptible to any degradation process in a compartment, it is persistent and the concentration of the substance will increase with ongoing emissions. Compartments where degradation processes are active provide sinks for the substance under investigation. In the following sections, the focus shall be on degradation processes.

4.1.1 Biodegradation: general aspects and ready biodegradability

Biodegradation is a process where microorganisms use the substance as a food source. Bacteria and mold fungi are present in almost all environmental compartments: water, sediment and soil, and degradation processes can take place under aerobic conditions (oxidation of molecules) as well as under anaerobic conditions (fermentation). Organic substances are changed due to the introduction of functional

https://doi.org/10.1515/9783110618952-004

groups, and due to enzymatic cleavage of the molecules (see Section 3.2.3). Under aerobic conditions, organic molecules may finally become mineralized to carbon dioxide and to water.

Waste water is to be submitted to wastewater treatment plants (WWTP), also called sewage treatment plants (STP), where bacteria "eat up" the organic material. Due to this process, in most instances the treated water is far less critical to aquatic organisms than the original waste water. There are different main methods to investigate the extend of biodegradation of organic substances:

- Decay of the parent substance, to be followed by a specific analytical method
- Reduction in dissolved organic carbon (DOC)
- Formation of CO_2
- Consumption of O_2

Assume your wastewater contains toluene as ingredient, and toluene shall be decayed. The stoichiometric equation for the complete oxidative decay due to respiration is

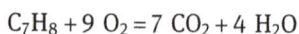

$$C_7H_8 + 9\ O_2 = 7\ CO_2 + 4\ H_2O$$

If the decay was followed by – for example – gas chromatography (GC) only, you could observe that toluene is disappearing, but you cannot be sure whether or not a substance is formed that slips your GC analysis. A metabolite could be formed which might even be more critical than toluene. Looking after DOC-decay, CO_2 formation or O_2 consumption is looking after the elimination (DOC) or mineralization of the organic molecule.

Depending on test conditions and results, the substance is rated as readily biodegradable (with or without limitations), not readily biodegradable, inherently biodegradable or not biodegradable. The OECD test guideline 301 series addresses ready biodegradability, the 302 series the inherent biodegradability. In inherent tests, the conditions for biodegradation are more favorable than in the ready tests (p.e., higher sludge to substance ratio). As source for degrading bacteria, activated sludge or effluent from municipal STP is used.

4.1.1.1 DOC decay

DOC by definition is organic, combustible material that passes a 0.42 μm filter. The water phase is combusted and the formed CO_2 detected. The DOC die-away test is run by purging air through the mixture of activated sludge (from the sewage plant) and test substance. The test substance must not be too volatile so it would be purged out, indicating a false-positive DOC decay. Actually, this test is not fit for toluene, but it may be used for diethylene glycol. Another false-positive decay would be indicated if the substance is adsorbed on sludge. Therefore, the first data point for DOC decay is taken 3 h after the start of the test. Any loss of DOC within 3 h is attributable to adsorption. It may also happen that a metabolite of the mother compound shows strong adsorption onto sludge; in that case, DOC decay is not mineralization. The total time period for these biodegradation tests is 28 days (Fig. 61).

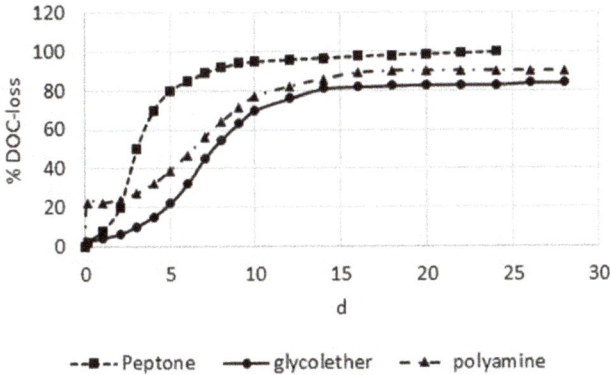

Fig. 61: Biodegradation followed by DOC loss.

Figure 61 shows the fictive DOC decay for three different substances. Peptone is pre-digested meat protein, and the bacteria can mineralize it immediately at a high rate. The glycol ether is less easily digestible, but like the peptone it does not adsorb strongly to the sludge. For the amine, the 3-h value shows adsorption to the sludge of about 22%. Further, degradation starts after a lag time of about 4 days. The lag time is the period bacteria need to accommodate to the "new" food source. By defi-nition, it is over when a 10% net decay due to microbial activity is achieved. After the lag time at day 4, DOC removal is more than 90% at day 10. The definition for ready biodegradability for DOC decay is:

– **Readily biodegradable**: within the 28-day study period, at least 70% DOC is removed without adsorption; this value has to be reached within a 10-day window that starts after the adaptation period, that is, 10% DOC removal (without adsorption).

Referring to Fig. 61, for the polyamine the interpretation runs as follows: 20% DOC loss is attributable to adsorption (3-h value). At about day 4, an addi-tional 10% DOC loss is achieved, so at day 4, the lag-time is over, and the 10-day window starts. At the end of this 10-day window, day 14, DOC loss is 85%. However, as 20% DOC loss have to be attributed to adsorption, the loss due to assimilation is (85–20)% = 65%, which is below the required 70%, and the sub-stance would not be rated as readily biodegradable.

When a TOC/DOC apparatus is not available, the DOC can be measured with the help of chromate oxidation; in this case, a generated result is termed "chemical oxygen demand" (COD). Organic carbons are oxidized by chromate in sulfuric acid and the presence of silver ions as catalysts at 140 °C; for toluene, the equation is:

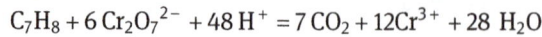

$$C_7H_8 + 6\,Cr_2O_7^{2-} + 48\,H^+ = 7\,CO_2 + 12Cr^{3+} + 28\,H_2O$$

Oxidation is performed with a surplus of chromate, and the non-consumed chromate can be titrated with Fe (II) solution, or it can be quantified by a photometric method. Some substances are not completely oxidized by the chromate method, for example nitrogen-containing polycondensated heterocycles or fluorinated carbons. The COD value may be lower than the corresponding DOC value.

4.1.1.2 CO₂ production or O₂ consumption

During biodegradation, oxygen is consumed and carbon dioxide is produced. Carbon dioxide production can be measured by precipitation of carbonates. In a closed system at constant temperature and a headspace, absorption of CO_2 would result in a decrease of pressure which can be translated into CO_2 production or O_2 consumption. If the closed system is completely filled with liquid, measurement of dissolved oxygen over time provides data on oxygen consumption.

A substance is rated as readily biodegradable if within the 28-day test either oxygen consumption or carbon dioxide production achieves 60% of the theoretical value within a 10-day window starting after the lag time. Figure 62 shows data for the ready biodegradation test for 2-methyl-2-heptanol. The positive control – a mixture of glucose and glutamate – shows the viability of the test system; 2-methyl-2-heptanol shows a lag time of 13 days and is not readily biodegradable in the test system chosen.

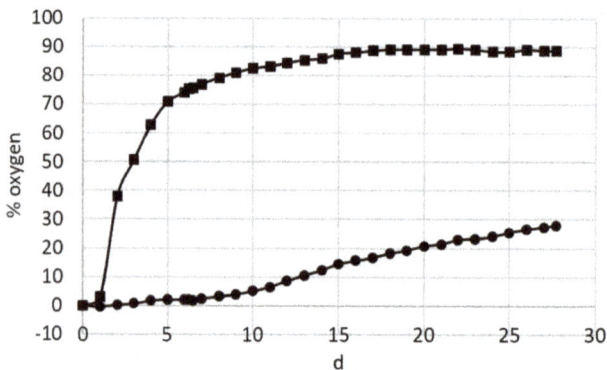

Fig. 62: Oxygen consumption (of theoretical oxygen demand) for 2-methyl-2-heptanol (circles) and glucose/glutamate as positive control (squares).

If the activated sludge has had time to adapt to the substance, degradation rates may increase significantly. This can be an advantage for on-site STPs at chemical production sites. Such an effect was observed for the industrial chemical methylene-4,4'-dianiline [23]; while almost all tests for ready and inherent biodegradability showed only limited biodegradation, a test with adapted sludge demonstrated a rapid decay of this substance.

Transformation to CO_2 is an indication that the substance under investigation is mineralized. Whether or not it is a detoxification depends on the composition of the parent compound and potential metabolites. This shall be illustrated with a theoretical exercise. Two commercially important di-organotin-compounds, dimethyltin-bis-decanoate $\{(CH_3)_2Sn^{2+}(C_9H_{19}CO_2^-)_2\}$ and di-n-octyltin-bis-decanoate $\{(C_8H_{17})_2Sn^{2+}(C_9H_{19}CO_2^-)_2\}$, are assumed to show ready biodegradability, say 65% CO_2 evolution within the 10-day window. If the di-alkyl-tin moiety is the toxicological critical part of the molecule, would you expect detoxification? You may wish to find it out yourself before you read the next paragraph!

As worst case, it might happen that only the carboxylate anion is digested. For $\{(CH_3)_2Sn^{2+}(C_9H_{19}CO_2^-)_2\}$, the theoretical yield in CO_2 is

$$C_{22}H_{44}SnO_4 + n\ O_2 = 22\ CO_2 + \cdots$$

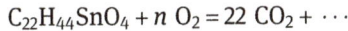

If now only the carboxylate anions are degraded, 20 equivalents of CO_2 would be formed, which makes a CO_2 yield of 90%, and the $\{(CH_3)_2Sn^{2+}(C_9H_{19}CO_2^-)_2\}$ would be rated as "readily biodegradable." However, the toxicological critical dimethyltin moiety is left unchanged!

4.1.2 Further tests on biodegradability

The OECD 302 series covers biodegradation tests providing a more favorable condition for substance decay. There are higher sludge-to-substance ratios used, and test conditions are more comparable to conditions in STPs compared to the OECD 301 tests.

The OECD 303 test guideline test actually simulates STP conditions on a laboratory scale (303A), and biodegradation in biofilms (303B); biofilms are formed by bacteria with extracellular matrix and create a special, bacteria-favoring matrix.

Somewhat outdated are the tests for biochemical oxygen demand (BOD) within a 5-day window (BOD_5). These tests range back into the 1960s and 1970s, when household detergents created foam piles on creeks and rivers every Saturday (the traditional day for laundry). Regulations were introduced demanding that detergents had to be biodegradable. A test system was developed where the oxygen consumption during substance incubation over 5 days was compared to the COD or the theoretical oxygen demand. Biodegradation was rated as quotient:

$$\text{Biodegradability} = \frac{BOD_5}{COD}$$

The OECD test guidelines, 307, 308 and 309, deal with the aerobic/anaerobic degradation in soil, in sediments and in surface waters. For OECD 308 and 309, the main target is the primary decay, that is, the disappearance of the test compound under environmental realistic conditions. Water and sediment samples may be taken from different freshwater, seawater or estuarine compartments which had not been adapted to the substance under investigation. Test substance loadings are (realistically!) very small and, therefore, the use of radiolabeled material might be required to follow the formation of metabolites and to check for mineralization. Though the technical performance of these tests is more challenging than test for ready biodegradability, the test conditions are more comparable and, therefore, more relevant to environmental conditions. Decay kinetics can be followed, and results from these tests do allow to draw the ultimate whether or not a substance is persistent in the tested compartment (see also Section 4.1.5).

4.1.3 Hydrolysis

Hydrolysis is a primary degradation pathway as the disappearance of the substance under interest is followed. For being a candidate for hydrolysis testing, the substance needs to contain hydrolysable groups, as esters, amides and epoxides. Certain halogenated compounds may also be sensitive against hydrolysis (Fig. 63).

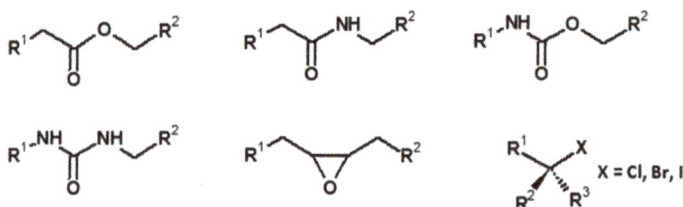

Fig. 63: Functional groups that are in principle susceptible to hydrolysis.

Hydrolysis testing is run at least at three different pH values, which represent the pH values that may be found in the environment: pH = 4, 7 and 9 (OECD test guidelines No. 111). The test may also be performed at pH 1.2 to check whether the substance is subject to hydrolysis in the stomach of mammals. The test is performed in the dark and at different temperatures, including 298 K. All products achieving a yield of 10% or more need to be identified. The need for different pH values for testing the hydrolysis is explained by the fact that protons as well as hydroxide ions catalyze the hydrolysis of, for example, carbonic acid esters (see textbooks for organic chemistry). Although water is a reactant, it is present in abundance, so its concentration does barely change and is incorporated in the rate constant. Reaction rates can be written as

$$\text{acid catalysis:} \frac{d[S]}{dt} = k_a \times [H^+] \times [S]$$

$$\text{neutral reaction:} \frac{d[S]}{dt} = k_n \times [S]$$

$$\text{base catalysis:} \frac{d[S]}{dt} = k_b \times [OH^-] \times [S]$$

In these equations, k_a and k_b are the rate constants [L/(mol × s)] for acid- and base-catalyzed hydrolysis, respectively, and k_n [1/s] is the rate constant for the non-catalyzed hydrolysis. If the hydrolysis rate constant data are taken at different pH values for a substance where hydrolysis is catalyzed by protons as well as by hydroxide ions, the following picture may be generated (Fig. 64).

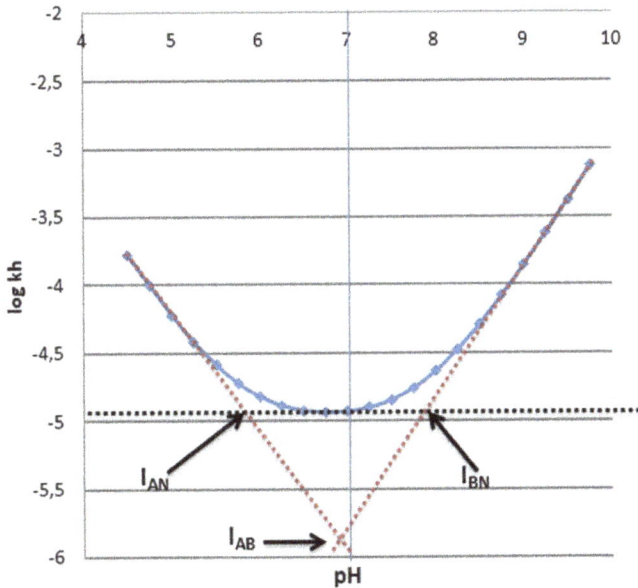

Fig. 64: Hydrolysis of a hypothetical substance at different pH values.

At intercept I_{AN}, $k_a \times [H^+] = k_n$; at the intercept I_{BN}, $k_b \times [OH^-] = k_n$; at intercept I_{AB}, $k_a \times [H^+] = k_b \times [OH^-]$. These relations allow to derive the individual rate constants [24]. For example, the hydrolysis rate constant for 2-chloro-2-methylpropane (t-butyl chloride) at 287 K is $k_n = 6.4 \times s^{-1}$; for ethyl acetate at 298 K, the rate constants are $k_a = 1.1 \times 10^{-4}$ L/(mol × s), $k_n = 1.53 \times 10^{-9}$ s^{-1} and $k_b = 0.11$ L/(mol × s) [24]. Hydrolysis rate is dependent on temperature, and measurements at fixed pH values and different temperatures generate data to derive the Arrhenius parameters for the rate constants, which then allow to extrapolate to other temperatures. The OECD test guideline 111 proposes to run a screening test at 50 °C first to check whether hydrolysis may at all play a significant role in the environmental decay of the substance.

Computer programs are available in the public domain that allows to calculate hydrolysis. See, for example, EPISUITE [5].

4.1.4 Direct and indirect photolysis

The wavelength of the daylight spectrum reaching the earth surface ranges from about 300 nm to more than 800 nm. At higher wavelengths, the radiation usually is no longer strong enough to trigger a noticeable transformation rate of exposed molecules. From the UV spectrum of the substance, you can get a rough estimate whether or not direct photolysis may play a role. If there is no absorption above 300 nm, you can neglect direct photolysis as a means for substance decay. OECD test guideline No. 316 describes a procedure to investigate direct photolysis in water. The decay of the substance is expected to follow first-order reaction kinetics, that is.

$$C_t = C_0 \times e^{-(k \times t)}$$

The rate constant k is dependent on the wavelength and the decadic absorption coefficient at the given wavelength and is calculated as

$$k = \varphi \times \sum_{290 \text{ nm}}^{800 \text{ nm}} \{\varepsilon_\lambda \times L_\lambda\}$$

with φ being the quantum yield (0–1), ε_λ as decadic molar adsorption coefficient [L/(mol × cm)], and L_λ as average daily solar photon irradiance at the wavelength λ [mmol/(cm^2 × day)]. L is dependent on the latitude and the season, and values are tabulated (Annex VII OECD test guideline 316). The molar adsorption coefficient ε_λ is derived from the UV–VIS spectrum taken in water, according to the Lambert–Beer law,

$$\log\left\{\frac{I_0}{I}\right\} = \varepsilon(\lambda) \times C \times D,$$

the concentration C given as mol/L and D as path length in cm. As a first approach, assume that $\varphi = 1$; with this assumption you can calculate whether a significant direct photodecay can be expected at all at the relevant latitude (50° latitude for middle Europe). Only in case the calculated half-life is less than 30 days, efforts need to be taken to establish the direct photodecay by experiment. This shall be demonstrated with an example. Figure 65 shows the UV–VIS spectrum of 4-nitrophenol in water.

As there is good absorbance in the region above 290 nm, the substance is a potential candidate for direct photolysis. According to the OECD guideline No. 316, k should be evaluated for every nm interval and sum up for the interval 290–380 nm.

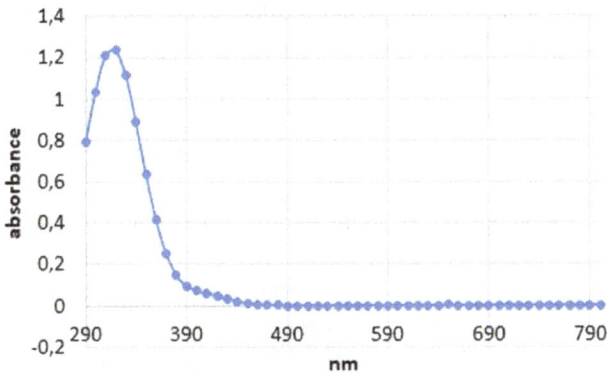

Fig. 65: UV-VIS spectrum of 4-nitrophenol in water (7.14 × 10^{-5} mol/L).

Here, we follow a simplified approach. The summation is performed for every 10 nm interval, and for the solar radiance in summer, the average of the 10 nm intervals is taken. Data are shown in the following table (Tab. 4). If every photon was successful in inducing a decay of one molecule, then the rate constant can be calculated as

$$k \ [\text{day}^{-1}] = \text{absorption coefficient} \left[\frac{L}{\text{mol} \times \text{cm}} \right]$$

$$\times \text{daily solar photon irradiance} \left[\frac{\text{mmol}}{\text{cm}^2 \times \text{day}} \right]$$

Tab. 4: Absorbance of a 4-nitrophenol solution in water (7.14 × 10^{-5} mol/L), corresponding extinction coefficients and theoretical maximum primary decay rate constant.

Wavelength [nm]	Absorbance (average)	Coefficient ε [L/(mole × cm)]	L in summer, 40° latitude [mmol/(cm^2 × day)]	First-order rate constant, k [1/day]
290	0.793	1.11E + 04	6.20E-05	6.88E-01
300	1.032	1.44E + 04	1.70E-04	2.46E + 00
310	1.210	1.69E + 04	2.50E-03	4.24E + 01
320	1.239	1.74E + 04	1.23E-02	2.13E + 02
330	1.114	1.56E + 04	5.60E-02	8.73E + 02
340	0.886	1.24E + 04	1.46E-01	1.81E + 03
350	0.635	8.89E + 03	1.62E-01	1.44E + 03
360	0.413	5.78E + 03	1.79E-01	1.04E + 03
370	0.251	3.52E + 03	1.91E-01	6.72E + 02
380	0.146	2.05E + 03	2.04E-01	4.18E + 02

If now the values for k are summed up, the result is $k = 6{,}512$/day. The corresponding half-life, $t_{0.5} = \ln(2)/k$, is 1.06×10^{-4} days. Therefore, direct photodecay may be very relevant, and a definite test should be run. Solar irradiation for middle Europe in dependence on the month is published by Frank and Klöppfer [25].

Xenon arc lamps provide a spectrum nearly identical to the solar spectrum. If sunlight is used, decay of control substances shall be investigated in parallel to identify the L_λ. Pure buffered water (pH = 4, 7 and 9), free of photosensitizers, has to be used as solvent. Temperature shall be 25 +/−2 °C.

Calculation programs for the atmospheric photodegradation are publicly available, for example, EPISUITE [5].

Irrespective to their own UV-VIS absorption spectrum, all organic substances can suffer indirect photolysis. Indirect photolysis is driven by reactive oxygen species. The most important one is the hydroxyl-radical, OH*, which is formed whenever the sun shines. As its half-life is very short (1 s), it is no longer present after sunset. Ozone and nitrous acid are precursors for OH* in the air:

$$O_3 + h\nu = O_2 + O^*$$

$$O^* + H_2O = 2\ OH^*$$

$$HONO + h\nu = NO + OH^*$$

In water, nitrate ions are the precursors for OH*:

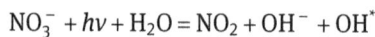

$$NO_3^- + h\nu + H_2O = NO_2 + OH^- + OH^*$$

Indirect decay in presence of nitrate in water can be quantified using substances with known degradation rate constants as probes; for example, 1-octanol has a second-order reaction rate constant of 6×10^9 L/(mol × s) [26]. When investigating the indirect photolytic decay of a substance j in water in presence of nitrate, 1-octanol can be used as a marker. Nitrate is added in great surplus, so the reaction becomes pseudo-first order. For the decay of octanol and substance j, the initial kinetic expression is

$$\ln\left(\frac{[X]_0}{[X]_t}\right) = k'_X \times \Delta t$$

where X stands for either octanol of compound j, and $k' = k \times [NO_3^-]$. Dividing the results gained with j by data for 1-octanol delivers k_j:

$$\ln\left\{\frac{[j]_0}{[j]_t}\right\}/\ln\left\{\frac{[oct]_0}{[oct]_t}\right\} = \frac{k_j}{k_{oct}} \Leftrightarrow$$

$$k_{oct} \times \ln\left\{\frac{[j]_0}{[j]_t}\right\}/\ln\left\{\frac{[oct]_0}{[oct]_t}\right\} = k_j$$

$[X]_0$ and $[X]_t$ can be measured, k_{oct} is known. Therefore, k_j can be calculated for the irradiation conditions chosen.

Dissolved organic matter (DOM) in water quenches OH*, and it also reduces the irradiation intensity. However, DOM can also sensitize oxygen to the more reactive singlet oxygen. Photodecay in water bodies has to be measured with realistic water samples to conclude on a degradation rate constant.

4.1.5 Assessment of persistency, bioaccumulation potential and toxicity (PBT)

The more persistent substances are in the environment, and the more bioaccumulative they are, the more care has to be taken they are not released into the environment. This is even more important if these substances are on top classified as toxic for any organism. A guidance document issued under Regulation (EC) No. 1907/2006 describes the persistency, bioaccumulation potential and toxicity (PBT) assessment [27]. Criteria are summarized in table R.11–1 of that document, and they are summarized in the following table (Tab. 5).

Tab. 5: Criteria for PBT and vPvB according to [27].

Property	PBT	vPvB
Persistence	Half-life >40 days in freshwater or >60 days in marine water or >120 days in freshwater sediment or soil or >180 days in marine sediment	Half-life >60 days in water or >180 days in any sediment or soil
Bioaccumulation	Bioconcentration factor in aquatic species >2,000	Bioconcentration factor in aquatic species >5,000
Toxicity	EC_{10} or NOAEC for aquatic species <0.01 mg/L, or Mutagen or carcinogen category 1, or Toxic to reproduction, or Specific Target Organ Toxicity on Repeated Exposure category 1 or 2	

Previous chapters have dealt with the degradability of substances and toxicity, and ecotoxicity is subject of the following chapters. A few words need to be mentioned concerning the criterion bioaccumulation.

The bioconcentration factor (BCF) is the equilibrium concentration of a substance in an aquatic species (mostly fish) and the surrounding water:

$$BCF = \frac{C_{\text{substance in fish}}}{C_{\text{substance in water}}} = \frac{k_{\text{uptake}}}{k_{\text{excretion}}}$$

OECD test guidelines describe the procedures of testing. For example, tests may be run either by checking the equilibrium concentrations, or they may be run establishing the first-order uptake and excretion constants in the organism. A first indication of bioaccumulation behavior is indicated by the octanol–water partition coefficient, the K_{OW}. For log $K_{OW} < 4.5$, the substance is unlikely to achieve a BCF of 2,000 or higher. A log K_{OW} larger than 4.5 is not the ultimate proof for high bioaccumulation potential as it does not cover any metabolism which usually will turn the substance into a more hydrophilic, less lipophilic compound.

4.2 Introduction into ecotoxicology

Basic biological and toxicological principles discussed in the previous chapter of course are valid for ecotoxicity as well. The distinction between toxicology and ecotoxicology is, therefore, somewhat artificial. One of the main differences is that of the perspective chosen.

Risk Management in Toxicology: make sure not a single individual human being is harmed.
Risk Management in Ecotoxicology: make sure a species does not get extinct in a habitat where it is normally present.

For toxicology, the prime target is to understand the hazardous properties of a substance to human health, so its benefits can be used while nobody gets harmed. For ecotoxicology, the main target can be described as making use (or no use) of substances in such a way that no species becomes extinct in any exposed ecosystem. The dilemma with ecotoxicology testing is that there are about 6,000 mammalian species, belonging to the about 44,000 species of vertebrate animals, not to mention the huge amount of invertebrate species, plants and microorganisms. Simply for physical viability (as well as for animal welfare and economic reasons), it is impossible to perform tests with every single species. Tests have to be performed with species which are selected in such a way that they are representative for a certain trophic level in an ecosystems, easy to handle in the lab, have short (at least not too long) generation times for rapid breeding, deliver robust test results and – hopefully – are sensitive enough to derive data which allow to set protective limits for the environment.

There are certain, distinct compartments in the environment which deserve attention in terms of ecotoxicological risk assessment: the aquatic environment (freshwater and seawater), the sediments at the ground of oceans, lakes and rivers and the terrestrial compartment. There may be certain special areas of high importance like the activated sludge basin in WWTPs. Tests can be run as acute tests and chronic tests. Chronic tests mostly focus on reproduction of the species investigated.

An area of increasing relevance and interest over the last 30 years is the interaction of chemicals with the hormone system. Sexual hormones being present or absent during the development can dictate the gender ratio in fish populations; chemicals interacting with thyroid gland hormones can result in disturbed development in tadpoles, to name just two examples. Changes introduced may not necessarily result in overt adverse effects in standard tests for chronic toxicity, but nevertheless might have a severe impact on the population fitness and survival and the ecosystem function. Meanwhile, there are several OCED test guidelines for in vitro and in vivo tests and tests in several species available to address the interaction of chemicals with the endocrine system.

Since their development, single species tests in defined, artificial environments were criticized for their low relevance for ecosystems. However, scientific tests need to show a certain robustness, comparability and repeatability and, therefore, standardization is required. It is not always the most sensitive species, that is tested but rather species which can be handled easily. In addition, test results on single species provide a first insight of the substance ecotoxicological potency. Based on data gained in single species tests, ecosystem studies in so-called mesocosms may be performed. Examples are artificial creeks installed in greenhouses where different aquatic organisms are exposed together (but some predators need to be separated from their prey), or mesocosm ponds of 1 m³ volume.

4.2.1 Testing for aquatic toxicity

Toxicity to aquatic organisms was the most frequent effect of ecotoxicity that became aware to the general population. Untreated wastewater delivered directly into creeks and rivers with upcoming industrialization caused piscicides, and the dead fish could be seen and smelled by everybody near the rivers. Testing for aquatic toxicity became a topic of high interest, and meanwhile internationally agreed test guidelines as well as rules for classification and labeling were enacted.

Aquatic toxicity testing is done along the trophic level of food chain algae – daphnia – fish. Certain bacteria toxicity testing is done as well, not leading to classification and labeling, but delivering important information which is then picked up in the environmental risk assessment. Bacteria testing is not confined to the aquatic environment. Depending on the distribution of the substance under investigation, bacteria toxicity tests can be run with soils and sediments as well.

Tests can be run static, semistatic or dynamic. In the static test, the test substance is added at the beginning of the test and the medium is not renewed. Concentrations have to be measured at the beginning and at the end of the test (and probably in between), and the mean of the concentrations have to be matched against effect. In semistatic tests, there is occasional renewal of the water, and dynamic tests are run as flow-through systems which require a continuous addition of the test substance. Water quality parameters need to be kept within certain boundaries, which covers temperature, pH, salinity and oxygen saturation. Tests are run with different dosages, negative control (no substance) and a positive (toxic) control to check the validity of the test system.

4.2.1.1 Acute and chronic aquatic toxicity

Tests for acute aquatic toxicity are run with comparatively short exposure times. They are aiming at incidences, and they provide a first insight into the aquatic toxicity of the test substance and help to define concentrations for chronic tests. The need for testing several species becomes clear as the most sensitive species does not exist (see Tab. 6).

Tab. 6: Acute toxicity data (mg/L): differences between species for similar aromatic compounds.

	2,4-Dinitrotoluene	2,4-Diaminotoluene	3,4-Dichloroaniline
Fish, LC_{50} (96 h)	6–40	0.5	2–9
Daphnia, EC_{50} (48 h)	23–35	1.6	0.33
Algae, IC_{50} (72 h)	1.8–3.5	126	1.0
Luminescent bacteria, IC_{50} (0.5 h)	45	95	0.65

The *activated sludge respiration inhibition test* according to OECD test guideline No. 209 [13] is run with activated sludge of a municipal WWTP. Active sludge consumes dissolved oxygen to decompose organic material by bacterial activity. The sludge is amended with readily biodegradable material (glucose, pre-digested protein), and different concentrations of the substance under investigation are added. After an equilibration time of 3 h, the oxygen decline over 30 min is monitored and compared to controls. IC_{50} and IC_{20} values can be derived. Discharge of the substance shall not result in killing the respiration activity of the receiving WWTP. The data generated do not result in classification and labeling, but they are important for risk assessment. As toxicity of substance to one special strain of bacteria may be compensated by higher metabolism turnover of other bacteria strains, the activated sludge respiration inhibition test is not very sensitive. Tests with *luminescent bacteria (Aliivibrio fischeri)* are attractive as they are easy to perform, require little time and costs and provide a first orientation on aquatic toxicity. More specific are tests checking the inhibition of nitrification: in the environment, proteins are degraded to ammonia which is transformed to nitrite and finally nitrate by specific bacteria. Nitrification inhibition tests can check for the decay of ammonium ions and/or the formation of nitrite and nitrate ions.

Algae tests are run with unicellular algae. They are exposed for 72 h under controlled pH and illumination conditions while the culture is on the exponential growth curve (starting with about 10,000 cells per milliliter). After the exposure time, the 50% effective concentration for the biomass (E_bC_{50}) and the growth rate (E_rC_{50}) are estimated by matching dose groups against controls. For each dose, the average specific growth rate is calculated as,

$$\mu = \frac{\ln(N_t) - \ln(N_0)}{t - t_0}$$

with N_0 and N_t being the number of cells at time points 0 and t, and growth inhibition is calculated as.

$$I = \frac{\mu_{control} - \mu}{\mu_{control}} \times 100\%$$

The NOEC or the EC_{10} of the test are used to evaluate the chronic toxicity of the substance to algae. Besides algae, the duckweed *Lemna minor* is another freshwater plant species that may be used in tests for freshwater toxicity. Plants are allowed to grow in test water containing different concentrations of the test substance for 7 days. After this period, the surface of fronts (the leaves), the biomass and the growth rate are measured (Fig. 66).

Fig. 66: Duckweed (*Lemna minor*) in a beaker, an example for higher aquatic plants in testing for ecotoxicity.

The water flea *Daphnia magna* (crustacean) feeds on algae and is a food source for fish. In *daphnia toxicity tests*, juvenile daphnia are exposed for 48 h. At the end of the test, immobile daphnia are counted and the EC_{50} is evaluated (Fig. 67).

For these small animals, it is tricky to decide what is "dead" and "not dead." Therefore, immobile daphnia which do not move any more are regarded as "affected," and a 50% effect concentration (EC_{50}), not a 50% lethal concentration (LC_{50}) is reported; unaffected daphnia move permanently.

For the *chronic daphnia toxicity test*, daphnia are exposed for 21 days. Every second day, daphnia are fed with algae. Immobile daphnia are removed daily, and at the end of the period, reproduction success is measured by counting juvenile daphnia. The

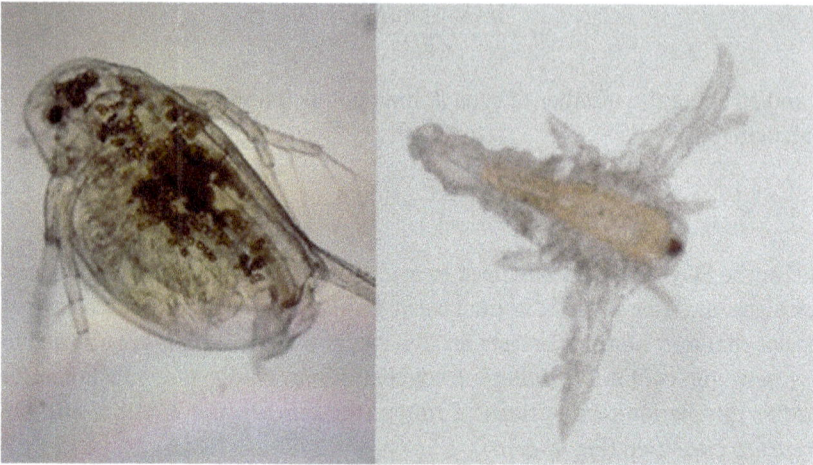

Fig. 67: Crustacean for aquatic toxicity testing (magnification 40x): *Daphnia magna* (left; freshwater) and *Artemia franciscana* (right; seawater).

highest test concentration should preferably show more than 50% effect, but without affecting survival of the adult daphnia. Oxygen content, hardness and pH of the water have to be checked at least every week.

In *acute fish toxicity tests*, fish of a certain species are exposed for 96 h against different concentrations of the test chemicals, and the number of dead fishes is counted at the end of the exposure period (every day, dead fish are removed; Fig. 68).

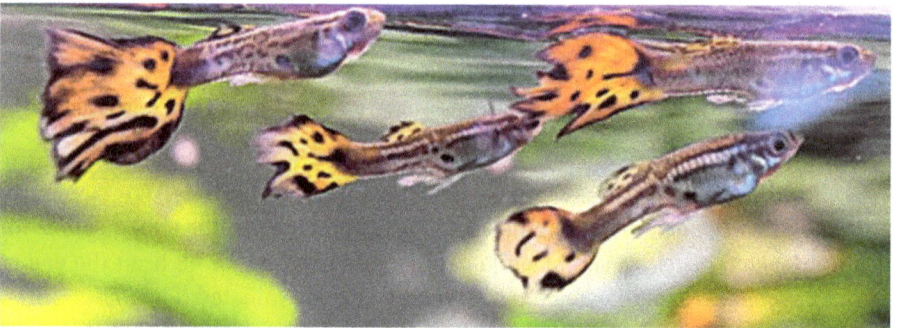

Fig. 68: *Poecilia reticulata* (Guppy, males) is one of several freshwater fish species that may be used for toxicity testing.

Test on early life stages of fish (p.e., zebrafish) use exposure of fresh fertilized eggs and the first larval stages. These tests allow to estimate the LC_{50} and may substitute traditional acute toxicity tests on fish.

Chronic tests on fish are run over 28 days with juvenile fish, or the early life stage test is performed: fresh fertilized eggs are exposed and observed over time until the hatched fish have achieved a juvenile status (21 days for zebra fish), The NOAEC of EC_{10} of these tests can be taken forward for risk assessment and classification and labeling.

4.2.2 Testing for sediment toxicity

Some substances released into water may show a tendency to adsorb to the sediment. Poor water solubility and high K_{OW} values are an indicator for distribution toward the sediment. Larvae of chironomids (midges), the sediment worm *Lumbriculus variegatus* and the plant *Myriophyllum sopicatum* may be chosen for acute and chronic tests in sediment-freshwater settings, as these species live on and in freshwater sediments. Acute tests focus on survival, and chronic tests address the reproduction of the species (Fig. 69).

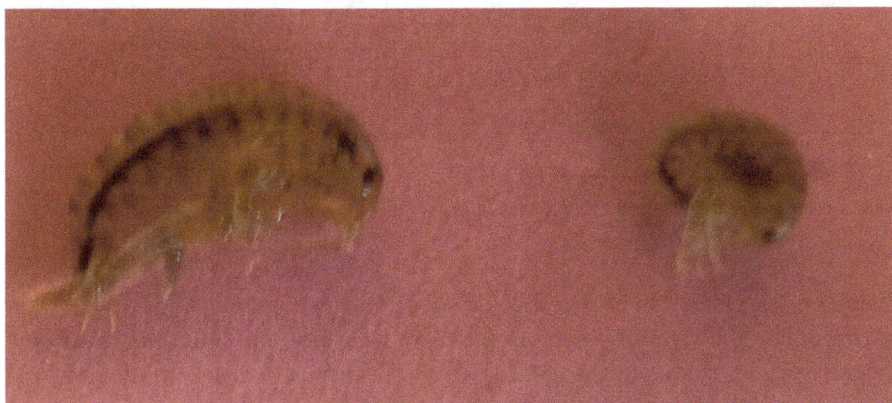

Fig. 69: A freshwater shrimp (*Gammarus* sp.) may be used for sediment toxicity testing. This example has a size of about 1 cm.

4.2.3 Testing for terrestrial toxicity

The terrestrial compartment (soil) is target compartment only if substances are designed to be used on/into soil, p.e., plant protection products or biocides. Industrial chemicals may be relevant for the soil compartment, especially if their degradability and vapor pressure is low. For the soil compartment, the influence of a substance to microorganisms, different species of invertebrates and plants can be investigated.

On a bacterial level, substances may interfere with the soil respiration, namely the transformation of nitrogen species and the release of carbon dioxide. In the nitrogen transformation test, soil is spiked with different concentrations of the test substance and plant powder. The plant material is normally degraded, and nitrogen in proteins is first released as ammonia and then oxidized to nitrite and finally nitrate. Over the 28-day test period, soil samples are extracted and analyzed for NH_4^+, NO_2^- and NO_3^-.

The *earthworm* (*Eisenia fetida*) is the species to test a substance for acute and chronic effects against oligochaetes.

Springtails (e.g., *Folsomia candida*) represent the arthropods species (hexapoda) and feed on soil algae and bacteria in humic layers of soil. Reproduction performance is tested over a period of 28 days.

With *Hypoaspis* species, *predatory soil mites* can be investigated for chronic effects of the test substance (Figs. 70 and Fig. 71).

Fig. 70: Eight day germination test with cress (*Lepidium sativum*) in presence of 5-mL copper solution (0, 17, 108, 270 and 680 mg/L; from left to right.

Effects on soil plants can be investigated concerning germination and early growth, or concerning the vigor of plants. In the plant vigor test, the substance is sprayed on young plant species in the 2–4 leaves stage, and the increase in biomass and plant appearance is controlled over a few weeks. Presence or absence of soil can make a huge difference in plant toxicity tests as soil may determinate the availability of a substance to plants.

Some OECD test guidelines address species of special interests, for example, birds, bees and bumble-bees. These tests need to be performed for plant protection products as they are dispersed over farmland, and significant exposure of these species is to be expected.

Fig. 71: Five-day germination test with radish (*Raphanus sativus*). Ten seeds in 70 g commercial potting soil, wetted with 20 mL water containing 1,020, 510, 205, 102, 51, 25, 12.5 and 0 mg Cu^{2+}/L.

4.3 Environmental modeling and risk assessment

The purpose of testing for ecotoxicity is the identification of ecotoxicological hazards. These may result in classification and labeling (currently only for endpoints in the aquatic compartment), so users can take care not to release the substance to the environment. Another, very important purpose of these tests is the environmental risk assessment. For the risk assessment, ecotoxicological data are required for all compartments that can be exposed to the substance. From the ecotoxicity data, the "Predicted No Effect Concentration" (PNEC) is calculated. For a certain compartment (comp.), the PNEC is defined as

$$PNEC_{comp} = \frac{(\text{Lowest ecotoxicological endpoint})_{comp}}{(\text{safety factor})}$$

For the compartment, the "Predicted Environmental Concentration" (PEC), a result of substance release rate, substance distribution and environmental degradation, is matched against the PNEC. A result of $PEC_{comp}/PNEC_{comp} \geq 1$ calls for remediate action.

4.3.1 Estimation of the predicted no effect concentration

The European Chemicals Agency has published a guidance document how to derive PNECs for the different compartments [28]. For inland ecosystems, PNECs are derived for microorganisms (target: STP), aquatic, sediment (benthic) and terrestrial organisms (agricultural soil). PNECs can be expended to fish- or worm-eating predators, as substances may accumulate in the food chain and, therefore, pose a specific risk to predators. For the marine environment, the focus is on aquatic and benthic organisms, fish-eating predators and top predators (see eagles, seals, etc.).

As a general rule, for the derivation of a PNEC, the lowest available endpoint is divided by a safety factor. The safety factor is reduced the more different types of species of the compartment were tested (Tab. 7).

Tab. 7: Safety factors for the derivation of a PNEC for freshwater organisms.

Point of departure	Safety factor
Lowest LC_{50} or EC_{50} from acute fish, daphnia and algae data	1,000
At least one EC_{10} or NOAEC from a chronic test with the most sensitive species in acute tests	100
Two EC_{10} or NOAEC from chronic tests, the most sensitive species from acute tests is included	50
EC_{10} or NOAEC available for fish, daphnia and algae	10

In general, EC_{10} data are preferred over the NOAECs. Imagine a substance where you have the following acute data points: LC_{50}/fish/96 h = 120 mg/L; EC_{50}/daphnia/48 h = 60 mg/L; IC_{50}/algae/72 h = 95 mg/L. These are acute data points from three different aquatic species – the minimum requirement to be able to derive a PNEC – and the daphnia value is the lowest one. In this case, the $PNEC_{aqua}$ is

$$PNEC_{aqua} = \frac{60}{1,000}\frac{mg}{L} = 0.06\frac{mg}{L}$$

If now data for a chronic fish study become available with $EC_{10,\,fish} = 15$ mg/L, this would not help much in terms of safety factor as the most sensitive species in acute tests – daphnia – is not covered. Assume in the next step an $EC_{10,daphnia}$ of 20 mg/L is generated in a chronic study. You now have two chronic data points covering the most sensitive species in acute tests. The safety factor reduces to 50, and the most sensitive endpoint is 15 mg/L. Therefore,

$$PNEC_{aqua} = \frac{15}{50}\frac{mg}{L} = 0.3\frac{mg}{L}$$

For marine organisms, due to their more homogenous environmental conditions, it is assumed that in general they are more sensitive than freshwater organisms. For that reason, safety factors are in general 10 times greater than for freshwater ecosystems. If there are only three acute test data for three different freshwater species available, the safety factor is 10,000. The factor is the same in case you have acute data from three different marine species. Reduction of this safety factor is possible if chronic data become available for the most relevant species from acute tests; however, step-wise an increased number of data points from marine organisms is required.

For microorganisms, the PNEC for the STP (PNEC$_{STP}$) is derived from the EC$_{50}$ or the NOEC of the activated sludge respiration inhibition test by applying a safety factor of 100 or 10, respectively. If an EC$_{10}$ or NOEC is available from a nitrification inhibition test in sewage, the safety factor is reduced to 1. In absence of data for bacteria toxicity testing, the NOEC from a toxicity control in ready biodegradability tests can be used, and a safety factor of 10 is applied to derive the PNEC$_{STP}$.

For the sediment, a PNEC can be estimated using data from freshwater organisms. Based on substance data as LOG K_{OW}, solubility and log K_{OC} (see Sections 2.2.4 and 2.2.6), the equilibrium concentrations of the substance in water and sediment can be calculated. $C_{sediment}/C_{water}$ is the additional safety factor to be applied. However, if the sediment is a compartment likely to show high concentrations of the substance, tests with sediment organisms should be run. The minimum requirement is one chronic test with a sediment species, and a safety factor of 100 is applied to the EC$_{10}$ (NOAEC). For two species chronic data, the factor reduces to 50, and for chronic data from three different sediment species goes down to 10.

As for sediment, the equipartitioning calculation may be applied for soil. When testing soil organisms, the logic for the derivation of the PNECs and safety factors is quite identical to the algorithm for freshwater organisms. The three different species may be, for example, soil microorganisms, plants and earthworms. Safety factors are 1,000, 100, 50 or 10, depending on the availability of relevant chronic tests.

Compounds with a high BCF and some metals (Pb, Cd, Hg, etc.) may enrich in the food chain and intoxicate a predator species which does not take up the substance directly (or there is only insignificant direct uptake). That process is called secondary poisoning. The Minimata catastrophe can be cited as a historical case of secondary poisoning (see mercury, Hg). Concerning the PNEC$_{oral}$ for secondary poisoning, interested readers are referred to the European Chemicals Agency (ECHA) document [28].

4.3.2 Predicted environmental concentration

ECHA issued a guidance document how to predict environmental concentrations [29]. Freeware programs are available to calculate the environmental concentrations of chemicals, for example, the EU program EUSES (European Union System for the Evaluation of Substances) [30, 31], and the Level III fugacity model from Trent University, Canada [32].

For environmental models, several physical–chemical parameters of the substance are required, at minimum molecular mass, water solubility, melting point and log K_{OW}. Other values can be extrapolated from these few constants, but the calculation results of the models become more robust the more measured data are available. For the environmental behavior, data on abiotic and biotic degradation are required. These degradation data are transferred to first-order rate constants for degradation. If there are no measured rates available for biodegradation in sediment and soil, a factor 10 lower degradation rate compared to surface waters can be assumed as default. In anaerobic soils and sediments degradation is assumed to be negligible, and the rate is set to zero. However, degradation rate constants or their corresponding half-lives generated in higher tier tests (OECD 308, OECD 309) can be used to overwrite the constants listed in Tab. 8 as these are a more realistic representation of the environmental behavior.

Tab. 8: Half-lives and first-order rate constants allocated to biodegradability results [31].

Test on ready biodegradability	Test on inherent biodegradability	Half-life surface waters (rate constant)
OECD 301 passed, including 10-day window criterion	–	15 days (0.047 × day^{-1})
OECD 301 passed, failing 10-day window criterion	–	50 days (0.014 × day^{-1})
Not readily degradable	No data	Infinite (0.0 × day^{-1})
Not readily degradable	>70%	150 day (0.0047 × day^{-1})
Not readily degradable	<70%	Infinite (0.0 × day^{-1})

The environmental models models estimate a steady-state equilibrium that is established between release of the substance and degradation rates in different compartments where to substance ends up. For release estimates, measured data can be used, if they are representative. The models have embedded reasonable worst-case data for release. In the EUSES program [31], different industrial, professional and private uses of the substance can be allocated to so-called Environmental Release Categories, which link the activity to a reasonable worst-case release rate [29].

Simple models (Fig. 72) assume a steady state between release, distribution and decay processes. Distribution constants for substances under consideration are taken in laboratory tests in non-saturated media. However, in the environment, a compartment may become saturated; consequently, it will no longer take up substance,

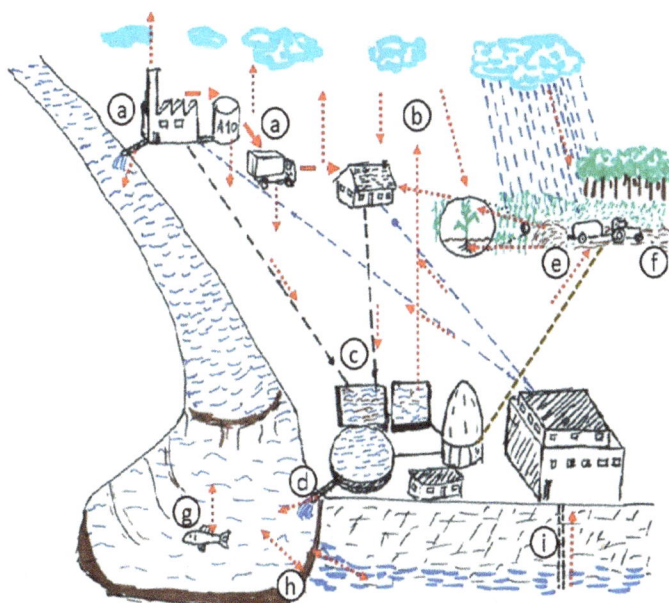

Fig. 72: Environment model. During production, storage, transport and use (large red arrows), emissions (small red arrows) to water, air and soil may occur (a). From the air, organisms may be exposed by inhalation as well as dry and wet deposition (b). Sewage is subjected to the sewage treatment plant (c), where nondegraded substance is emitted to the air, receiving surface waters (d) and/or sewage sludge (e), the latter either incinerated or spread on farmland. Plants may take up the substance by above-earth parts and via roots. Contaminated plants and fish are routes for human exposure. From soil, the substance may evaporate or irrigate into the ground, perhaps down to groundwater levels (f). In surface waters, organisms may take up the substance (g), and the substance can equilibrate to sediment and into soil/groundwater (h). Via drinking water based on groundwater, consumers might be exposed (i).

although the equilibrium is not reached. To account for this situation, Mackay et al. [33] introduced the terminus "fugacity" and defined a unit region for modeling. Fugacity, f, can be regarded as an equivalent to the vapor pressure, a tendency to leave a compartment. At equilibrium, f has the same value in all compartments.

In a box scheme, the unit region and exchange processes can be sketched as shown in Fig. 73.

Concentration and fugacity are linked to each other by the fugacity capacity, Z, which is a measure for the compartments capacity to take up the substance:

$$C \left[\frac{\text{mol}}{\text{m}^3}\right] = Z \left[\frac{\text{mol}}{\text{m}^3 \times \text{Pa}}\right] \times f \text{ [Pa]}$$

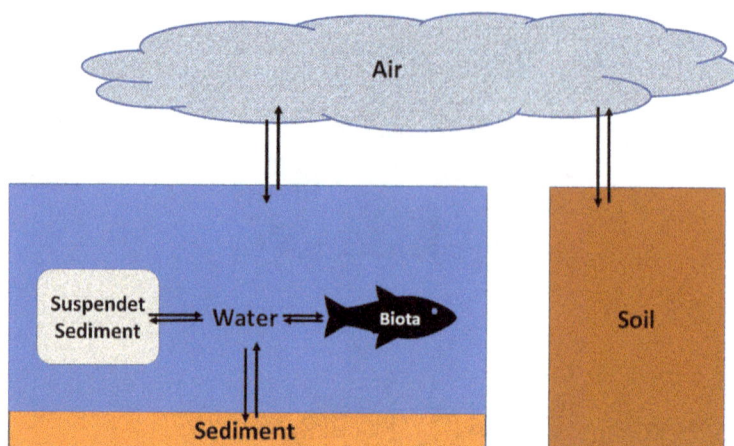

Fig. 73: Simplified environmental model and exchange processes. The water compartment is in equilibrium (is in exchange) with the sediment, suspended sediment, biota (fish and other organisms) and the air compartment; the air compartment is in equilibrium (exchange) with the soil compartment.

Based on the physical data of a substance, the fugacity capacity can be calculated for the different compartments, as listed in Tab. 9. Again, the importance of accurate measurement of the physical constants becomes obvious.

Tab. 9: Unit region for environmental modeling according to [33].

Compartment	Volume [m³]	Depth [m]	Surface [m²]	Fraction organic carbon	Density [kg/m³]
Air	1.00E14	1.00E03	1.00E11	–	1.2
Water	2.00E11	20	1.00E10	–	1.00E03
Soil	9.00E09	0.1	9.00E10	0.02	2.40E03
Sediment	1.00E08	0.01	1.00E10	0.04	2.40E03
Suspended sediment	1.00E06	–	–	0.2	1.50E03
Biota (fish)	2.00E05	–	–	–	1.00E03

Table 10 summarizes the algorithms for the calculation of the fugacity capacities for the six compartments for the environmental modelling. In that table, C = concentration [mol/m³]; n (mol); R = universal gas constant [8.314 J/(mol × K)]; T = temperature [K]; H = Henry's constant; K_d, soil-water distribution coefficient coefficient; $K_{OC} = K_d/f_{OC}$, the soil–water distribution coefficient normalized for the content or organic carbon in soil, f_{OC}.

Tab. 10: Fugacity capacities (Z) for the six compartments.

Compartment (no.)	Definition of Z
Air (1)	$C = n/V = (\text{vapor pressure})/(RT) = f/(RT) = Z \times f \Leftrightarrow Z_1 = 1/(RT)$
Water (2)	$Z_2 = (\text{Water solubility})/(\text{vapor pressure}) \Leftrightarrow Z_2 = 1/H' = (RT)/H$
Soil (3)	$K_d = C_{soil}/C_{water} = K_{oc} \times f_{oc} = Z_3/Z_2 -> Z_3 = K_d \times Z_2 \Leftrightarrow Z_3 = (K_{oc} \times 0.02)/H'$
Sediment (4)	$Z_4 = (K_{oc} \times 0.02)/H'$
Suspended sediment (5)	$Z_5 = (K_{oc} \times 0.2)/H'$
Biota (6)	$BCF = C_{fish}/C_{water} = Z_6/Z_2 \Leftrightarrow Z_6 = BCF/H'$

As an example, we will calculate the data for the chemical acetophenone (CAS-No. 98-86-2). Data for the substance are taken from [34] and are summarized in Tab. 11.

Tab. 11: Data for acetophenone.

Property	Value
Melting point (MP)	293 K
Boiling point (BP)	475 K at 101,300 Pa
Density	1,030 kg/m^3
Vapor pressure (VP)	45 Pa at 298 K
log K_{ow}	1.65
K_{oc}	50
Water solubility (SOL)	6.2 kg/m^3 at 298 K
Henry constant H'	0.766 Pa × m^3/mol
BCF (calculated)	0.47
Half-life surface waters	360 h (readily biodegradable)
Half-life soil and sediment	3,600 h (extrapolated from half-life in surface waters)
Half-life air	144 h

If these data are used in the algorithms given in Tab. 10, the fugacity capacities for acetophenone in the different compartments can be calculated, and Tab. 10 is converted to Tab. 12.

Tab. 12: Z-values for acetophenone at 293 K.

Compartment	Z-value
Air	$Z_1 = 4.11\text{E-}04$
Water	$Z_2 = 1.305$
Soil	$Z_3 = 1.305$
Sediment	$Z_4 = 2.611$
Suspended sediment	$Z_5 = 13.06$
Biota	$Z_6 = 0.614$

At static equilibrium, fugacities in the different compartments all have the same values, and distribution constants are given by the ratio of the Z-values.

The fugacity model level I assumes a distribution of a certain amount of a substance emitted in the environment, and degradation processes are neglected. When equilibrium is established, the fugacity f has the same value in all compartments. Assume an amount of 100,000 kg acetophenone was emitted in an environment as defined in Tab. 9. This total amount is to be found in:

$$100,000 \text{kg} = C_{air} \times V_{air} + C_{wate} \times V_{water} + C_{soil} \times V_{soil}$$

$$+ C_{sed} \times V_{sed} + C_{suspsed} \times V_{suspsed} + C_{biota} \times V_{biota}$$

Now, with the fugacity (f) being the same in all compartments, because of

$$f = \frac{C_{water}}{Z_{water}} = \frac{C_{air}}{Z_{air}} \leftrightarrow C_{air} = C_{water} \times \frac{Z_{air}}{Z_{water}}$$

the mass balance over all compartments j can be expressed in terms of C_{water}:.

$$100,000 \text{kg} = \sum_j \left\{ V_j \times C_{water} \times \frac{Z_j}{Z_{water}} \right\}$$

With the different Z-values given in Tab. 12 and volumes listed in Tab. 9, the result is

$$C_{water} = 4 \times 10^{-7} \frac{\text{kg}}{\text{m}^3} = 400 \frac{\text{ng}}{\text{L}}$$

This exercise is to be repeated for the other compartments. Such calculations can be done with the pocket calculator if necessary, but freeware programs are available. For example, the level I fugacity model of Trent University [35] delivers the following graphical output for acetophenone (Fig. 74).

The models become more complicated but also more realistic, if degradation processes are included. With degradation processes, it may have a great influence on final concentrations in which compartment the substance is emitted first. In addition, gain of substance from neighbor regions and losses due to transport to neighbor regions may be included. The time-dependent exchange between compartments now needs to be covered, and in the fugacity models a transfer rate is introduced, D [mol/ (h × Pa)]. For the mass transport rate, N' between the boundary layer of the two compartments A and B, we can write

$$N'_{A,B} = -D_{A,B} \times (f_A - f_B)$$

and with contact surface $S_{A,B}$ and transfer rates r_A and r_B, $D_{A,B}$ is

$$D_{A,B} = \frac{S_{A,B}}{\frac{1}{r_B \times Z_B + \frac{1}{r_A \times Z_A}}}$$

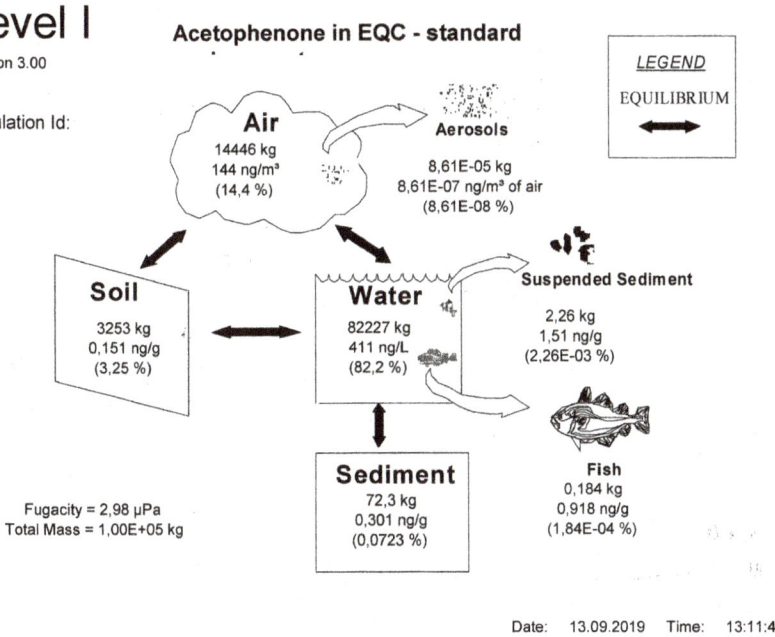

Fig. 74: Graphical output of the level I fugacity model simulation for 100 t acetophenone.

For most substances, r_A and r_B are not available and have to be approximated with data from simple molecules like CO_2. Degradation rates are to be introduced as first-order rate constants or as half-lives. If available, metabolic decay by fish may be introduced. Approximate data may be generated with fish hepatocytes in vitro, via estimation of the Michaelis–Menten kinetics parameter V_{max} and K_M and using them in PBTK modeling (see Section 3.2.5); PBTK models may be adapted to fish [36]. Other than the fugacity model level I, level III requests an immission input of the chemical per hour for each compartment. For a certain compartment i being in contact with other compartments j, the time-dependent amount in that compartment is modeled as

$$\frac{dN_i}{dt} = E_i = \text{Input} + V_i \times C_i \times k_i + \sum_j \{D_{ij} \times (f_i - f_j)\}$$

In this equation, "Input" represents an immission source, and k_i is the sum of all first order rate constants for that compartment.

Coming back to our example, acetophenone, in the first case, only emissions into air and water were assumed, namely 10 kg/h into air and 1,000 kg/h into water. Results are shown in Fig. 75.

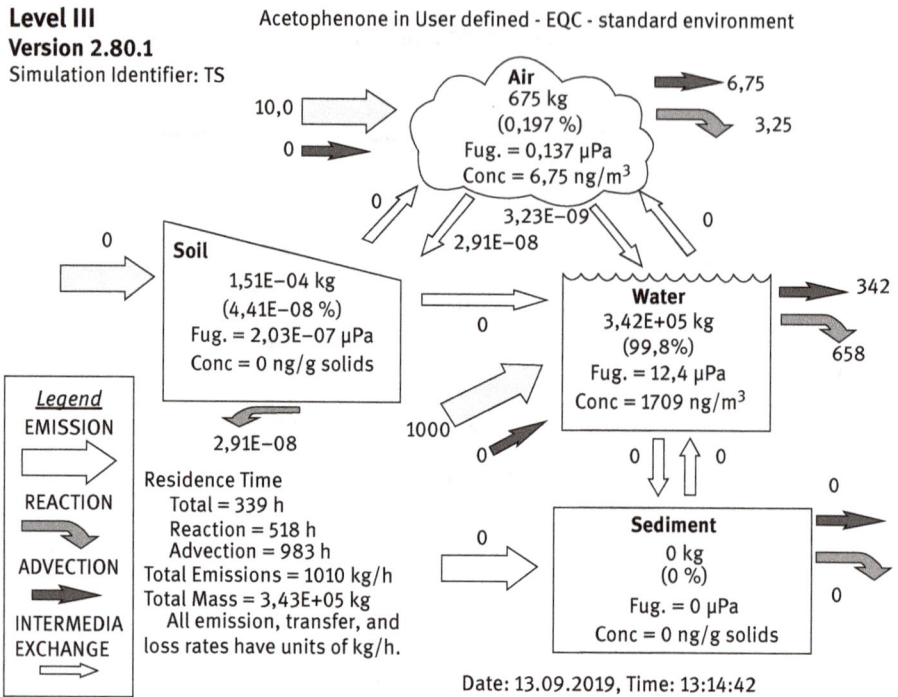

Fig. 75: Copy of level III fugacity model data graphical output for acetophenone. 10 kg/h and 1,000 kg/h emission into air and water, respectively.

Compared to the level 1 simulation, the substance is nearly exclusively to be found in water, and the air as well as the soil compartment have become negligible. However, the picture changes if half of the amount emitted into water is redirected into soil. Results are shown in Fig. 76.

The fugacity models experienced further developments, and ionic substances can be incorporated in an activity model, taking into account the pH values of the environmental compartments [37].

With the aid of modeling, the PEC can be estimated. Manipulation of estimated parameters can reveal the robustness (or weakness) of certain assumptions, and a prioritization for the generation of further substance related data may be possible. The scale of the model can be changed to investigate either more local or rather regional effects. In the EUSES model, the local and regional scale is predefined [31]. Concerning the first-order rate constants for substance decay, it shall be memorized that tests for ready and inherent biodegradability typically look after the minerali-zation of the substance, but environment simulation tests primarily look after the primary decay. The potential formation of environmentally more stable compounds has to be considered.

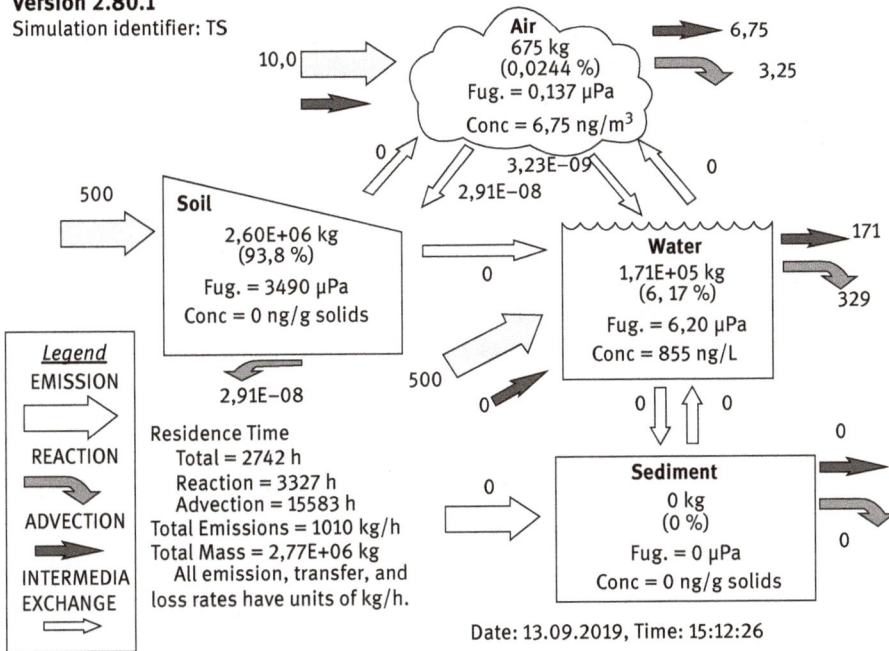

Level III
Version 2.80.1
Simulation identifier: TS

Acetophenone in User defined - EQC - standard environment

Air
675 kg
(0,0244 %)
Fug. = 0,137 µPa
Conc = 6,75 ng/m³

10,0

6,75
3,25

0
3,23E−09
2,91E−08
0

Soil
2,60E+06 kg
(93,8 %)
Fug. = 3490 µPa
Conc = 0 ng/g solids

500

Water
1,71E+05 kg
(6, 17 %)
Fug. = 6,20 µPa
Conc = 855 ng/L

0

171
329

2,91E−08

500

0
0   0

0

Legend
EMISSION

REACTION

ADVECTION

INTERMEDIA
EXCHANGE

Residence Time
Total = 2742 h
Reaction = 3327 h
Advection = 15583 h
Total Emissions = 1010 kg/h
Total Mass = 2,77E+06 kg
All emission, transfer, and
loss rates have units of kg/h.

0

Sediment
0 kg
(0 %)
Fug. = 0 µPa
Conc = 0 ng/g solids

0
0

Date: 13.09.2019, Time: 15:12:26

Fig. 76: Copy of level III fugacity model data graphical output for acetophenone, emissions of 10, 500 and 500 kg/h into air, water and soil, respectively.

For environmental risk assessment, the quotient PEC/PNEC is generated for the different compartments. For PEC/PNEC ≥ 1, further action is required. This may cover the generation of more (up to now missing) data to refine the PEC and/ or PNEC, or to reduce emissions.

At the end of this chapter, an illustrative experience from the 1990s shall be mentioned. At a production site, an incident in the middle of a wonderful sunny and warm summer resulted in the release of several tons diethylene glycol (DEG; 2-(2-hydroxy)ethoxy-ethanol) into a nearby creek. From the safety data sheet, the following data were available:

- water solubility: totally miscible at 293 K;
- acute 96-h toxicity fish: LC_{50} > 32,000 mg/L;
- acute 48-h toxicity daphnia magna: >10,000 mg/L;
- 7 day-IC_{10} green algae: 2,700 mg/L;
- 30-min EC_{20} activated sludge: >2,000 mg/L;
- OECD 301 A, 28 days: > 90% DOC removal; readily biodegradable.

Samples were taken from the creek at several locations and analyzed by GC. The highest concentrations found were about 100 mg DEG per liter water. Due to the hot summer, water flow was low, but the leak had been stopped and there was no more increase in DEG concentration but rather a slow dilution. What was to be expected?

With such data demonstrating low to negligible ecotoxicity, nobody expected something serious to be going on. Ready biodegradability of the DEG raised the expectation that bacteria will eliminate the substance quickly. After 2 days, many dead trout were observed in the creek. What has happened?

Several hypotheses were brought forward, for example that something else was emitted together with the DEG which has significant fish toxicity, or that trout was especially sensitive against DEG (the acute fish toxicity data from the safety data sheet were generated with another fish species). Finally, the following explanation was accepted: trout is more sensitive to low oxygen levels than many other freshwater fish species. As it was a hot summer, the comparatively high water temperature resulted in a critically low oxygen concentration, already. The release of the readily biodegradable DEG gave the bacteria food, and while at daylight photosynthesis of green water plants might have compensated the oxygen demand by bacteria, photosynthesis stops in the dark while biodegradation goes on. The drop in dissolved oxygen below viable levels for trout caused the high number of dead Trout.

4.4 Exercises

(1) Biodegradation tests look after the formation of CO_2 or consumption of O_2. Why is a substance rated as readily biodegradable, if in a limited time the CO_2 production has arrived at 60% of the theoretical yield? Shouldn't it be 80% or – better – 100%?

(2) When a substance was investigated for biodegradability, test A delivered the result 75% DOC decay, but test B reported 50% O_2 – consumption, only. (a) What errors in performing/evaluating the tests may have occurred? (b) Could there be an explanation under condition that both tests were run properly?

(3) Pentachlorophenol is a compound which is very toxic to aquatic life. Hexaethylene glycol-mono-(pentachlorophenyl)-ether $(C_6Cl_5-O-(C_2H_4O)_6-CH_3)$ was shown to be readily biodegradable, as 65% CO_2 were produced within a 10-day window in an OECD 301B test. Therefore, is everything o.k. provided emissions occur only toward STPs?

(4) What is the theoretical oxygen demand for 1 g of 1-chloro-2-propanol, if the full oxidation delivers carbon dioxide, water and the chloride anion? Write down the complete stoichiometric equation. Use molecular weights for $C = 12$, $H = 1$, $O = 16$, $Cl = 35.5$ g/mol.

(5) Evaluate the following data (Tab. 13) generated in a test for ready biodegradability with a substance of only poor volatility:

Tab. 13: Data for DOC die-away test (OECD 301A).

Time	3 h	1 day	3 days	7 days	10 days	12 days	20 days	28 days
Δ DOC	15%	19%	25%	40%	78%	87%	90%	92%

(6) In Fig. 65, the UV-VIS spectrum of 4-nitrophenol is given, and in Tab. 4 the maximum direct photolysis rate constants are calculated. These data are for summer, 40° latitude. Under assumption of a maximum quantum yield, calculate the minimum half-life for 4-nitrophenol using data for middle Europe for January and June, respectively (Tab. 14).

Tab. 14: Irradiation density (mmol photons/cm^2/day).

Nm	300	310	320	330	340	350	360	370	380
Jan.	7.43E-09	8.83E-05	1.90E-03	7.01E-03	1.23E-02	1.39E-02	1.52E-02	2.05E-02	2.34E-02
June	8.88E-05	5.06E-03	2.39E-02	4.93E-02	6.89E-02	7.52E-02	8.00E-02	1.06E-01	1.17E-01

(7) For 1,2-dichloro benzene (ODB), the atmospheric rate constant for the reaction with OH molecules is k_{OH} = 4.2E-13 cm^3/(molecule × s). If reaction with OH radicals is the only pathway for ODB decay in the atmosphere, what is the half-life at the 50th latitude in June ([OH*] = 1.5E + 06 molecules/cm^3, constant) and January ([OH*] = 5.0E + 04 molecules/cm^3, constant)?

(8) For an ester, the following half-lives were reported in dependence in the pH value: $t(1/2)$ = 3.285 h (pH = 4.0), 1.59 h (pH = 7.0) and 0.03 h (pH = 9.0). Calculate the rate constants for acid catalyzed (k_a), neutral (k_n) and base catalyzed (k_b) hydrolysis.

(9) Why does a high log K_{OW} indicate, that a substance may be bioaccumulative? Is there a need for checking the BCF by exposing fish?

(10) What does the assessment of ecotoxicity aim at, and what is the problem of ecotoxicity assessment?

(11) In testing for aquatic ecotoxicity, why are tests with fish evaluated on the basis of an LC$_{50}$, tests with daphnia an EC$_{50}$ and tests with algae an IC$_{50}$?

(12) Which different environmental compartments can be addressed by ecotoxicity testing?

(13) What is the purpose of investigating whether a substance can inhibit activated sludge, and which kind of organisms are tested?

(14) What is meant with PNEC, how is it derived?

(15) What is the minimum data requirement to derive a PNEC for freshwater organisms?

(16) For a substance, the acute toxicity data for freshwater organisms are LC_{50}/fish/96 h = 2.0 mg/L, EC_{50}/daphnia/48 h = 12 mg/L and IC_{50}/algae/72 h = 8 mg/L. Further, the IC_{10}/algae/72 h is 2.0 mg/L. If one further test shall be performed to arrive at a lower assessment factor, which species should be selected?

5 Dose–response in toxicology and ecotoxicology

5.1 Introduction

Tests in toxicology try to find out the response of organisms to exposure to a substance. According to the "old" paradigm of Paracelsus, "the dose decides whether a substance is non-toxic," the effect provoked by a substance is dependent on the intrinsic property "toxicity" and the exposure:

$$\text{Effect} = \text{Toxicity} \times \text{Exposure}$$

Toxicity is an intrinsic property of the substance like vapor pressure, melting point and so on. Exposure is dictated by physical–chemical properties of the substance and handling conditions. It describes how far the sensitive organ system comes into direct contact with the toxin. For a corrosive acid, the sensitive target system is any biological tissue, and in first instance unprotected skin. The intrinsic property "corrosion" may generate damage only if there is direct contact between the acid and skin for a sufficient time. Another substance may be toxic if swallowed, but "only" harmful by skin contact due to the fact that the target organ, that is, the liver, receives a higher load of the substance by swallowing than by skin contact.

Increase of exposure is achieved by extending the contact time and the concentration of the substance at the target tissue. In (eco)toxicology testing, the exposure time is standardized according to the type of study (acute, subacute, subchronic or chronic), and different dosages are given to the organism under scrutiny.

Today, computer programs are available for free for nonlinear curve analysis, and these programs deliver the parameters of the functions, averages, standard deviations and confidence intervals [38, 39]. Because of this fact, the following sections about the "old" tools of logit analysis or probit analysis seem to be little bit overdone and outdated. However, in lab courses, acute ecotoxicity tests with invertebrates are easy to be performed. Analyzing the data statistically in a more "manual" way helps to solidify the acquired knowledge and to generate a feeling for the performance of toxicity tests. The chemical engineer will remember the technique of data analysis by linear regression that is underlying the logit analysis.

5.2 Response to exposure: dichotomous, quantal and continuous outcomes

The response to the exposure may be dichotomous, quantal or continuous:
- Dichotomous response: yes/no; dead/alive; tumor/no tumor
- Quantal response: no, low, medium, high and very high damage
- Continuous response: increase in blood pressure, body weight depression

https://doi.org/10.1515/9783110618952-005

At low doses, there may be no response of the exposed population. The highest dose tested that does not trigger an adverse response is the no observed adverse effect level (NOAEL); the lowest dose tested, that provokes an adverse reaction, is the lowest observed adverse effect level (LOAEL). To define what is adverse has to be decided by toxicologists, pathologists and/or veterinarians. Just as an example, a dose that increases the liver weight in exposed rats by 10% although statistically significant is not necessarily adverse if the behavior and performance of the rats is not changed, and other effects are not detectable. However, if fatty deposits are observable in the livers of the rats, this is adverse as this finally may end in liver function failure. As another example, at dose 1 the blood clotting time may by slightly extended, but not statistically significant to the unexposed control group, and this is probably not adverse. At dose 2, the difference to the control group is statistically significant, and there are indications of intestinal hemorrhage (bleeding): this is adverse. At increasing dose, the damage increases, and dose–response curves may be drawn. On the ordinate, for dichotomous data the % response is plotted, and for continuous data the difference to the control group may be sketched (Fig. 77).

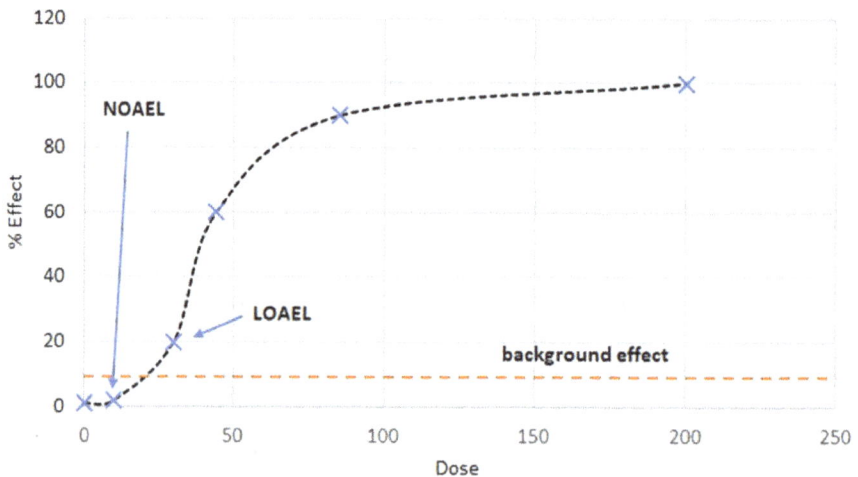

Fig. 77: Illustrative dose-response graph. Crosses represent the dosages tested and the% effect triggered, for example, mortality. For the species tested, 10% effect is within the normal range that can be observed without any external influence (summary of experience over many tests). As a result, the dose at the value 10 is not adverse (not damaging) and is, therefore, the NOAEL. The dose at 30, in contrast, causes damage above background; this is LOAEL.

5.3 Dose–response for dichotomous data

Tests for mortality were most likely the first that were done in (eco)toxicological testing. A group of test organisms is exposed to different dosages or concentrations of the test substance, and after a certain time period the dead individuals were counted. As a working example, see the following data for acute toxicity testing with fish (Tab. 15):

Tab. 15: Data from a fish acute toxicity test.

Dose [mg/L]	Number exposed	Number dead	% mortality
0.107	10	0	0
0.222	10	1	10
0.444	10	6	60
0,849	10	9	90
1.98	10	10	100

To find values for 50% mortality or 10% mortality, you may try to read it from the graph, but this is only a crude estimate (Fig. 78). Traditionally, people looked for a way to transform data in such a way that a plot delivered a straight line where the intercept and the slope could be derived as substance specific parameters. This was an appropriate and feasible way for data interpretation in terms of averages and standard deviations, before nonlinear data analysis became an easy task to be run by programs on readily accessible computers. Two of the "traditional" methods for the linearization of data for regression analysis are the logit analysis and the probit analysis. Both methods are demonstrated in the following section.

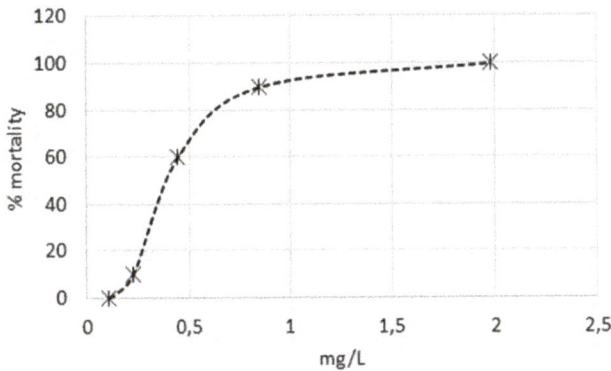

Fig. 78: % mortality against dose in an acute fish toxicity test.

5.3.1 The logit transformation

To transform quantal data into a linear equation of the form, the logit transformation allows to draw a straight line and derive the slope and intercept. The concentration is taken as the natural logarithm, LN, and the response is translated into the so-called logit:

$$\text{logit}(j) = \text{LN}\left\{\frac{d_j + w_j}{n_j - d_j + w_j}\right\}$$

where d_j = number of dead fish at dose j, n_j = number or exposed fish at dose j and w_j = weight of the dose group j:

$$w_j = \frac{n_j}{\sum n_j}$$

Adding w_j causes a small distortion of the data but allows to include the two nearest data points with $d = 0$ and $d = n$, each, which would otherwise deliver undefined expressions for the logit calculation. That is, you can make use of the data at $c = 0.107$ mg/L and 1.98 mg/L; imagine there would be another data point, say $c = 1.5$ mg/L, which would cause 100% mortality, that is, 10 dead fish. In this case, the data point $c = 1.5$ mg/L would be included, but the data point for $c = 1.98$ mg/L would be discarded. If you have several data points with a probability $P = 0\%$ or $P = 100\%$, only make use of the highest dose with $P = 0\%$ and the lowest dose with $P = 100\%$.

In case there are sufficiently data available so that data points with $d = 0$ and $d = n$ are not needed, the logit can be expressed in terms of likelihood P (probability for dead), and we can write

$$\text{logit}(j) = \text{LN}\left\{\frac{d_j}{n_j - d_j}\right\} = \text{LN}\left\{\frac{P}{1-P}\right\}$$

with $0 < d_j < n_j$ and $P = (d_j / n_j) \times 100$.

Before electronic calculators became available, people preferred to use the decadic logarithm instead of the natural logarithm; you can do so as well, but be careful to be consistent throughout; do not mix. The data given in Tab. 16 are transformed accordingly, and LN (concentration) is plotted against the logit (Fig. 79).

Tab. 16: Logit transformed data from Tab. 15.

Dose [mg/L]	LN(dose) (= x)	Number exposed	Number dead	W	Logit (= y)
0.107	−2.2349	10	0	0.2	−3.9318
0.222	−1.5051	10	1	0.2	−2.0369
0.444	−0.8119	10	6	0.2	0.3895
0,849	−0.1637	10	9	0.2	2.0369
1.98	0.6831	10	10	0.2	3.9183

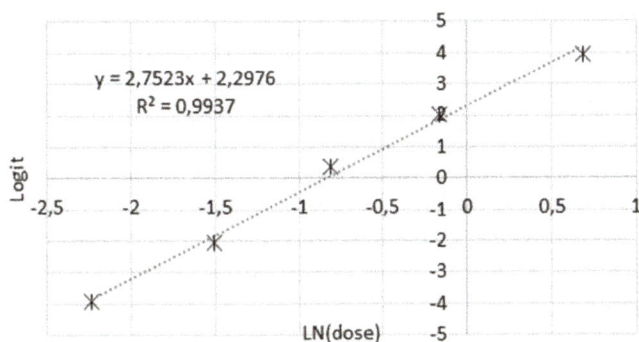

Fig. 79: Plot of LN(dose) against logit.

For acute toxicity, and for comparison of substances, the values with 50% effect are taken as benchmarks. These values are used to allocate substances to categories of acute toxicity. As the 95% confidence interval is smallest around the LC_{50} values, they are best suited to compare compounds to each other. At the LC_{50}, $P = 0.5$, and

$$\text{logit} = LN\{1\} = 0 = a + b \times LN(LC_{50})$$

$$LC_{50} = e^{\left(-\frac{a}{b}\right)}$$

and with $a = 2.2978$, $b = 2.7523$, the average lethal concentration is $LC_{50} = 0.43$ mg/L.

Note: if not all dose groups have the same number of exposed individuals, take the weighting factor into account to derive the mean values of the logit and LN(concentration), that is,

$$\mu \, (\text{mean}) = \sum w_j \times n_j$$

Linear regression analysis allows to calculate 95% confidence interval, which is smallest around the average of the respective variable (see textbooks of statistics). In acute toxicity testing – especially if different tests with the same substance and the same species show diverging results – the derivation of the predictive interval for the LC_{50} can be useful so diverging test results can be compared statistically. To do so, the standard deviation of the "residuals, E," $E = y - (a + bx)$ is required.

For the calculation of the 95% confidence interval, the following parameters need to be calculated (data see Tab. 17):

$$SD_E = \sqrt{\frac{\sum e^2}{\sum E^2}}$$

$$SD_p = SD_E \times \sqrt{1 + \frac{1}{n} + \frac{(LN(LC_{50}) - AV_x)^2}{(n-1) \times SD_x^2}}$$

$$\text{lower } 95\% \text{ logit}(LC_{50}) = \text{logit}(LC_{50}) - t_{n-2;0.975} \times SD_p$$

$$\text{upper } 95\% \text{ logit}(LC_{50}) = \text{logit}(LC_{50}) + t_{n-2;0.975} \times SD_p$$

Note: logit(LC_{50}) = 0. With $n = 5$ (number of dose groups that could be analyzed), the results are $SD_E = 0.2864$, $SD_p = 0.3137$. For $t_{n-2;0.975}$, data have to be taken from t-tables of statistics books. For $n = 3–6$, t-values are 12.706, 4.303, 3.183 and 2.7776, respectively. With $t_{3;0.975} = 3.183$, the 95% confidence-interval for the logit (LC_{50}) is [−0.9986; 0.9986], and the 95% confidence interval for the LC_{50} is exp [(logit(LC_{50})$_{\text{low/high}}$ − a)/b] = [0.303 mg/L; 0.625 mg/L].

Tab. 17: Logit data transformation for confidence intervals.

	LN(dose) (= x)	Logit (= y)	$E = y - (a + bx)$	E^2
	−2.2349	−3.9318	−0.0783	0.0061
	−1.5051	−2.0369	−0.1921	0.0369
	−0.8119	0.3895	0.3265	0.1066
	−0.1637	2.0369	0.1898	0.0360
	0.6831	3.9183	−0.2459	0.0605
Sum				0.2461
Average, AV	−0.8065	0.0779		
SD	1.1358	3.136		

5.3.2 Probit analysis

The probit analysis assumes that the response to an exposure follows the normal-distribution function. The logit analysis does not have this precondition and because of the biological background explained below might be preferred. However, in pre-electronic calculator times probit analysis was convenient, as you could solve the dose–response curve by looking up tabulated values (Tab. 18). The response P is transformed into the probit first by looking up the appropriate z-value in the normal distribution table (i.e., for 40% mortality, $P = 0.4$ and $z = -0.253$), and add the integer 5. That is, for 40% mortality, probit = 4.747 (Tab. 19). The probit is matched against ln (concentration), and the straight line allows to calculate slope and intercept (Fig. 80).

The probit is not defined for $P = 0$ and $P = 1.0$; for $P = 0$ and $P = 1$, use $P = 0.01$ and $P = 0.99$, respectively. At LC_{50}, we have probit = 5. Therefore:

$$\text{Probit} = 5 = 1.6483 \times LN(LC_{50}) + 6.3801 \Leftrightarrow LC_{50} = 0.43 \text{ mg/L}$$

Tab. 18: Probit transformation table.

P-value	0	0.01	0.02	0.03	0.04	0.05	0.06	0.07	0.08	0.09
0		2.6737	2.9463	3.1192	3.2493	3.3551	3.4452	3.5242	3.5949	3.6592
0.1	3.7184	3.7735	3.8250	3.8736	3.9197	3.9636	4.0055	4.0458	4.0846	4.1221
0.2	4.1584	4.1936	4.2278	4.2612	4.2937	4.3255	4.3567	4.3872	4.4172	4.4466
0.3	4.4756	4.5041	4.5323	4.5601	4.5875	4.6147	4.6415	4.6681	4.6945	4.7207
0.4	4.7467	4.7725	4.7981	4.8236	4.8490	4.8743	4.8996	4.9247	4.9498	4.9749
0.5	5.0000	5.0251	5.0502	5.0753	5.1004	5.1257	5.1510	5.1764	5.2019	5.2275
0.6	5.2533	5.2793	5.3055	5.3319	5.3585	5.3853	5.4125	5.4399	5.4677	5.4959
0.7	5.5244	5.5534	5.5828	5.6128	5.6433	5.6745	5.7063	5.7388	5.7722	5.8064
0.8	5.8416	5.8779	5.9154	5.9542	5.9945	6.0364	6.0803	6.1264	6.1750	6.2265
0.9	6.2816	6.3408	6.4051	6.4758	6.5548	6.6449	6.7507	6.8808	7.0537	7.3263

Tab. 19: Probit transformed data.

Dose [mg/L]	LN(dose)	Exposed	Dead	% Dead	P	Probit
0.107	−2.2349	10	0	0	0.01	2.6737
0.222	−1.5051	10	1	10	0.1	3.7184
0.444	−0.8119	10	6	60	0.6	5.2533
0.849	−0.1637	10	9	90	0.9	6.2818
1.98	0.6831	10	10	100	0.99	7.3263

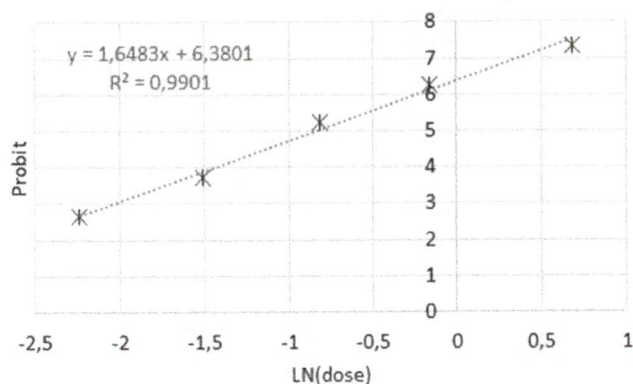

Fig. 80: Probit plot of the data.

The probit analysis delivers the same result as the logit analysis. Slight deviations may occur.

As a first rough estimate, the LC_{50} could be extrapolated from a straight line
drawn from the next LN(dose) value below the LC_{50} to the next LN(dose) value

above the LC_{50}; these two data points allow to set the slope and intercept of the strait line and to estimate the LC_{50} from this equation, % mortality = $A \times$ LN(dose) + B. In our case, the slope A is

$$A = \frac{60 - 10}{-0.8119 - (-1.5051)} = 72.129$$

If we take now the data for 60% mortality and introduce the value for A, the equation becomes

$$60 = 72.127 \times (-0.8119) + B \leftrightarrow B = 118.56$$

With this simple estimate, we would calculate for 50% mortality dose of 0.38 mg/L; in our example, this estimate is just a factor ≤ 1.2 different form the logit and probit results; this is not too bad and would not change classification and labelling for acute aquatic toxicity (see later sections).

5.3.3 Cancer data

Cancer is a disease with a high burden to affected persons and their relatives. Regulations addressing marketing and use of carcinogenic substances aim at tight control of the risk posed by these substances. A definition of risk and appropriate descriptors are required to quantify the risk and risk management measures. Several descriptors exist to describe the risk posed by carcinogens.

Relative risk (RR): this is the risk in relation to an unexposed population; divide probability of disease for exposed ($P1$) by that for nonexposed ($P0$):

$$RR = P\,(1)/P\,(0)$$

Excess risk (ER): describes additional cases due to exposure.

$$ER = P\,(1) - P\,(0)$$

Attributable risk (AR): describes the proportion of diseases that are attributable to the exposure.

$$AR = \{P\,(1) - P\,(0)\}/P\,(1)$$

Cancer prevalence: at a given time, you check a population for the presence of cancer. The result is called prevalence.

Cancer incidence: over an observation period, you count every new cancer case. This is set against exposure-man-years.

$$\text{Incidence} = \frac{\text{number of new cancer cases over } t \text{ years}}{\sum_0^t (\text{person} \times \text{exposure time})}$$

For cancer risk evaluation, the probability P for the incidence of cancer in dependence on exposure dose is modeled. There are several dose–response functions available, and only two shall be briefly presented here. For cancer probability, the dose and the time contribute to the manifestation of disease, so the likelihood of acquiring cancer can be described as

$$P(t, D) = 1 - \exp(-a \times D^b \times t^k)$$

with D = dose, t = time elapsed until the tumor is diagnosed and k = number of stages a cell has to pass before it is a malignant cancer cell (k = 3,. . ., 6).

A function can be based on the multistage cancer model, where P is modeled by a polynome. In the past, the polynome was limited to second degree, which could be handled without the use of computers, and parameters were found by quadratic regression:

$$P = 1 - \exp - (q_0 + q_1 \times D + q_2 \times D^2)$$

Alternatively:

$$P = \{1 + \exp - (q_0 + q_1 \times D + q_2 \times D^2)\}^{-1}$$

The factor q_0 stands for the background incidence of cancer.

The Weibull function is another way to describe dose–response for dichotomous data:

$$P = B + (1 - B) \times (1 - \exp[-a \times D^b])$$

B stands for the background incidence. If the background incidence is treated as a constant and P is the probability of acquiring cancer for those who otherwise would have been left out, the equation becomes simplified:

$$P = 1 - e^{-(a \times D^b)}$$

the probability of not getting cancer although being exposed is

$$1 - P = e^{-(a \times D^b)}$$

the ratio of these two probabilities, therefore, is

$$\frac{P}{1-P} = e^{(a \times D^b)} - 1 \Leftrightarrow$$

$$1 + \frac{P}{1-P} = \frac{1}{1-P} = e^{(a \times D^b)} \Leftrightarrow$$

$$LN\left\{\frac{1}{1-P}\right\} = a + b \times D$$

Plotting LN{1/(1–P)} against the dose D delivers "a" as intercept and "b" as slope. If the background B is maintained, a transformation resulting in an equation for a straight line delivers

$$LN\left\{\frac{1-B}{1-P}\right\} = a + b \times D \Leftrightarrow LN(1-P) = LN(1-B) - a - b \times D$$

The "classical" algorithm for cancer risk estimation based on animal models is the polynomial model cut at stage 2:

$$P = 1 - e^{\left(-a-b\times D-c\times D^2\right)} \text{ and } 1-P = e^{\left(-a-b\times D-c\times D^2\right)} . ->$$

$$\frac{1}{1-P} = e^{\left(a+b\times D+c\times D^2\right)} \Leftrightarrow LN\left\{\frac{1}{1-P}\right\} = a + b \times D + c \times D^2$$

This equation can be solved by quadratic regression.

The transformations shown so far allow to solve dose–response equations with the help of simple calculators. The United States Environmental Protection Agency (US EPA) holds available a free software program, "benchmark dose software" (BMDS) for performing dose–response analysis by nonlinear regression [38]. Another electronic tool for dose–response modeling of cancer data is the program PROAST, which is accessible for free [40].

Before calculation programs for non-linear dose-response fitting became readily available, the US EPA cancer risk analysis made use of the polynomial model. For the lowest dose applied in the animal experiment, the upper 95% confidence interval is chosen and extrapolated linearly to $D = 0$. From this linear straight line, the cancer risk for (very) low dose exposure is extrapolated. As an example, take data from a cancer study with oxy-4,4'-dianiline (ODA) in rats (Tab. 20).

Tab. 20: Thyroid gland tumors in male rats exposed to 4,4'-oxydianiline via feed.

Dose [ppm]	Animals exposed	Animals with tumors	P	LN(1/(1–P))
0	46	1	0.022	0.022
150	47	1	0.022	0.022
300	48	8	0.167	0.183
800	50	13	0.26	0.301

Analysis of these data with Microsoft Excel delivers the following graphics for a polynomial second degree (Fig. 81):

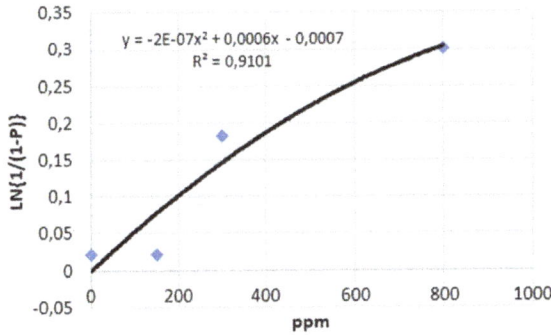

Fig. 81: Modeling thyroid tumor data in male rats as polynome of second degree.

The result shown in Fig. 81 needs to be regarded with care: the factor $-2E\text{-}07$ implies that if the dose is large enough, the substance will be cancer protective. This is nonsense (unless you regard preterm death caused by the substance as cancer-protective)! The BMDS version 3.1 makes use of several quantal dose–response models and performs a quality check automatically. As a result, model outputs may be rated as "viable – recommended," "viable – alternate" or "questionable." Actually, the polynomial second degree was rated as questionable and the linear model was rated as viable. The graphical output is shown in Fig. 82. The benchmark dose (BMD_{10}) causing a 10% extra risk is 288 ppm and its lower 95% confidence interval ($BMDL_{10}$) is 198 ppm. In the USA, the $BMDL_{10}$ is the starting point for the risk evaluation. From this value, a linear line is drawn to the data origin. This straight line has a slope factor, which can be used to calculate the excess cancer risk.

$$ER = (\text{slope} - \text{factor}) \times \text{dose}$$

The BMD_{10} is a more robust estimate than the $BMDL_{10}$, but the latter is better suited to address the power of the study. For example, old cancer studies made use of only 20–25 animals per gender and dose. The BMD_{10} may not necessarily change but due to the reduced number of exposed animals (and as a consequence reduced power of the test) the $BMDL_{10}$ will be lower.

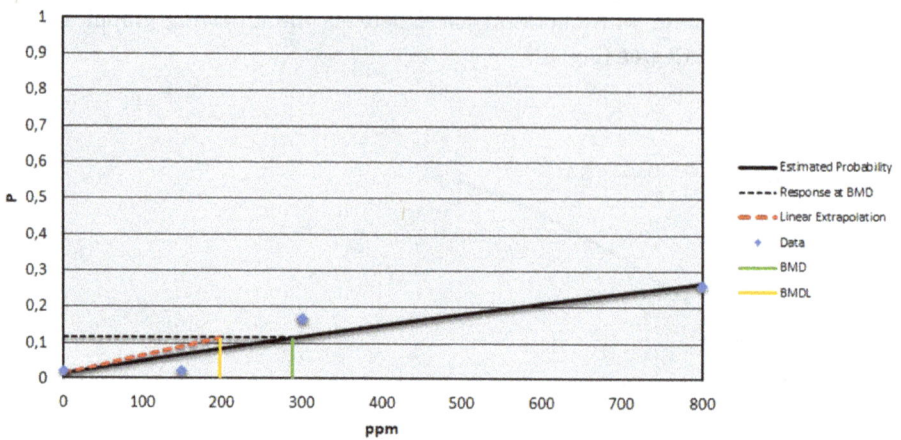

Fig. 82: Graphical output of the benchmark dose software v3.1 for oxydianiline data (Tab. 20), polynomial model first degree (= linear model).

In the European Union, another concept for cancer risk assessment is preferred: the T_{25} concept and its application will be demonstrated in the following section.

5.3.4 T_{25} concept for carcinogenic risk evaluation

The T_{25} concept runs as follows. From a given data set, you identify the lowest dose that induces a significant increase in cancer response, where significance is checked with the Fisher's test (look up textbooks on statistics). In the example with ODA in rats, 300 ppm is the dose that causes a significant increase in cancer incidence in the rat (error: $P < 0.05$). The net cancer incidence at 300 ppm is

$$I_{300} = 18.3\% - 2.2\% = 16.1\%$$

T_{25} is the dose that causes a net cancer incidence of 25%, and it is assumed that the ratio between dose and incidence is linear:

$$\frac{300 \text{ ppm}}{16.1\%} = \frac{T_{25}}{25\%} \leftrightarrow T_{25} = 466 \text{ ppm}$$

To evaluate a tolerable dose for human beings it is important to agree on a tolerable cancer risk. In many regulations, the tolerable risk is one additional case in 1 million people exposed over lifetime. As the T_{25} was derived from feeding data, an allometric scaling to man is not required (for dosages given as mg/kg body weight, a scaling factor of 4 would be required for rat to man transformation; that is $T_{25,\text{man}} = T_{25,\text{rat}}/4$). A dose of 446 ppm in food would cause 250,000 additional cancer cases in 1 million exposed people. For the tolerable risk, the tolerable concentration in food is

$$\frac{\text{Tolerable dose}}{1} = \frac{T_{25}}{250,000} = \frac{446\,\text{ppm}}{250,000} \Leftrightarrow$$

Tolerable concentration = $1.78E{-}03$ ppm = 1.78 ppb.

5.4 Biological background for dose–response models

Like in reaction kinetics in chemistry, the dose–response models showing the best fit to experimental data do not necessarily guide us to an unequivocal understanding of molecular processes driving the outcomes. However, they may provide an idea of what might happen at molecular level and can help to formulate theories which then can be checked in experiments. For the biological background for dose–response, a few ideas will be presented in the following section.

5.4.1 Receptor-transmitted toxicity

For this model, the effect is triggered by the occupation of a receptor on the surface of a cell, which triggers the cell to do something or stops the cell from doing something (p.e., blocking of nerve-cell transduction by neurotoxins) or blocks an important molecule (p.e., Fe^{2+} in hemoglobin by CO).

A ligand L binds to a receptor, R, and the dissociation constant is defined as

$$K_D = \frac{[R] \times [L]}{[RL]}$$

the total concentration of receptors is $[R]_t = [RL] + [R]$; this is introduced into the equation

$$K_D = \frac{([R]_t - [RL]) \times [L]}{[RL]} \Leftrightarrow \frac{K_D}{[L]} = \frac{[R]_t - [RL]}{[RL]} = \frac{[R]_t}{[RL]} - 1 \Leftrightarrow P = \frac{[RL]}{[R]_t} = \frac{[L]}{K_D + [L]}$$

P is the occupation of receptors and has values $P = 0,\ldots,1$.

Fig. 83 shows the plot of P against $[L]$ for $K_D = 5$. Maximum response may be achieved at comparatively high concentrations only, and the plot is not very easy to read, especially finding the P_{50} value (50% receptor occupation). A plot against $\log[L]$ changes the picture (Fig. 84). As P is the ratio of occupied receptors divided by the total number of receptors, we can write

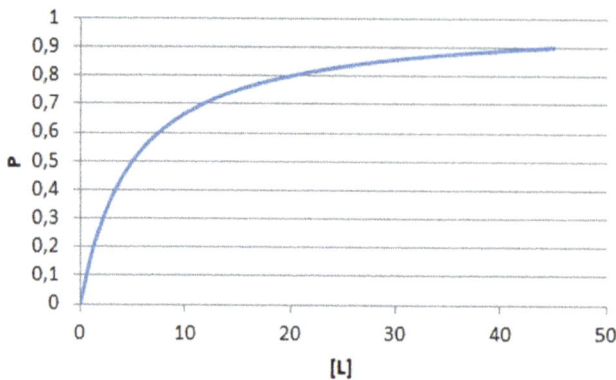

Fig. 83: Receptor occupation P in dependence on ligand concentration with $K_D = 5$.

$$P = \frac{[RL]}{[R]_t} = \frac{[L]}{K_D + [L]}, \quad -> 1 - P = 1 - \frac{[L]}{K_D + [L]} = \frac{K_D + [L] - [L]}{K_D + [L]} = \frac{K_D}{K_D + [L]}; => $$

$$\frac{P}{1-P} = \frac{[L]}{K_D} => LN\left(\frac{P}{1-P}\right) = LN[L] - LN(K_D)$$

if the reaction between ligand and receptor was a-order in L, the equation would read

$$LN\left(\frac{P}{1-P}\right) = a \times LN[L] - LN(K_D)$$

The similarity to the logit is obvious!

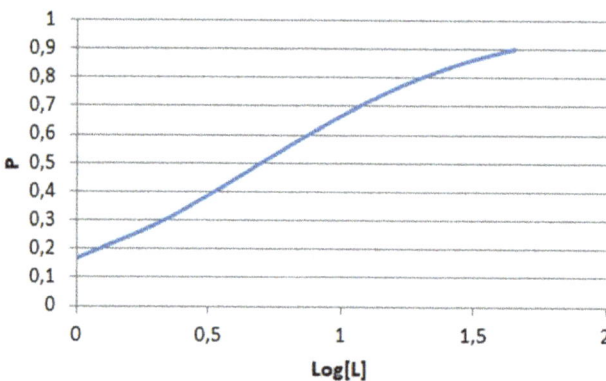

Fig. 84: Receptor occupation P in dependence on the logarithm of the ligand concentration.

Now assume that you have a concentration of the ligand that results in 50% occupation of the receptor:

$$P = 0.5 = \frac{[L]_{50}}{K_D + [L]_{50}} \Leftrightarrow K_D = [L]_{50}$$

Therefore,

$$P = \frac{[L]}{[L]_{50} + [L]}$$

If the receptor has more than one binding site for the ligand or a functional protein can catch more than one molecule, the equation and plot changes. An example is hemoglobin (R), which can bind four molecules of oxygen (L) before it is completely occupied:

$$RL_4 = RL_3 + L; K_1 = \frac{RL_3 \times L}{RL_4}; RL_3 = RL_2 + L; K_2 = \frac{RL_2 \times L}{RL_3}; RL_2 = RL + L; K_3 = \frac{RL \times L}{RL_2}$$

$$RL = R + L; K_4 = \frac{R \times L}{RL}; ->$$

Under assumption that all K's have the same value:

$$K_1 \times K_2 \times K_3 \times K_4 = \frac{RL_3 \times L}{RL_4} \times \frac{RL_2 \times L}{RL_3} \times \frac{RL \times L}{RL_2} \times \frac{R \times L}{RL} \Leftrightarrow K^4 = \frac{R}{RL_4} \times L^4$$

or more general:

$$K^n = \frac{R}{RL_n} * L^n$$

This equation is named Hill equation after the publication of Archibald Hill. Figures 85 and 86 show the change of the curve in dependence on the Hill parameters K and n.

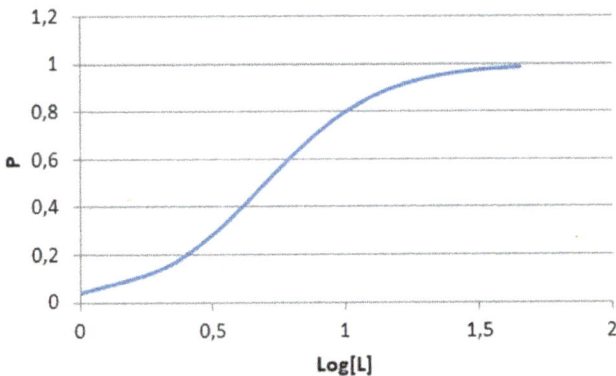

Fig. 85: Hill plot for $K_D = 5$ and $n = 2$.

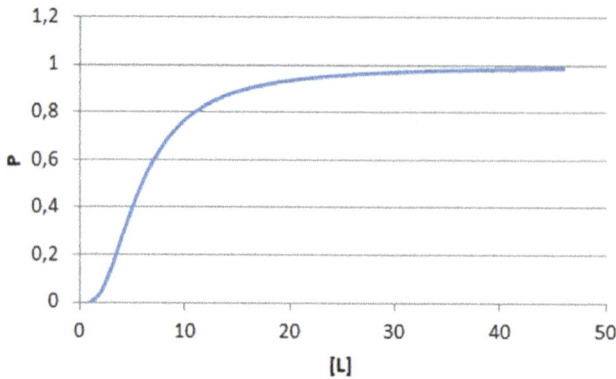

Fig. 86: Hill plot for $K_D = 5$ and $n = 4$.

Usually, an effect increases with increasing ligand (or toxin) concentration. In the debate of adequate NOAEL definition for endocrine-active substances, ideas of "nonlinear dose response" were presented. See, as two of many examples, the publication of Conolly and Lutz from 2004 [41] and the review from Vandenberg et al. from 2012 [42]. Examples for molecular mechanisms may be that an overload of ligand does deactivate the receptor; ideas behind this concept are that either the ligand bridging two receptor causes an activation and that more free ligands are going to split these bridges, or the ligand may act in a way similar to "substrate inhibition" in enzyme catalysis, that is, a non-competitive deactivation. The algebraic expression for the latter is

$$R + L = RL; \; K_1 = \frac{[RL]}{[R] \times [L]}$$

$$RL + L = RL_2; \; K_2 = \frac{[RL_2]}{[RL] \times [L]}$$

$$P = \frac{[RL]}{[R] + [RL] + [RL_2]} = \frac{K_1 \times [L]}{1 + K_1 \times [L] + K_1 \times K_2 \times [L]^2}$$

Here, only RL is the activated form of the receptor. If $K_1 = 0.1$ and $K_2 = 1.0$, a bell-shape curve results as shown in Fig. 87.

As another example for a bell-shaped curve, take the bacterial reverse mutation assay (Ames test) where increasing concentrations of a substance first increase the number of revertants, but higher concentrations exert cytotoxicity which results in an observable decrease in revertants.

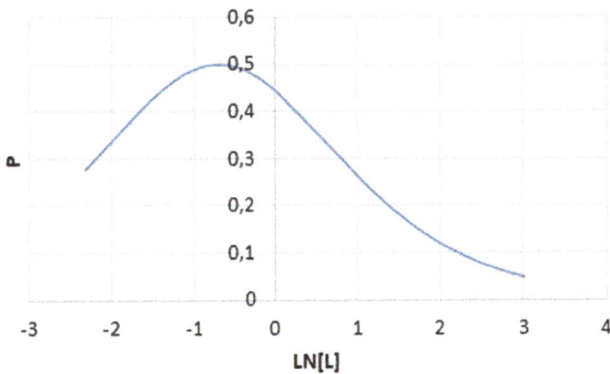

Fig. 87: Nonlinear dose–response curve.

5.4.2 E_{max} model

In this model, it is assumed that a substance may provoke an effect but the response cannot be higher than E_{max}. It is comparable to the Michaelis–Menten kinetics of enzymes:

$$E = \frac{E_{max} \times [X]}{E_{50} + [X]} \Leftrightarrow \frac{E}{E_{max}} = \frac{[X]}{E_{50} + [X]}$$

This expression is – on a mathematical basis – equivalent to the previous shown receptor model. It implies that such curves as shown above do not necessarily require the involvement of a receptor.

5.4.3 A few thoughts on "thresholds"

By definition, for stochastic processes like DNA damage there is no theoretical threshold, every DNA-reactive molecule can hit and change the DNA. However, whether this impact is turned into an adverse outcome depends on several factors: the hit has to be in a relevant region of the DNA, has to change the genetic code in such a way it is not a silent mutation and needs to escape repair processes. Therefore, although substances do not have a theoretical threshold, in toxicology studies they may show a practical threshold. In general, any efficient repair-mechanism of the organisms to impacts caused by external molecules will end up in an at least practical threshold. Only if the frequency of damages overcomes the repair process capacity, an adverse effect prevails.

Receptor-transmitted toxicity is a model that leads to the assumption of theoretical thresholds. For example, it is conceivable that a certain minimum of receptors must be occupied by their agonists before an electric pulse is triggered in a nerve cell. The model of a nerve pulse resembles a threshold mechanism without differentiation: the pulse is either released or not. How would a dose–response curve look like where a certain minimum dose is required to trigger an effect at all, but then the effect increases with increasing dose? Based on a publication of Spassova [43], a dose–response function may be extended by a factor $\{(x - x_0) - ((x - x_0)^2)^{0.5}\}/(2 \times ((x - x_0)^2)^{0.5})$, where x is the dose applied, x_0 is the threshold dose and A the background of the effect in a control group:

$$y = A + \frac{(x - x_0) + \sqrt{(x - x_0)^2}}{2 \times \sqrt{(x - x_0)^2}} \times f(x)$$

where $f(x)$ is any suitable dose–response function.

5.5 Questions

(1) What is meant by dose–response, and what is normally to be expected in toxicology testing?
(2) What is meant by dichotomous, quantal and continuous response?
(3) Please see the following data for a fish acute toxicity test with a liquid substance (Tab. 21). Find the LC_{50} by the logit and by the probit method.

Tab. 21: Acute fish toxicity data.

Concentration [mg/L]	Number of exposed fish	Number of dead fish
3.2	10	0
7	10	4
9	10	8
10	10	10

(4) You may have realized that in task 3, the value for 10 mg/L seems to be a little "out of range." If now it is considered that the water solubility of the compound tested is about 8–9 mg/L, what idea do you have to explain the logit and probit graphs?

(5) From an acute toxicity study with daphnia, you have the following data (Tab. 22); find the EC_{50} by the logit method and calculate its 95% confidence interval.

Tab. 22: Acute daphnia toxicity data.

Dose [ppm]	Number of daphnia exposed	Number of immobile daphnia
0.4	20	2
0.8	20	7
1.2	20	14
2.4	20	19

(6) The cancer incidence in a control population shall be 2%. In an exposed population, the incidence shall be 10%. Calculate the RR, the AR and the ER and explain their meaning.

(7) In a cancer study, the lowest dose that caused as significant increase in cancer in exposed rats shall be 10 mg/kg b.w., and 32% of rats show tumors; in the control group, the cancer incidence was 2% only. Calculate the T_{25} and the maximum exposure level for an excess risk of 1:1 million for human beings.

(8) Derive an equation for receptor occupation (P) and the logit if the receptor binds two ligands: $R + 2L = RL_2$.

6 Classification, labeling and packaging of chemicals

The classification of substances can result in a label carrying pictograms and information to inform the user about hazards, and the label will provide basic hints how to prevent risk. This is one of the main purposes of the classification and labeling (C&L) regulations. In addition to that, the classification may result in restricted access to, and use of a substance. As this may have a high impact on the marketing of a substance, common rules on how to perform C&L are required.

Basis for the European Union Regulation (EC) No. 1272/2008 and its amendments, the latest being COMMISSION REGULATION (EU) 2019/521 [44], is the implementation of the United Nations Globally Harmonized System (GHS) for the C&L of dangerous substances and mixtures [45]. It defines hazard classes, describes how test results lead to C&L, defines standards for labels and allocates responsibilities for C&L. The regulation is subdivided in several Titles and Annexes. Annex VI, part 3 lists substances with an EU harmonized C&L. First read chapters "Introduction into Ecotoxicology and Environmental Behavior" (Chapter 4), "Basics in Toxicology" (Chapter 3) and "Physical and Chemical Properties of Substances" (Chapter 2) before you go ahead with this chapter.

A guidance document concerning the labeling, and another document for the application of classification, labeling and packaging criteria is available on the European Chemicals Agency (ECHA) website [46, 47].

6.1 Title I: general issues

Title I of Regulation (EU) No. 1272/2008 describes the scope of the regulation and some general aspects. The following products are *exempted from the regulation 1272/2008*, most of them due to specific legislations:
- Radioactive substances
  - They are subject to directive 2013/59/EC.
- Products under custom supervision
  - They are not yet on the EU market.
- Nonisolated intermediates
  - There is no chance to investigate their properties.
- Substances in scientific research and development, provided strictly controlled handling is assured.
- Transport of dangerous goods
  - For transport, these are classified and labeled according to dangerous goods regulations. For road transport, for example, see [48].

https://doi.org/10.1515/9783110618952-006

- Pharmaceuticals in their finished form, ready for consumer use
  - They fall under directives 2001/82/EC and 2001/83/EC.
- Medical devices
  - Directives 93/42/EC and 98/78/EC and Regulation (EU) No. 2017/745 are applicable for medical devices.
- Cosmetic products in their finished form, ready for consumer use.
  - They fall under Regulation (EU) No. 1223/2009.
- Food- and feed-stuff in their finished form, ready for consumer use
  - These products are subject to regulation 178/2002/EC.

In article 2, some useful definitions can be found. For example, a *polymer* is a substance characterized by a sequence of one or more monomer types; it consists of a mixture where the components differ mainly due to the number of monomer units, and the mixture shows a molecular weight distribution. In this mixture, the weight majority is held by a molecule of at least three covalently bound monomer units in one chain, covalently bound to an initiator or another monomer unit. This majority weight molecule makes up less than 50% of the whole mixture.

In general, manufacturers, importers, traders and users have to classify and label a substance or mixture before it is placed on the market.

The harmonized C&L in Annex VI part 3 is binding. However, if hazards are known which are not covered by Annex VI, part 3, the C&L shall be extended accordingly.

6.2 Title II: hazard classification

Manufacturers, importers and downstream users are obliged to find out whether their substance/mixture is hazardous. This can be done by consulting information made available from suppliers, and also by looking up literature and databases. Two very helpful databases shall be mentioned here (but observe their disclaimers!): the Hazardous Substances Data Bank (HSDB) [49], the German GESTIS database (Gefahrstoff Informations-System) that can be accessed in German and in English [50], and the website of the ECHA [51]. In these databases, you can search the product by name or chemical abstracts system number (CAS-No.).

Depending on the market volume, tests according to article 8 have to be performed to check for hazardous properties. The regulation 1907/2006/EC on the Registration, Evaluation and Authorization of Chemicals (REACh) defines tonnage volumes which request the generation of data against chemical safety, toxicology and ecotoxicology.

Experience from occupational hygiene (exposed workforces: incidents, medical surveillance) and epidemiological data provide an important insight into hazardous properties of substances and need to be incorporated into the substance evaluation.

For mixtures, the regulation provides calculation methods for toxicological and ecotoxicological hazards. Calculation methods are attributable to the fact that it is physically not possible to test all mixtures, and the main target is to avoid unnecessary animal testing. Nevertheless, it may happen that for certain applications, mixtures need to be tested in animal tests to be able to define the hazards and risks posed by the mixture. However, test results cannot change the mixture calculation results for the endpoints carcinogenicity, mutagenicity and reproduction toxicity (CMR).[1]

Whenever possible and acceptable in terms of risk, animal tests must be avoided. Tests with primates are prohibited for the purposes of this regulation. Tests with volunteers shall not be performed to disqualify animal test data. However, it may happen that animal test data leave contentious points for discussion in terms of transferability to human beings. If, for example, it is unclear whether an occupational exposure limit (OEL) based on animal data is sufficiently protective, tests with volunteers may be proposed. An ethics committee needs to endorse these tests which are run under close medical surveillance. One example from the past is aniline. One of the most sensitive acute effects caused by aniline in human beings is the formation of methemoglobin; for this endpoint, man is more susceptible than the rat. In addition, it was uncertain in how far dermal uptake via the air adds to the body burden caused by inhalation. Based on studies with exposed volunteers, the German MAK value (Maximale Arbeitsplatz Konzentration; German for Occupational Exposure Limit) of 2 ppm was maintained.[2]

If tests are to be run, they have to comply with standards referred to in regulation 1907/2006/EC. That is, performing labs need to demonstrate their competence (staff, equipment, experience, etc.), and test guidelines as laid down in Regulation (EC) No. 440/2008 have to be followed; these test guidelines are nearly identical to the OECD test guidelines for the testing of chemicals [13]. Animal tests need to be run under "Good Laboratory Practice" (GLP); labs need to be certified for running such tests, only animals from certified breeders can be used for the tests. Tests shall be run under consideration of the physical form of the substance as it is brought on the market and the form it is likely to be used by downstream users.

For the calculation of toxicological and ecotoxicological hazards of mixtures, for acute toxicity, irritation/corrosion and acute aquatic toxicity, ingredients are added up for every toxicological endpoint and exposure route separately. Every ingredient shall be respected with its specific concentration limit (SCL). If SCLs are not allocated, generic concentration limits (GCL) have to be used as given in the tables of Annex 1. Cut off levels of components in the mixture for incorporation into the calculation methods are as follows:
- the lowest SCL of the substance; if the substance has no SCL, then
  - for substances of acute toxicity category: 1–3: 0.1%
  - for sensitizers category 1 or 1A: 0.1%

---

1 A reason is that these endpoints are regarded as specific critical, and the sensitivity of the test methods is limited while a huge protection level is desired. A dilution would just reduce the detection sensitivity, but not inactivate the mutagenic principle.
2 See: https://onlinelibrary.wiley.com/doi/pdf/10.1002/3527600418.mb5263d0064.

- for sensitizers category 1B: 1.0%
- for substances of acute toxicity category 4: 1.0%
- for substances of aquatic chronic toxicity category 2–4: 1.0%
- for substances causing skin and eye irritation or corrosion: 1.0%
- for substances with aspiration risk: 1.0%
- for carcinogens and mutagens category 1: 0.1%
- for carcinogens and mutagens category 2: 1.0%
- for substances toxic to reproduction category 1: 0.3%
- for substances toxic to reproduction category 2: 3.0%
- for substances of aquatic acute and chronic toxicity category 1: (0.1%)/ (M-Factor)
- substances hazardous to the ozone layer category 1: 0.1%

The multiplication factor (M-factor) is introduced for substances with very high aquatic toxicity, which means very low LC_{50} levels. An LC_{50} below 1 mg/L would result in an acute aquatic toxicity classification category 1; if the LC_{50} is below 0.1 mg/L, the substance is ten times more toxic; therefore, the M-factor is 10; for an LC_{50} below 0.01 mg/L, M = 100 and so on.

Deviation from the calculated C&L of mixtures is possible if adequate and reliable data support this. For example, a component of the mixture was classified as very toxic by inhalation as mist, but the high viscosity of the mixture does no longer allow the generation of respirable mists, the mixture may no longer be classified according to acute inhalation toxicity. Alternatively, hazardous components in the mixture may be biologically no longer available (additive in polymer matrix), or synergistic or antagonistic effects could be demonstrated.

A mixture need not be classified as explosive, oxidizing or flammable
- if none of the components shows these properties.
- if a mixture is supplied in a ready-to-use aerosol dispenser compliant with directive 75/324/EEC.

The manufacturers, importers and users of a substance have to update the information about hazardous properties of their substances on a regular basis, as new data may become available in the scope of national and international chemical programs. If necessary, the C&L has to be updated accordingly.

6.3 Title III: hazard communication

The first information on hazardous properties made visible for the user is the label on the packaging. The label covers the following, compulsory pieces of information, in the language of the region where the product is placed on the market:
- name, full address and telephone number of the supplier.

- packages for the general public: nominal quantity (unless this is given some-where else on the package).
- product identifiers
  - name of the substance; it has to be identical to the name given in the safety data sheet.
  - if the substance is listed in Annex VI, part 3, that name has to be used, or, if not listed in part 3, the substance name given somewhere else in Annex VI; or, if the substance is not listed in Annex VI at all, the CAS-number with IUPAC name or another, internationally used name, or – in case of absence of a CAS-number – the IUPAC name shall be printed on the label.
  - for mixtures, the trade name or the purpose of the mixture has to be printed on the label and has to be identical to the name given in the safety data sheet.
- identity of all substances of a mixture that
  - contribute to C&L as acute toxic, corrosive, serious eye damage, carcinogenic, mutagenic, toxic to reproduction, sensitizers, specific target organ toxicity, as-piration hazard,
    - a total of four substances may be sufficient, if all hazard categories are covered.
- if applicable, hazard pictograms with signal words
- if applicable, hazard statements (H-statements);
- if the mixture is hazadous and sold to professional or private users, the Unique Formula Identifier (UFI) has to be added. This is an alpha-numric code which puts poison centers in the position to get access to the precise composition of the mixture.
- if applicable, precautionary statements (P-statements)
  - a total of six should be sufficient.
- if applicable, additional information
  - p.e., EU-hazard phrases; name and identity of sensitizers being present be-tween 10% and < 100% of the respective SCL or GCL; "Safety data sheet available on request"; "For professional use only."

H- and P-statements are organized as
- H200-series: physical–chemical hazards.
- H300-series: toxicological hazards.
- H400-series: environmental hazards.
- P100-series: general recommendations (p. e. P102: keep away from children).
- P200-series: prevention (e.g., P280: Wear protective gloves/protective clothing/ eye protection/face protection).
- P300-series: remediation measures (p. e. P330: rinse mouth with water).
- P400-series: storage (p. e. P405: Store locked up) .
- P500-series: disposal.

On request and for a defined fee, alternative chemical names may be used on the label to keep the product identity confidential, if the substance has no OEL, is not classified as carcinogen, mutagen or toxic to reproduction, not corrosive, not causing serious eye damage, not acute toxic categories 1–3, no specific target organ toxicity (STOT) category 1 and not toxic to aquatic organisms category 1 (acute/chronic) or 2 (chronic). Statements like "nontoxic" and "ecological," which provoke the impression that the hazardous substance/mixture is not critical, are forbidden.

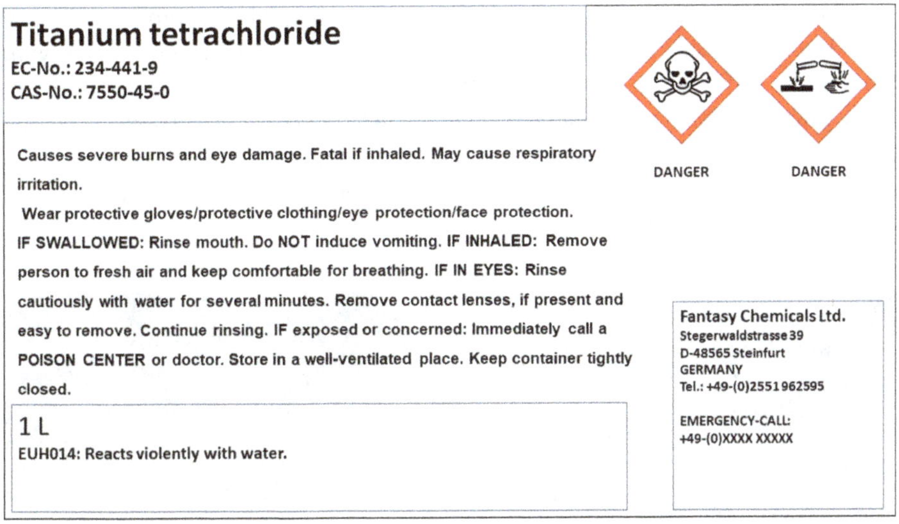

Titanium tetrachloride
EC-No.: 234-441-9
CAS-No.: 7550-45-0

Causes severe burns and eye damage. Fatal if inhaled. May cause respiratory irritation.

DANGER   DANGER

Wear protective gloves/protective clothing/eye protection/face protection.
IF SWALLOWED: Rinse mouth. Do NOT induce vomiting. IF INHALED: Remove person to fresh air and keep comfortable for breathing. IF IN EYES: Rinse cautiously with water for several minutes. Remove contact lenses, if present and easy to remove. Continue rinsing. IF exposed or concerned: Immediately call a POISON CENTER or doctor. Store in a well-ventilated place. Keep container tightly closed.

1 L
EUH014: Reacts violently with water.

Fantasy Chemicals Ltd.
Stegerwaldstrasse 39
D-48565 Steinfurt
GERMANY
Tel.: +49-(0)2551 962595

EMERGENCY-CALL:
+49-(0)XXXX XXXXX

Fig. 88: Example label for titanium tetrachloride.

Fig. 89: GHS-pictograms (symbols) 1–9.

In case of many labels, GHS 02 and 03 may be dropped if the label GHS 01 has to be applied. If GHS 05 or 06 or GHS 08 has to be used, GHS 07 must not appear if it was warning against the same hazard category.

Labels should be fixed on the packaging in such a way that they are inedible and clearly legible if the package is stored in the appropriate position. The label has to have a specific minimum size, depending on the size of the packaging, and symbols need to have at least 10% of the label size. For containments up to 3 L, the label should have it lest a size of 52 mm × 74 mm (if feasible), and the pictogram shall have a minimum size of 10 mm × 10 mm.

In case of package inside a package, each containment has to have the appropriate labeling. Only in case of dangerous goods, it is sufficient that the outer,

transport-package carries the labeling according to the dangerous goods regulations [48], provided the inner packaging carries the label according to Regulation
(EC) No. 1272/2008.

A hazardous product may be packed in such a small entity, that a full label is
no longer applicable. At minimum, the smallest package shall carry the product
identifier, the pictogram and the telephone number of the supplier. The outer packaging shall carry the full label. In case of difficulties in applying the full label, use
shall be made of tie-out tags or foldout labels.

Water-soluble packaging for single use may be left unlabeled, provided the volume does not exceed 25 mL, the small units are delivered in an outer packaging
that has the full label, the content is neither acute toxic category 1–3, nor STOT category 1 (*and not sold to the general public*), nor skin corrosive or eye damaging category 1, nor flammable, self-heating or oxidizing category 1, nor self-reactive or an
peroxide class A or B, and not carcinogenic, mutagenic or toxic to reproduction,
and it is not a biocidal product.

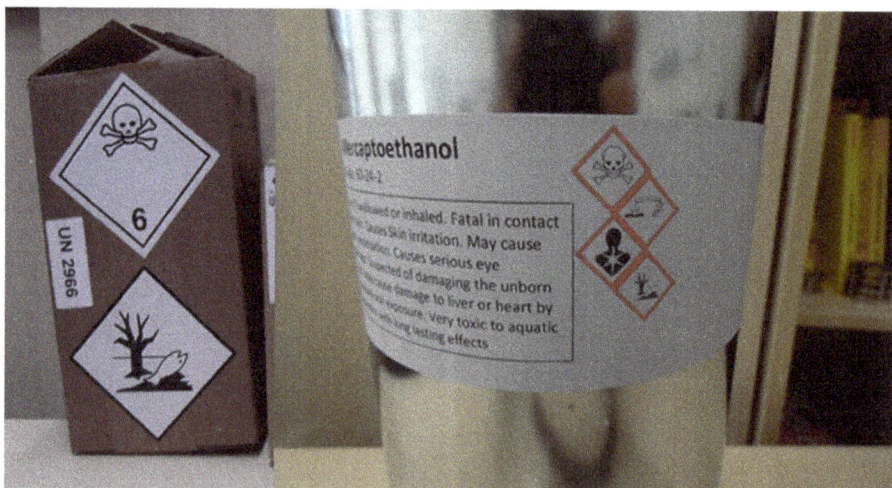

Fig. 90: Labeling as dangerous good on the outer (transport-) package (left), GHS-labels on the
inner (handling-) package (right).

6.4 Title IV: packaging

The title IV describes requirements for the packaging and is comparatively short.
Generally, the packing shall be robust so that no product escapes under normal use
and storage conditions. The material of the packaging and fastenings shall not be
cracked by the content and shall not undergo any (hazardous) reactions. If the hazardous product is sold to the general public, the packaging shall not have a shape

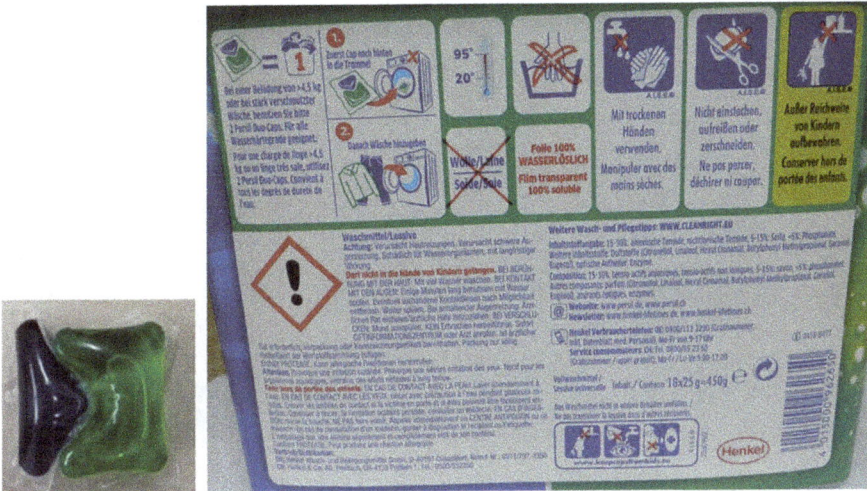

or form that arises or attracts the curiosity of children, and in general it shall not be misleading so it can be mixed up with packaging for food, beverages or pharmaceuticals.

Products for the general public containing acute toxic substances category 1–3, corrosive substances or STOT category 1 substances must be equipped with *child-resistant fastenings*; *tactile warnings and pictograms* have to be applied for products falling into acute toxicity categories 1–4, eye damage or skin corrosion category 1, STOT category 1, CMR category 2 or flammability category 1 or 2 or presenting an aspiration hazard. Note that products classified as carcinogen, mutagen of toxic to reproduction category 1 must not be sold to the general public, exempted such products that have a specific derogation, for example gasoline containing benzene.

If a substance is also classified as a dangerous good, the packaging fulfilling the requirements of the dangerous goods regulations are regarded as appropriate for dangerous substances. The regulations on dangerous goods have detailed entries on the quality of packaging and how to check it against cracking, fire, pressure, shock impact, etc.

6.5 Title V following

The next titles deal, to a large extend, with harmonized C&L, the role of the agency and of the manufacturers/importers/users. Manufacturers and importers shall propose a harmonized C&L to the agency. Harmonization is explicitly desired for sensitizers and CMR substances. After finalization of the consultation process, the agency

can conclude on a harmonized C&L, and the corresponding substance will then be listed in Annex VI, part 3.

Member states shall appoint bodies responsible for emergency medical help for intoxications/accidents with hazardous substances and mixtures (poison centers). These centers need to receive all relevant information about the products on the market, including complete disclosure of formulations. Information to be submitted to poison centers is laid down in annex VIII which enters into force on first January 2021. The Unique Formula Identifier (UFI) is a code that needs to be made available to poison centers, and which shall guarantee the unambiguous identification of the mixture by every poison center in the EU. These centers keep the information provided confidential, as far as this is in agreement with emergency medical treatment.

Any *advertisement for hazardous products* has to inform the addressee about the hazardous properties of the product (article 48).

All information that was used to conclude on an appropriate C&L for a product has to be stored and held retrievable *up to 10 years* after the last shipment of the product (article 49).

If a member state has justifiable reasons for concern that the current classification, labeling and packaging of a product poses a considerable risk against health and environment – although C&L is compliant with regulation (EC) No. 1272/2008 – it can take appropriate measures; this may include a request for a callback. In that case, this member state has to inform the ECHA and the other member states immediately, and the commission needs to take a decision within 60 days whether or not the measures are justified.

6.6 Annex I, part I: deviation from C&L requirements

Certain products may be placed on the market with simplified labeling:
- Gas cylinders up to 150 L: commercial name possible, provided the name of the hazardous substance in mentioned on the cylinder.
- Cartridges containing propane, butane or liquefied petrol gas: pictogram, H- and P-statement for flammable gases are sufficient.
- Massive metals, alloys and polymers do not need to be labeled if they do not present a hazard by inhalation, swallowing of skin contact.
- Explosives and pyrotechnics placed on the market to be used as such need only to be labeled as explosive.

Packaging not exceeding a volume of 125 mL can be equipped with a reduced labeling, where H- and P-statements may be omitted for certain hazardous products, p. e. if they are flammable liquids category 2 or 3, oxidizers category 2 or 3, skin or eye irritants category 2, to name a few. For STOT after single exposure (SE) category 2 and 3 and STOT after repeat exposure (RE) category 2 and acute toxic substances category 4 the H- and P-statements may be omitted, if these products are not sold to the general public. Products hazardous to the aquatic environment category 1 or 2 do not need H- and P-statements, but category 3 and 4 do need them: category 3 and 4 do not have the pictogram GHS 09.

6.7 Annex I, part II: physical hazards

6.7.1 Explosion hazard

Fig. 93: GHS01.

Explosion hazard is defined as a rapid decomposition of material where a huge volume of gas is formed suddenly, that exceeds the original volume of the material manifold. If the material decomposes in a confined space, a strong and rapid build-up of pressure results (until the confinement ruptures). Explosion is, typically associated with the release of heat and generation of fire (although the latter is not always the case). Explosive substances are sensitive to temperature, mechanical shock and/or friction. The test methods are standardized under the scope of guidelines for dangerous goods. Depending on the outcome of the tests, the explosive substances are allocated to one of the total six classes. The EU guidance allows an upstream theoretical assessment whether a substance may have explosive properties. There are several suspect groups that may raise concern and usually would trigger testing; in presence of these groups, the oxygen balance should be calculated to decide on the need for further testing (see Chapter 2):

Tab. 23: Categories of explosives.

Class	Symbol	Signal word	Hazard phrase	Example
1	GHS01	Danger	H200. Unstable explosive	Trinitro glycerol; dry hydroxylamine
1.1	GHS01	Danger	H201. Explosive; mass explosion hazard	Ammonium perchlorate, lead azide
1.2	GHS01	Danger	H202. Explosive; severe projection hazard	
1.3	GHS01	Danger	H203. Explosive; fire-, blast- or projection hazard.	Dry hydroxy benzotriazole
1.4	GHS01	Warning	H204. Fire or projection hazard	
1.5	---	Danger	H205. May mass explode in fire.	
1.6	---	---	---	

Typical precautionary statements allocated to, for example, class 1.2 explosives are
- P210: Keep away from heat, hot surfaces, sparks, open flames and other ignition sources. No smoking.
- P230: Keep wetted with . . .
- P234: Keep only in original container.
- P240: Ground/bond container and receiving equipment.
- P250: Do not subject to grinding/shock/ . . . /friction.
- P280: Wear protective gloves/protective clothing/eye protection/face protection.
- P370 + P372 + P380 + P373: In case of fire: Explosion risk! Evacuate area! DO NOT fight fire when fire reaches explosives.
- P401: Store . . . (*list appropriate conditions*).
- P501: Dispose of contents/container to . . . (*specify*).

Explosive substances may be mixed with desensitizers when brought on the market. These mixtures are no longer explosive, but they will become explosive again when the desensitizer is removed. An example is wetted picric acid, or wetted *N*-hydroxybenzotriazole. If the decomposition energy of the substance is at least 300 J/g, the desensitized mixture may fall into one of the four categories of desensitized explosives.

Tab. 24: Classification of desensitized explosives.

	Category 1	Category 2	Category 3	Category 4
Pictogram	GHS02	GHS02	GHS02	GHS02
Signal word	Danger	Danger	Warning	Warning
Hazard statements	H206: Fire, blast or projection hazard: increased risk of explosion if desensitizing agent is removed.	H207: Fire or projection hazard: increased risk of explosion if desensitizing agent is removed.	H207: Fire or projection hazard: increased risk of explosion if desensitizing agent is removed.	H208: Fire hazard: increased risk of explosion if desensitizing agent is removed.

Mixtures of flammable substances with inorganic oxidizers may be explosives (gun powder). In any case of doubt, the mixture should be tested. Inorganic oxidizers are allocated to classes 1 (very strong) to 3 (rather weak). If the content of an oxidizer class 1 or 2 is below 15%, or a class 3 oxidizer is below 30%, the mixture is not expected to be explosive.

There are some extra phrases in the EU beyond the global GHS:

- EUH001: explosive when dry (picric acid and 1-hydroxy benzotriazole are brought on the market as mixtures with water).
- EUH006: explosive with or without contact to air (ethyne).
- EUH044: risk of explosion when heated under confinement (if the substances or mixture is not classified as explosive already).

An organic substance or mixture of organic substances shall *not* be classified as explosive, if

- there are no chemical groups associated with explosive properties (see Section 2.3.1), or
- the oxygen balance is below −200 or
- the decomposition energy is below 500 J/g (calorimetric measurement) and the onset temperature for decomposition is above 500 °C.

A mixture of an organic substance with an inorganic oxidizer shall not be classified as explosive if

- the oxidizer is allocated to category 1 or 2 and makes up not more than 15% of the mixture or
- the oxidizer is category 3 and makes up not more than 30% or the mixture.

6.7.2 Flammability

Fig. 94: GHS02.

Flammability is the property of a substance to catch fire. Gases are classified as extremely flammable, category 1 (H220) if a content of 13% or less in air can be ignited or if the ignitability range stretches over at least 12% points. Other flammable gases are category 2, H221 (flammable gas).

A *pyrophoric gas* is a flammable gas that ignites spontaneously in contact with air at temperatures of 54 °C or below.

A *chemically unstable gas* is a flammable gas that is able to react explosively even in the absence of air and oxygen.

Tab. 25: Labeling for flammable gases.

	Category 1A	Gases 1a which meet pyrophoric or unstable criteria			Category 1B	Category 2
		Pyrophoric	Chemically unstable			
			Category A	Category B		
Pictogram	GHS02	GHS02	GHS02	GHS02	GHS02	None
Signal Word	Danger	Danger	Danger	Danger	Danger	Warning
Hazard Statement	H220: Extremely flammable gas.	H220: Extremely flammable gas. H232: May ignite spontaneously if exposed to air.	H220: Extremely flammable gas. H230: May react explosively even in the absence of air.	H220: Extremely flammable gas. H231: May react explosively even in the absence of air at elevated temperature or pressure.	H221: Flammable gas.	H221: Flammable gas.

Typical precautionary statements for flammable gases are
- P210: Keep away from heat, hot surfaces, sparks, open flames and other ignition sources. No smoking.
- P377: Leaking gas fire – do not extinguish unless leak can be stopped safely.
- P381: Eliminate all ignition sources if safe to do so.
- P403: Store in a well-ventilated place.

For pyrophoric gases, P280 (Wear protective gloves/protective clothing/eye protection/face protection.) and P222 (Do not allow contact with air), and for chemically instable gases P202 (Do not handle until all safety precautions have been read and understood) is used.

Substances brought on the market as aerosol cans need to be tested concerning their flammability, irrespective whether they release gas dispersed liquids or solids or foams, pastes or powders. Based on the test results belong to flammable aerosols category 1 (H222: extremely flammable aerosol) or category 2 (H223: flammable aerosol).

Flammable liquids have frequently been the reason for major accidents. They are classified in dedendence on their boiling point and their flash point. The *flash point* is that temperature at which the liquid generates enough vapor so that a flame brought to the surface of the liquid spreads over the liquid. The substance may be heated in an open cup (open cup test) or – typically preferred today – in a closed cup, were the lid of the cup is briefly removed while the flame is brought near the substance surface whenever a temperature is reached where the flammability shall be tested. The flash point defines to what category of flammable liquids the substance belongs. Category 1 (H224: extremely flammable liquid and vapor) has a flash point below 23 °C and a boiling point below 35 °C (hydrogen cyanide; 2-methyl-1,3-butadiene; diethyl ether). Category 2 (H225: Highly flammable liquid and vapor) are flammable liquids with a flash point below 23 °C, but the boiling point is above 35 °C (tertiary-butyl alcohol; ethanol; propanol). Category 3 liquids (H226: Flammable liquid and vapor) have a flash point of at least 23 °C and not higher than 60 °C (*n*-butanol; bromo benzene). Diesel oils, gas oils and light heating oils with a flash point up to 75 °C are exemptions in so far as they also belong to category 3 flammable liquids although the flash point might be higher than 60 °C.

Liquid mixtures brought on the market may contain flammable components, but due to other ingredients added, the freshly prepared mixture may have flash points higher than 60 °C. If the mixture contains at least 5% of a flammable or highly flammable components, and the flash point is at or below 93 °C, EU-hazard (EUH) phrases have to be added:
- EUH209: Can become highly flammable in use.
- EUH209a: Can become flammable in use.

Tab. 26: Categories of flammable liquids.

	Category 1	Category 2	Category 3
Pictogram	GHS02	GHS02	GHS02
Signal word	Danger	Danger	Warning
Hazard Statement	H224. Extremely flammable liquid and vapor.	H225. Highly flammable liquid and vapor.	H226. Flammable liquid and vapor.
Flash-point	< 23 °C; initial boiling point < 35 °C.	< 23 °C; initial boiling point ≥ 35 °C.	23 °C ≤ flash point ≤ 60 °C; OR: Diesels, gasoil, light heating oil, 55 °C ≤ flash point ≤ 75 °C.

Typical precautionary statements allocated to flammable liquids are

– P210: Keep away from heat, hot surfaces, sparks, open flames and other ignition sources. No smoking.
– P233: Keep container tightly closed.
– P240: Ground/bond container and receiving equipment.
– P241: Use explosion-proof electrical/ventilating/lighting/ . . . /equipment.
– P242: Use only nonsparking tools.
– P243: Take precautionary measures against static discharge.
– P280: Wear protective gloves/protective clothing/eye protection/face protection.
– P303 + P361 + P353: IF ON SKIN (or hair): Take off immediately all contaminated clothing. Rinse skin with water/shower.
– P370 + P378: In case of fire: Use . . . to extinguish. (*specify*).
– P403 + P235: Store in a well ventilated place. Keep cool.
– P501: Dispose of contents/container to . . . (*specify*).

Flash points of liquid mixtures containing flammable liquids may be calculated based on a method published by Rasmussen and Gmehling [1]. If the calculated flash point is at least 5 K higher than the classification limit, the mixture is not to be allocated to that class. For example, a calculated flash point of 30 °C would justify to classify the mixture not as category 1 or 2 flammable liquid; however, it has to be classified as category 3 flammable liquid.

For plant safety and safe handling there are other properties connected to flammability which do not have directly an influence to C&L. These are the *Explosion Limits* and the *Self-ignition Temperature*. Both data points have to be submitted to the ECHA for substance registration.

The upper and lower explosion limits are the concentration of a gas or a vapor in air which will cause an explosion if ignited. The self-ignition temperature is that temperature of a surface which sets the substance under investigation under fire.

To test this, an empty vessel is heated up in a heating block. At certain temperatures, a drop of the test-substance is added, releasing an audible "plop" when the self-ignition temperature is reached or exceeded.

Flammable solids are classified according to their burning behavior, that is the spread velocity of a flame when set on fire and the capability to jump over a defined wetted zone. Both categories are allocated the H228: Flammable Solid (phosphorus pentasulfide; red phosphorus).

6.7.3 Pyrophoric and self-heating substances

Pyrophoric substances (GHS02, Danger) are those which cause fire within 5 minutes when exposed to air, for example, white phosphorus, butyl lithium solution and are labeled as

- H250: Catches fire spontaneously if exposed to air.
- Typical precautionary statements are
- P210: Keep away from heat, hot surfaces, sparks, open flames and other ignition sources. No smoking.
- P222: Do not allow contact with air.
- P231 + P232: Handle under inert gas. Protect from moisture.
- P233: Keep container tightly closed.
- P280: Wear protective gloves/protective clothing/eye protection/face protection.
- P302 + P334: IF ON SKIN: Immerse in cool water/wrap in wet bandages.
- P370 + P378: In case of fire: Use . . . to extinguish. (*specify*)

Some substances heat up when exposed to air without immediately catching fire.

Tab. 27: Classification of self-heating substances.

	Category 1	Category 2
Pictogram	GHS02	GHS02
Signal word	Danger	Warning
Hazard Statement	H251: Self-heating; may catch fire.	H252: Self heating in large quantities; may catch fire.

Examples for category 1 are sodium dithionite and sodium methanolate; magnesium powder is an example for category 2. For both categories, typical precautionary statements are

- P235: Keep cool.
- P280: Wear protective gloves/protective clothing/eye protection/face protection.

- P407: Maintain air gap between stacks/pallets.
- P413: Store bulk masses greater than ... kg/ ... lbs. at temperatures not exceeding ... °C/ ... °F. (*specify*)
- P420: Store away from other materials.

6.7.4 Water reactive substances

Substances reacting dangerously with water deserve extra attention. They are allocated to three categories. Category 1 substances react vigorously with water at ambient temperatures; the gas evolved ignites spontaneously, or flammable gas is emitted at a rate of 10 or more liters per kg substance per minute. If the substance does not meet category 1 criteria, but flammable gas is emitted at a rate of 20 liters per kg substance per hour, it is category 2. Any other substance reacting with water at ambient temperatures emitting flammable gas belongs to category 3.

Tab. 28: Classification of water reactive substances.

	Category 1	Category 2	Category 3
Pictogram	GHS02	GHS02	GHS02
Signal word	Danger	Danger	Warning
Hazard Statement	H260: In contact with water releases flammable gases which may ignite spontaneously.	H261: In contact with water releases flammable gases.	H261: In contact with water releases flammable gases.

Magnesium alkyls; calcium phosphide; phosphorus pentasulfide are examples for category 1 substances, aluminum powder and calcium are examples for category 2. Typical precautionary statements in combination with category 1 and 2 substances are
- P223: Do not allow contact with water.
- P231 + P232: Handle under inert gas. Protect from moisture.
- P280: Wear protective gloves/protective clothing/eye protection/face protection.
- P302 + P335 + P334 + P378: IF ON SKIN: Brush off loose particles from skin. Immerse in cool water/wrap in wet bandages. Use ... to extinguish. (*specify*)
- P402 + P404: Store in a dry place. Store in a closed container.
- P501: Dispose of contents/container to ... (*specify*)

6.7.5 Oxidizers

Fig. 95: GHS03.

Oxidizers may easily release oxygen or oxygen equivalents, p. e. active halogens, which make flammable substances highly flammable. The categories are allocated by comparison to defined standards.

For *oxidizing* liquids, Category 1 is an oxidizer where the 1:1-mix with cellulose shows an ignition time less then, or an increase in pressure equivalent or more than a 1:1-mix of cellulose/50% perchloric acid (H270: May cause fire or explosion; strong oxidizer). Examples are chlorine dioxide and compressed oxygen. Category 2 is a substance when mixed 1:1 with cellulose causes a pressure increase equivalent to 40% $NaClO_3$ solution mix with cellulose, but the reaction does not justify category 1 (H271: May intensify fire; oxidizer). For category 3, a 1:1 mix with cellulose is equivalent to a pressure increase caused by a 1:1-mix 65% HNO_3/Cellulose (H272: May intensify fire; oxidizer.).

Oxidizing solids are classified in three categories as well having the same H-phrases, but the categories 1 to 3 are compared cellulose/$KBrO_3$ mixtures. *Category 1*: a 4:1 to 1:1 mix with cellulose has a faster burning rate than 3:2 $KBrO_3$/Cellulose. *Category 2*: a 4:1 to 1:1 mix with cellulose has a faster burning rate than 2:3 $KBrO_3$/Cellulose. *Category 3*: a 4:1 to 1:1 mix with cellulose has a faster burning rate than 3:7 $KBrO_3$/Cellulose.

Tab. 29: Classification of oxidizers.

	Category 1	Category 2	Category 3
Pictogram	GHS03	GHS03	GHS03
Signal word	Danger	Danger	Warning
Hazard statement	H271: May cause fire or explosion; strong oxidizer	H272: May intensify fire; oxidizer	H272: May intensify fire; oxidizer

Precautionary statements for category 1 oxidizers are
- P210: Keep away from heat, hot surfaces, sparks, open flames and other igni-
  tion sources. No smoking.
- P220: Keep/Store away from clothing/ . . . /combustible materials. (*insert addi-
  tions as appropriate*).
- P280: Wear protective gloves/protective clothing/eye protection/face protection.
- P283: Wear fire/flame resistant/retardant clothing.
- P306 + P360: IF ON CLOTHING: Rinse the contaminated clothing immediately
  and skin with plenty of water before removing clothes.
- P371 + P380 + P375: In case of major fire and large quantities: Evacuate area.
  Fight fire remotely due to the risk of explosion.
- P370 + P378: In case of fire: Use . . . to extinguish. (*specify*)
- P420: Store away from other materials.
- P501: Dispose of contents/container to . . . (*specify*)

6.7.6 Self-reacting substances and organic peroxides

Self-reacting substances do not fulfill definitions of explosives, but they may either
release a critical amount of heat at decomposition (300 J/g or more), or a 50 kg package (or more) suffers self-accelerating decomposition at temperatures of 75 °C or less.

Organic peroxides – though showing a comparable behavior – are dealt with
separately. A reason may be that organic peroxides may as well act like explosives,
oxidants and irritant/corrosive materials. They tend to explosive decomposition,
burn rapidly and/or are sensitive to shock-impact and friction. A common feature is
the C-O-O-R element in the structure. Dibenzoyl peroxide and di-tertiary-butyl peroxide are radical initiators as the disproportionate at elevated temperatures.

Hydroperoxides are typically produced and used on-site, like cumene hydroperoxide. Hydroperoxide formation occurs during the curing of linseed oil. Some organic
substances are sensitive to unintended hydroperoxide formation, namely diethyl
ether, tetrahydrofurane, dioxane and tetraline (bi-cyclo[4.4.0]decane). In presence of
air and sunlight, omnipresent OH* radicals initiate the activation of these molecules,
which than undergo a chain reaction leading to an enrichment of hydroperoxides.
When the original compound, having a higher vapor pressure and lower boiling
point, evaporates, the concentration of the hydroperoxide increases. This makes the
mixture more and more heat sensitive, and violent explosions may follow. It is utmost important to check such solvents on hydroperoxides. This can be done with potassium iodide paper, which turns blue on the presence of hydroperoxides, or Fe^{2+}
solutions which change color from greenish to pale yellow.

Tab. 30: Types of self-reacting substances/peroxides.

	Type A	Type B	Type C + D	Type E + F	Type G
Properties	May detonate when heated.	Not A, but pressure buildup when heated may cause explosion.	Not A or B, deflagration possible when heated.	No or only partial deflagration when heated under confinement.	None of A–F, SADT* 60 °C or more.
Pictogram	GHS01	GHS01 + GHS02	GHS02	GHS02	–
Signal word	Danger	Danger	Danger	Warning	–
Hazard Statement	H240. Heating may cause an explosion.	H241. Heating may cause a fire or explosion.	H242. Heating may cause a fire.	H242. Heating may cause a fire.	–
Examples	1-Hydroxy-1'-hydroperoxydicyclohexylperoxide	Dibenzoyl peroxide	Dilauryl peroxide	di-tert-butyl peroxide	

*self-accelerating decomposition temperature

Some typical precautionary statements allocated to these substances are
- P210: Keep away from heat, hot surfaces, sparks, open flames and other ignition sources. No smoking.
- P234: Keep only in original container.
- P235: Keep cool.
- P240: Ground/bond container and receiving equipment.
- P280: Wear protective gloves/protective clothing/eye protection/face protection.

Whenever a molecule has a peroxy group (R-O-O-R), it is to be classified as peroxide, unless:
- its available oxygen from organic peroxide is not higher than 1%, and the content of hydrogen peroxide equivalents is not higher than 1%, or
- its available oxygen from peroxide is not higher than 0.5%, and the content of hydrogen peroxide equivalents is more than 1%, but not more than 7%.

The available oxygen in a mixture is calculated as

$$AO = 16 \times \sum_j \left(\frac{n_j \times c_j}{m_j} \right)$$

with n_j = number of peroxide groups in molecule j, c_j its mass concentration in the mixture (%) and m_j the molecular mass of the organic peroxide j (g/mol). Hydrogen peroxide may be present on purpose (mixtures) or as a residual from the manufacturing process.

6.7.7 Compressed gases

Containments with compressed gases pose a hazard in so far as that overheating or mechanical stress may end up in heavy rupture of the containment. The label and hazard phrases shall ensure that the containments are prevented from overheating and secured against mechanical shock. The hazard symbol is the gas bottle, and the hazard and precautionary statements are
- H280: Contains gas under pressure; may explode if heated,
  - P410 + P403: Store in a well-ventilated place.
- H281: Contains refrigerated gas; may cause cryogenic burns and injury
  - P282: Wear cold insulating gloves/face shield/eye protection.
  - P336: Thaw frosted parts with lukewarm water. Do not rub affected areas.
  - P315: Get immediate medical advice/attention.
  - P403: Store in a well-ventilated place.

It should be noted that refrigerated gas or expanding compressed gas may be very cold and can cause burns on direct contact. Violent evaporation of deep freeze, liquefied gas indoors may expel oxygen in air so people inside are in danger of suffocation.

6.7.8 Corrosive to metals

Corrosivity was originally a hazardous property in the area of toxicology, only. However, corrosion to materials can pave the way to major incidents by failure of piping and storage tanks. Further, corrosivity to metals may cause release of hydrogen

Fig. 97: GHS05.

which may cause rupture of the containment and poses a fire and explosion hazard. A corrosion rate of at least 6.25 mm on either steel or aluminum per year at 55 °C results in a classification.

The Hazard Statement is H290: May be corrosive to metals. The symbol is the same as for substances that cause severe eye damage of chemical burns to the skin. Precautionary statements are

- P234: Keep only in original container.
- P390: Absorb spillage to prevent material damage.
- P406: Store in a corrosive resistant/ . . . container with a resistant inner liner.

In case a substance is classified as corrosive to metals but is not corrosive to the skin nor does it cause serious eye damage, the package ready for consumer use does not need to carry the GHS05 label.

6.8 Annex I, part III: health hazards

6.8.1 Acute toxicity

Testing of acute toxicity is the typical first toxicological endpoint that is identified. It allows to estimate the acute risk that can be triggered by a single exposure to the substance. International testing guidelines describe how these tests shall be performed. For acute toxicity testing, a single dose of the substance exerts its activity on the test animals. The general testing regime for acute tests is as follows.

- single exposure of a given number of test animals to the substance.
- after exposure animals have to be observed for a certain time period.
- counting of animals that died during the observation period.
    - looking for irregularities in organs and tissues by macroscopic visual inspection.

In previous times much emphasis was laid on exact detection of the dose (oral or dermal exposure) or the concentration (inhalation exposure) that causes 50% fatalities (50% lethal dose (LD_{50}) or 50% lethal concentration (LC_{50})). Since the 1990s, it was generally accepted that for protection of human health, it would be sufficient to know the range in which the LD_{50} or LC_{50} falls. Adapted tests need fewer animals, and results are not delivered as precise LD_{50} or LC_{50} values, but in form of, for example, 50 mg/(kg b.w.) < LD_{50}/oral/rat ≤ 300 mg/(kg b.w.). For the calculation of the toxicity of mixtures (see Section 6.8.1.1), LD_{50} or LC_{50} values are used in algorithms; however, for substances were only an interval was reported, the acute toxic estimate (ATE) is used instead (Tab. 31–35). Under the more recent testing regimes, not only death but also morbidity of test animals may be sufficient to categorize the substance as acute toxic.

Tab. 31: Classification based on acute oral toxicity: LD_{50} intervals and their ATE of the acute toxicity categories.

Category	LD_{50} [mg/kg]	ATE	Hazard Statement	Pictogram	Signal word
1	≤ 5	0.5	H300. Fatal if swallowed	GHS06	Danger
2	5 < LD_{50} ≤ 50	5.0	H300. Fatal if swallowed	GHS06	Danger
3	50 < LD_{50} ≤ 300	100	H301. Toxic if swallowed	GHS06	Danger
4	300 < LD_{50} ≤ 2,000	500	H302. Harmful if swallowed	GHS07	Warning

LD_{50}: lethal dose (oral or dermal) that kills 50% of the exposed animals.
LC_{50}: lethal concentration (in air of water) that kills 50% of the exposed animals.
ATE: acute toxic estimate, an estimated LD_{50} or LC_{50} for the dose or concentration interval that covers 50% mortality

For oral exposure, a single dose is instilled into the stomach of the animal (typical rat, sometimes mouse). For dermal exposure, the test substance is applied on the shaved skin and occluded for 24 h to make sure the animal cannot lick the substance; this test is typically performed with rats or rabbits. After 24 h, the exposed skin is rinsed. After oral or dermal acute exposure, the animals are observed for a certain time period (14 days). The cut-off values for the acute toxicity categories after dermal exposure are shown in Tab. 32.

Tab. 32: Classification based on acute dermal toxicity: LD_{50} intervals and their ATE of the acute toxicity categories.

Category	LD_{50} [mg/kg]	ATE	Hazard Statement	Pictogram	Signal word
1	≤ 50	5	H310. Fatal in contact with skin	GHS06	Danger
2	50 < LD_{50} ≤ 200	50	H310. Fatal in contact with skin	GHS06	Danger
3	200 < LD_{50} ≤ 1,000	300	H311. Toxic in contact with skin	GHS06	Danger
4	1000 < LD_{50} ≤ 2,000	1,100	H312. Harmful in contact with skin	GHS07	Warning

In the GHS, a category 5 is defined with an LD_{50} between 2,000 and 5.000 mg/kg. In the EU, this category is not used. Recommended precautionary statements for category 1 and 2 substances are
- P264: Wash . . . thoroughly after handling. (*hands, body,* . . . *:specify*)
- P270: Do not eat, drink or smoke when using this product.
- P301 + P310: IF SWALLOWED: Immediately call a POISON CENTER/doctor/ . . .
- P321: Do NOT induce vomiting.
- P330: Rinse mouth.
- P405: Store locked up.
- P501: Dispose of contents/container to . . . (*specify*)

Precautionary statements for category 1 and 2 substances for acute dermal toxicity are
- P262: Do not get in eyes, on skin, or on clothing.
- P264: Wash . . . thoroughly after handling. (*hands, body,* . . . *:specify*)
- P270: Do not eat, drink or smoke when using this product.
- P280: Wear protective gloves/protective clothing/eye protection/face protection.
- P302 + P352: IF ON SKIN: Wash with plenty of water/ . . .
- P310: Immediately call a POISON CENTER/doctor/ . . .. (p. e. the supplier).
- P321: Do NOT induce vomiting.
- P361 + P364: Take off immediately all contaminated clothing and wash it before reuse.
- P405: Store locked up.

For inhalation exposure, the animals have to inhale the substance over a 4 h period. The preferred test species is the rat. For inhalation tests it has to be decided/respected in what form the substance is likely to appear during normal handling and use (Tab. 33–35). It is conceivable that gas and gas phases above liquids with a comparatively low boiling point (vapors) will be inhaled as gas. Liquid materials, however, can be sprayed (spray aerosols), or they may form condensation aerosols when they are handled at elevated temperatures (condensation aerosols). Finely dispersed dust is a form which allows inhalation of solid materials. Dusts and mists are summarized as aerosols, and the aerosol diameter may have a very strong influence on the outcome of acute inhalation toxicity. Aerosols means mists and/or dusts. They will more or less rapidly deposit on surfaces over time. Testing aerosols is a technical challenge and interpretation of results may be complicated. To create robust experimental animal test data which are reproducible, typically aerosols of very small diameters have to be generated. The reasons are that these aerosols need to be stable over the time period between generation and inhalation, and they have to pass the nose "labyrinth" of the rat which is a compulsory nose breather; the median aerodynamic diameter needs to be 4 μm and below. If the rat nose is seen as a maze, the human nose is a straight highway into the lung, and particles with a median aerodynamic diameter (MAD) of less than 10 µm can reach the alveoli. These differences sometimes make translation of rat experimental data to human risk challenging. For inhalation toxicity, guidelines for dangerous goods (transport regulations) require acute inhalation toxicity data for 1 h exposure. The EU regulation allows to transform such data by using the simple form of Haber's rule:

- $LC_{50} \times (\text{exposure-time})^n = \text{constant}$,
  - with $n = 1$ as default.

Tab. 33: Gas classification based on acute inhalation toxicity: LC_{50} intervals and their ATE of the acute toxicity categories.

Category	LC_{50} [ppm]	ATE	Hazard statement	Pictogram	Signal word
1	≤ 100	10	H330: Fatal if inhaled.	GHS06	Danger
2	$100 < LC_{50} \leq 500$	100	H330: Fatal if inhaled.	GHS06	Danger
3	$500 < LC_{50} \leq 2{,}500$	700	H331: Toxic if inhaled.	GHS06	Danger
4	$2{,}500 < LC_{50} \leq 20{,}000$	4,500	H332: Harmful if inhaled.	GHS07	Warning

Tab. 34: Vapor classification based on acute inhalation toxicity: LC_{50} intervals and their ATE of the acute toxicity categories.

Category	$LC>_{50}$ [mg/L]	ATE	Hazard statement	Pictogram	Signal word
1	≤ 0.5	0.05	H330: Fatal if inhaled.	GHS06	Danger
2	$0.5 < LC_{50} \leq 2$	0.5	H330: Fatal if inhaled.	GHS06	Danger
3	$2 < LC_{50} \leq 10$	3	H331: Toxic if inhaled.	GHS06	Danger
4	$10 < LC_{50} \leq 20$	11	H332: Harmful if inhaled.	GHS07	Warning

Tab. 35: Dust and mist (aerosol) classification based on acute inhalation toxicity: LC_{50} intervals and their ATE of the acute toxicity categories.

Category	LC_{50} [mg/L]	ATE	Hazard statement	Pictogram	Signal word
1	≤ 0.05	0.005	H330: Fatal if inhaled.	GHS06	Danger
2	$0.05 < LC_{50} \leq 0.5$	0.05	H330: Fatal if inhaled.	GHS06	Danger
3	$0.5 < LC_{50} \leq 1.0$	0.5	H331: Toxic if inhaled.	GHS06	Danger
4	$1.0 < LC_{50} \leq 5.0$	1.5	H332: Harmful if inhaled.	GHS07	Warning

Reports from older toxicity studies deliver distinct LD_{50} and LC_{50} values. If an oral LD_{50} of 1,300 mg/kg was reported, the substance has to be classified as H302, harmful if swallowed, the GHS07 pictogram and the signal word "Warning" have to be added. In more recent reports, substances are allocated to categories. That is, if the substance is rated as category 3 for acute dermal toxicity, the dermal LD_{50} is somewhere between 200 and 1,000 mg/kg. For the calculation of the acute toxicity of mixtures, LD_{50} values have to be used if available. For substances without a precise LD_{50} (whatever "precise" means in biological tests), the ATE has to be used.

A special case are substances which can be swallowed, but due to their low viscosity they may enter the lung and can cause fatal lung damage. Lamp oils are a sad example that has caused fatalities in toddlers in the EU. Surface active agents (detergents) are another example. The kinematic viscosity, η, is critical. If $\eta \le 20.5$ mm^2/s at 40 °C, the liquid substance is classified and labeled with GHS symbol No. 8 and H304: may be fatal if swallowed and enters airways.

Fig. 98: GHS symbol 08 for substances with hazard statement H304: may be fatal if swallowed and enters airways.

Mixtures containing 10% or more of such H304 substances – where the addition of concentrations is performed over all aspiration hazard substances present at 1% or more – and with a kinematic viscosity of $\eta \le 20.5$ mm^2/s at 40 °C, are to be classified. If the mixture separates into distinct layers, and at least one of these layers contains 10% or more substances with aspiration hazard and has a viscosity of $\eta \le$ 20.5 mm^2/s at 40 °C, the whole mixture is classified as aspiration hazard. Sometimes, the kinematic viscosity is not available, but it may be calculated from the dynamic viscosity, given in milli-pascal seconds (mPas), and dividing the dynamic viscosity by the density of the liquid (g/cm^3).

Special substance properties, but also results of acute toxicity testing, can call for additional hazard statements beyond those given in the GHS. For such cases, the EU allocates EUH statements. Some examples with a link to acute toxicity are:
- EUH070: Toxic by eye contact.
  - For a substance that is neither fatal nor toxic in contact with skin.
- EUH071: Corrosive to the respiratory tract
- EUH029: Contact with water liberates toxic gas
  - Aluminum phosphide, AlP

- EUH031: Contact with acid liberates toxic gas
    - NaOCl decomposes to chlorine in the presence of acid.
- EUH032: Contact with acid liberates very toxic gas
    - KCN

6.8.1.1 Acute toxicity: mixtures

For animal welfare reasons, for financial reasons and because of mere lab capacities, it is impossible to test every mixture of substances that is brought on the market. Experience gained in toxicology over decades has shown that acute toxicity of mixtures can be calculated by the dose addition of the ingredients. According to GHS, the mixture toxicity is calculated as

$$\frac{100\% - \left(\sum(\text{unknown}), \text{if} \geq 10\%\right)}{\text{ATE}_{\text{mix}}} = \sum \frac{C_i}{\text{ATE}_i}$$

ATE is the acute toxicity estimate for the mixture (mix) and for an individual component in the mixture (i). C_i is the concentration in percentage by weight. If the mixture contains at least 10% components of unknown toxicity, this amount has to be subtracted from the 100% of the denominator on the left hand side. In the algorithm, ATE_i are to be replaced by $\text{LD}_{50,i}$ or $\text{LC}_{50,i}$, if these data points are available. ATE_i data are to be taken from the tables above (Tab. 31–35). If 1% or more of the mixture consists of unknown components or components without test data, the warning phrase *X percent of the mixture consist of component(s) of unknown toxicity*. In the safety data sheet, a more precise specification can be added, if data allow to do so, for example: "15% of the mixture consists of components with unknown acute dermal toxicity."

As an example, assume a mixture containing 8% substance A, acute oral toxicity category 2 and 80% of substance B, acute oral toxicity category 4. The rest of the mixture is not toxic. With the above algorithm we calculate

$$\frac{100}{\text{ATE}_{\text{mix}}} = \frac{8}{5} + \frac{80}{500} \quad \leftrightarrow \quad \text{ATE}_{\text{mix}} = \frac{100 \times 500}{880} = 57 \text{ mg/kg}$$

57 mg/kg is in the ATE range of acute oral category 3 (>50–300 mg/kg), so the mixture is classified as acute oral toxic category 3, label H301: Toxic if swallowed and the skull and crossbone label (GHS 5) with signal word "Danger."

If the remaining 12% of this mixture are components of unknown toxicity, the calculation has to be modified:

$$\frac{100 - 12}{\text{ATE}_{\text{mix}}} = \frac{8}{5} + \frac{80}{500} \quad \leftrightarrow \quad \text{ATE}_{\text{mix}} = \frac{88 \times 500}{880} = 50 \text{ mg/kg}$$

In this case, the mixture is acute toxic category 2 by oral exposure, pictogram GHS05 and H300: Fatal if swallowed.

Now imagine that for substance A you would have an LD_{50} value of 50 mg/kg. B still is acute oral category 4. The algorithm now is

$$\frac{100}{ATE_{mix}} = \frac{8}{50} + \frac{80}{500} \leftrightarrow ATE_{mix} = 312 \text{ mg/kg}$$

Now the mixture is in the acute oral toxicity category 4, H302: Harmful if swallowed, label GHS 07 (exclamation mark) and signal word WARNING.

In such calculations it may happen, under rare circumstances, that the calculation says the mixture is, p. e. category 3, but the most severe class the individual ingredients belong to is category 4. According to regulation 1272/2008/EC, a *mixture cannot belong to a more severe acute toxicity category than the ingredient with the most severe toxicity class.*

For these calculations, components of category 1–3 are included if they are present at least at 0.1%. Category 4 substances are included if they are present at least at 1%. Substances with SCL lower than 1.0% or 0.1% have to be included down to their SCL, respectively.

Calculating acute toxicity for mixtures with ingredients where the single components where tested in different physical states needs careful considerations. Assume an arbitrary mixture where you have the following composition (Tab. 36):

Tab. 36: Mixture, example for classification for inhalation toxicity.

Component, (molecular mass)	Content	Tested as	ATE	LC_{50}	Category
A (500 g/mol)	5%	Dust	0.5 mg/L/4 h	Not available	3
B (100 g/mol)	10%	Gas	100 ppm/4 h	200 ppm/4 h	2
C (200 g/mol)	5%	Vapor	3 mg/L/4 h	Not available	3
D (250 g/mol)	80%	Mist	1.5 mg/L/4 h	Not available	4

If this mixture is to be sprayed or dispersed, all ingredients may be inhaled as mist and, therefore, need to be covered for the calculation. Where LC_{50} values are available, they have to be used instead of the ATE. The ppm for component B needs to be transformed into mg/L. To do so, the ideal gas law is applied. 200 ppm in the air means that the partial pressure of the component in air is $0.0002 \times 101,300 \text{ Pa} = 20.26 \text{ Pa}$; this value is used to calculate the concentration in mg/L (or g/m^3):

$$\frac{n}{V} = \frac{p}{RT} = \frac{20.26}{8.314 \times 298} = 0.0082 \frac{mol}{m^3} = 0.8 \frac{mg}{L}$$

As a result, component B will be included with an LC_{50} of 0.8 mg/L, which is used instead of an ATE. Component C was tested as vapor, but the mixture may be applied as spray aerosol. It is assumed, that the ATE for dusts/mists category 3 is applicable, so the ATE for dusts/mists category 3, 0.5 mg/L will be used as ATE for component C, and not the ATE of 3 mg/L for vapors. The calculation for the spray-aerosol, therefore, runs as

$$\frac{100}{\text{ATE}_{\text{mix}}} = \frac{5}{0.5} + \frac{10}{0.8} + \frac{5}{0.5} + \frac{80}{1.5} \Leftrightarrow \text{ATE}_{\text{mix}} = 1.165\frac{\text{mg}}{\text{L}}$$

This value is between 1 and 5 mg/L (Tab. 35). For spraying, this mixture is harmful by inhalation, category 4.

The conversion of ppm into mg/m³, and vice versa, for $p = 101{,}300$ Pa and $T = 298$ K is

$$\text{ppm} = 24.454 \left[\frac{\text{L}}{\text{mol}}\right] \times \frac{\text{mg}}{\text{m}^3} \times \frac{1}{\text{MW}} \left[\frac{\text{mol}}{\text{g}}\right]$$

with MW as molecular weight in dimension g/mol.

6.8.1.2 Acute toxicity: STOT SE

Acute intoxication may cause adverse effects other than death which requires appropriate warning. Such effects may be organ damage – p. e. jaundice due to acute liver intoxication with methylene-4,4'-di-aniline, blindness caused by acute methanol intoxication – or less prominent effects like airway irritation, narcotic effects or skin effects due to degreasing.

Substances can be allocated to STOT SE categories 1, 2 or 3. If a single dose, either the Lowest Observed Adverse Effect Level (LOAEL) or the Lowest Observed Adverse Effect Concentration (LOAEC), is in the range of the acute toxicity categories 1–3 (skull & crossbones categories, that is, for example, ≤ 300 mg/kg by oral exposure) and results in organ dysfunction or failure, the substance is *STOT SE category 1*. If for organ dysfunction a dose is required as high as acute toxicity category 4 (p. e., higher than 300 mg/kg, but not greater than 2,000 mg/kg for oral dosages), the substance belongs to *STOT SE category 2*.

Tab. 37: LOAEL/LOAEC cutoffs for STOT SE categories.

Exposure route	Category 1	Category 2
Oral	LOAEL ≤ 300 mg/kg b.w.	2,000 ≥ LOAEL > 300 mg/kg b.w.
Dermal	LOAEL ≤ 1,000 mg/kg b.w.	2,000 ≥ LOAEL > 1,000 mg/kg b.w.
Inhalation, gas	LOAEC ≤ 2,500 ppm/4 h	20,000 ≥ LOAEC > 2,000 ppm/4 h
Inhalation, vapor	LOAEC ≤ 10 mg/L/4 h	20 ≥ LOAEC > 10 mg/L/4 h
Inhalation, dust or mist	LOAEC ≤ 1.0 mg/L/4 h	5.0 ≥ LOAEC > 1.0 mg/L/4 h

STOT SE category 1 or 2 is not additive. That is, a mixture is classified only if an individual ingredient reaches or exceeds the specific or generic concentration limit (SCL or GCL). Existing official SCLs are listed in the substance tables of Annex VI, regulation

1272/2008/EC. Generic concentration limits are given in Tab. 38. For STOT SE category 1 or 2, the signal is the GHS 08 label ("decaying chest"), but the signal word is "DANGER" for category 1, and "WARNING" for category 2.

Tab. 38: GCL for mixture classification STOT SE.

Component	Mixture	
	Category 1	Category 2
Category 1	≥10%	≥1%, <10%
Category 2		≥10%

If possible, specific concentration limits shall be derived for substances classified for STOT SE. The LOAEC or LOAEL of a study is divided by the lower cutoff for the category, the generic value (GV), and the result is reduced to the next lower integers, 5, 2 or 0, and then multiplied with 100%. For example, if a substance in an acute dermal study has caused severe liver effects at a dose of 60 mg/kg (LOAEL), the substance would belong to STOT SE category 1. The SCL is now calculated as

$$\frac{LOAEL}{GV_1} \times 100\% = \frac{60 \text{ mg/kg}}{1,000 \text{ mg/kg}} \times 100\% = 0.6\% \rightarrow 0.5\% = SCL(STOT \text{ SE } 1)$$

In this example, the SCL is 0.5% and is to be used instead of the GCL of 10%. A mixture containing 0.5% or more of this substance has to be classified as STOT SE category 1. The SCL for STOT SE category 2 is

$$\frac{LOAEL}{GV_2} \times 100\% = \frac{60 \text{ mg/kg}}{2,000 \text{ mg/kg}} \times 100\% = 0.3 \rightarrow 0.2\% = SCL(STOT \text{ SE } 2)$$

NOTE: if the calculation of the SCL results in a higher value then the GCL, the SCL shall be discarded, and the GCL shall be used for classification of mixtures.

For Specific Target Organ Toxicity, derive and use a Specific Concentration Limit (SCL) for the classification of preparations, if LOAEL/LOAEC data are available, and if the SCL is lower than the Generic Concentration Limit (GCL).
Calculated results are rounded down to integers 5, 2, 1 or 0.

Occupational experience is a frequent cause for the classification of a substance as *STOT SE category* 3. H335 – may cause respiratory irritation – is allocated to substances causing irritation and comparable sensations to the respiratory tract, and H336 – may cause drowsiness or dizziness – is frequently allocated to solvents with narcotic effects. Such properties may be found in exposed workforces even if animal experiments did not indicate them.

For STOT SE category 3, the mixture is classified with H335 or H336, if the sum of all ingredients with either H335 or H336 reaches or exceeds a content of 20%. This summation covers all ingredients present at or above 1.0% (or less, if there is a lower SCL). Examples for hypothetical mixtures of STOT SE category 3 substances are given in the following table. STOT SE category 3 substances and mixtures are labeled with the GHS07 symbol (exclamation mark) and the signal word "WARNING"!

Tab. 39: Classification of mixtures against STOT SE category 3, GCL.

Component	Hazard phrase	Mixture A	Mixture B	Mixture C	Mixture D
Toluene	H336	10%	10%	12%	10%
Diethyl ether	H336	5%	5%	12%	10%
Dichloro-dimethylsilane	H335	8%	18%	10%	17%
Chloro-trimehtylsilane	H335	8%	8%	0.1%	4%

Mixture A is not to be classified as STOT SE category 3, but the other mixtures are; mixture B gets H335, mixture C H336, and mixture D is to be classified with H335 and H336.

Tris-o-cresylphosphate is an example for a substance STOT SE category 1. In the past, it was used as plasticizer, solvent and lubricant. Single exposure can cause a die-back of neuronal axons with some delay, predominantly in legs, followed by the arms, and causing paralysis. Chicken are test animals which reacted more sensitive against this effect than the rat. The specific concentration limit for STOT SE category 1 is 1%, the specific concentration limit for STOT SE category 2 is 0.2%.

Tab. 40: Labeling for STOT SE.

	Category 1	Category 2	Category 3
Pictogram	GHS08	GHS08	GHS07
Signal Word	Danger	Warning	Warning
Hazard Statement	H370: Causes damage to organs (*state organ(s) and relevant route(s) of exposure*)	H371: May cause damage to organs (*state organ(s) and relevant route(s) of exposure*)	H335: May cause respiratory irritation; H336: May cause drowsiness or dizziness.

Typical precautionary statements for STOT SE category 1 substances
- P260: Do not breathe dust/fumes/gas/mist/vapors/spray.
- P264: Wash . . . thoroughly after handling. (*hands, body, . . . :specify*)
- P308 + P311: If exposed or concerned: Call a POISON CENTER/ doctor/ . . .
- P321: Do NOT induce vomiting.
- P405: Store locked up.
- P501: Dispose of contents/container to . . . (*specify*)

6.8.2 Irritation and corrosion to skin and eyes

Skin irritation and corrosion are local effects where the substance causes tissue damage at the side of contact. In general, corrosion causes an irreversible effect, while irritation is reversible. Classical animal tests were developed in the 1940's, typically using the rabbit as sensitive species. The test substances are applied on the shaved skin for a certain time (during which the substance is occluded so it cannot be wiped off) or as liquid or finely ground powder in the eye. After the prescribed contact time, the result is taken.

Corrosive and irritating substances are very painful for the experimental animals. Therefore, according to OECD test guidelines on corrosion/irritation testing and according to regulations 440/2008/EC and 1272/2008/EC, alternative methods have to be checked first. The algorithm is:
- substances/mixtures with a pH ≤ 2 or pH ≥ 11.5 are corrosive by definition, unless . . .
  - . . . the investigation of alkaline or acidic reserve indicates the substance is probably not corrosive (low reserve); in that case, proceed to in vitro testing.
- Hydroperoxides are corrosive.
- Peroxides are at least irritating.
- If in vitro testing shows the substance is neither corrosive nor irritating, proceed to in vivo testing.

Skin effects category 1 are corrosions. This category is sub-divided in categories 1A, 1B and 1 C, depending on the contact time needed to induce the irreversible effect (Tab. 41).

Tab. 41: Categories of corrosive substances.

Contact time. . .	. . . resulting in irreversible damage within. . .	Category
≤ 3 min	≤ 1 h	1A
> 3 min, ≤ 1 h	≤ 14 d	1B
≤ 4 h	≤ 14 d	1 C

Category 1 triggers the hazard phrase H314: causes severe skin burns and eye damage, and the pictogram GHS05. For skin irritation, the hazard phase is H315: causes skin irritation, and the pictogram GHS07 is used. Irreversible eye damage leads to H318: causes severe eye damage and GHS05 label (category 1), nonirreversible, but severe damage leads to H319: causes severe eye irritation and GHS07 label.

Mixtures are classified according to their composition. Mixtures with a pH not higher than 2 or not lower than 11.5 shall be considered as corrosive, unless the acid reserve/alkaline reserve is low. In case of low alkaline/acid reserve, the mixture may be tested with in vitro systems. However, if the pH of the mixture is not 'extreme' or the low alkaline or acid reserve indicates no corrosive properties, and if the ingredients of the mixture were tested for irritation/corrosion, the properties of the mixture may be calculated ('bridging principle').

The regulation 1272/2008/EC distinguishes between "additive" and "nonadditive" behavior for irritation and corrosion. Corrosion- and irritation-like damage can not only be triggered by pH shift; direct toxicity to skin cells (keratinocytes), other chemical reactivity, effects on the blood capillaries, or surface activity can also induce such local damage. Surface active agents may disintegrate the cell membrane; aldehydes (glutaraldehyde) can crosslink proteins, and phenols (2,4-dichloro phenol) can precipitate or inhibit proteins and cause cell death; skin absorptive compounds of heavy metals can inhibit enzymes in the keratinocytes; dimethyl sulfate inhibits proteins by alkylation; potassium dichromate, silver nitrate and hydrogen peroxide are examples of substances probably acting via oxidative damage. Substances which do not behave additively have typically very low specific classification limits of < 1%. The following listing may help to differentiate "classical" from "nonclassical" irritating and corrosive substances.

- strong influence on the ion product of water (pH-value) indicates classical behavior.
- non-classic behavior for corrosion/irritation is induced or indicated by, for example,
  - strong surface activity
  - classification as oxidant (hydrogen peroxide)
  - classification as acute toxic substance (category 1 > cat. 2 > cat. 3 > cat. 4; dermal > oral > inhalation)
  - cell damage by alkylation and/or crosslinking
    - Bis(2-chloroethyl)-sulfide (S-Lost) and Tris(2-chloroethyl)-nitrile (N-Lost) are alkylating, skin damaging warfare agents. Dimethyl sulfate is an alkylating substance classified as corrosive.

In case of additive behavior, the mixture is classified and labelled according to Tab. 42. If a mixture contains 3% of a substance corrosive 1A and 3% of a substance corrosive 1B and additivity can be assumed, the mixture has to be classified as corrosive category 1B as only the sum of both corrosives exceeds the generic concentration limit.

If a substance has a specific concentration limit, the specific limit has to be chosen. In such cases, for additive substances Tab. 42 is no longer useful. An algorithm is used instead:

$$\text{If } \sum \frac{C_i}{\text{concentration limit for } i} \geq 1 \text{: classify}$$

In that equation, use specific concentration limits whenever possible. For substances with no specific concentration limit, take the generic concentration limit.

If a mixture is to be classified as corrosive, the label GHS05 has to be applied; GHS07 (exclamation mark) must not be used for labelling irritating properties; however, if the mixture is – for example – also harmful if swallowed, then the label GHS07 has to be added.

Tab. 42: Classification of mixtures (2 < pH < 11.5) containing additive corrosive and/or irritating substances, generic concentration limits.

Components	Mixture	
	Category 1	Category 2
Sum of category 1 substances	≥ 5%	≥ 1%, < 5%
Sum of category 2 substances	–	≥ 10%
10 × {Sum of category 1 substances} + sum of category 2 substances		≥ 10%

How the subcategories work for mixtures shall be explained by an example. Assume a mixture is composed of components as shown in Tab. 43. The corrosive/irritating properties shall be additive.

Tab. 43: Sample mixture skin corrosion/irritation; effects are additive, and 2 < pH < 11.5.

Component	Classification (skin)	Content
Acid X	1A	3%
Acid Y	1C	3% (0%)
Additive Z	2	2%
Water	None	92% (95%)

Acid Y may either be present or absent. In presence of acid Y, the calculation runs as follows:

- 3% acid category 1A; < 5% => mixture is not category 1A.
- 3% acid category 1A + 3% acid category 1 C = 6% (of acid, at least 1 C) > 5% => mixture is category 1 C

This result would also be generated by the algorithm shown above. For corrosion, testing for 1 C:

$$\frac{3\% \ (\text{acid } X)}{5\%} + \frac{3\% \ (\text{acid } Y)}{5 \%} = \frac{6}{5} > 1$$

If we now omit acid Y and level up the content of water from 92 to 95%, the calculation is:
- 3% acid category 1A; < 5% = > mixture is not category 1A.
- 10 × 3% acid category 1A + 2% additive category 2 = 32% > 10% = > mixture is category 2.

With the algorithm mentioned above, the calculation would read, for category 1,

$$\frac{3\% \ (\text{acid } X)}{5 \%} < 1$$

that is: no classification as category 1. Checking for category 2:

$$\frac{3\% \ (\text{acid } X)}{1 \%} + \frac{2\% \ (\text{additive } Z)}{10 \%} = \frac{32}{10} > 1$$

and consequent the mixture has to be classified as category 2.

In case of non-additive components, Tab. 44 is to be used. Here, the irritating/corroding substances are not added up.

Tab. 44: Classification of mixtures (2 < pH < 11.5) containing non-additive corrosive and/or irritating substances, generic concentration limits.

Component	Concentration limit	Mixture
Alkaline with pH ≥ 11.5	1%	Category 1
Acidic with pH ≤ 2	1%	Category 1
Other category 1 corrosive, not additive	1%	Category 1
Other irritant category 2	3%	Category 2

Substances causing eye damage are allocated to category 1 (H318: Causes serious eye damage) or category 2 (H319: causes serious eye irritation). Concerning classification of mixtures against local effects on the eyes, there is also a distinction made between "additive" and "nonadditive" ingredients. To stay on the safe side, in case of any doubt whether ingredients are additive or not, calculate the mixture via both ways. An over-labelling is less likely causing problems at the workplace than a missing label.

Concerning eye-damage, it is expected that every substance corrosive to the skin can cause serious eye damage, so testing substances corrosive to the skin against eye effects is rejected.

Tab. 45: Classification of mixtures (2 < pH < 11.5) against local eye effects, additive components, generic concentration limits.

Component	Concentration limit	Mixture
Sum of category 1 substances (eye and skin)	≥ 1%, < 3%	Category 2
Sum of category 1 substances (eye and skin)	≥ 3%	Category 1
Eye effects category 2	≥ 10%	Category 2
10 × {category 1 substances} + category 2	≥ 10%	Category 2

Tab. 46: Classification of mixtures (2 < pH < 11.5) against local eye effects, non-additive components, generic concentration limits.

Component	Concentration limit	Mixture
Alkaline with pH ≥ 11.5	≥ 1%,	Category 1
Acidic with pH ≤ 2	≥ 1%,	Category 1
Other ingredient category 1	≥ 1%,	Category 1
Other ingredient category 2	≥ 3%	Category 2

For local effects on the skin as well as on eyes, ingredients of mixtures are only taken into account if they are present at 1% or more. Exempted are those substances which have lower, specific concentration limits (SCF), which will be covered down to their SCL. For illustration, a hypothetical mixture is listed in Tab. 47.

Tab. 47: Mixture (2 < pH < 11.5) with ingredients classified for eye effects, additivity is applicable.

Component	Classification	Concentration (%)	SCL
Component A	Eye damage	0.1	No
Component B	Eye damage	1.2	No
Component C	Eye damage	0.8	≥ 0.1% eye irritation, ≥ 1% eye damage
Component D	Eye irritation	2	≥ 5% eye irritation
Water	None	95.9	

In this mixture, component A can be neglected as it is below the consideration limit of 1%; component C cannot be neglected as it is present at /above its SCL. For category 1 (eye damage), the calculation runs as follows:

$$\frac{[B]}{GCL_B} + \frac{[C]}{SCL_C} = \frac{1.2}{3} + \frac{0.8}{1} = \frac{3.6}{3} = 1.2 > 1$$

so the mixture is classified as causing serious eye damage, category 1. If component B and D would be replaced by water, the mixture is no longer category 1, but it is still category 2 due to the lower SCL of component C.

As final example in this chapter, estimate the C&L with respect to irritation/corrosion of a mixture of 100 g of 30% hydrochloride acid with 56 g of 25% ammonia solution. Both substances are listed in Appendix 2, and the reader now may start his own calculations before further reading.

While looking up data in Appendix 2, the reader may have realized that both substances have specific concentration limits, and that mixing 100 g of a 30% hydrochloric acid with 56 g of something else makes a 19.2% hydrochloric acid, and 56 g of a 25% ammonia solution mixing with 100 g of something else makes a 16% ammonia solution, and these results might be matched against the SCLs. However, actually, in this mixture, 30 g HCl (M = 36.5 g/mol) = 0.82 mole HCl meet 14 g NH_3 (M = 17 g/mol) = 0.82 mol NH_3, and both react with each other forming 0.82 mol ammonium chloride, which results in 44 g/156 g = 28.2% NH_4Cl in water. Ammonium chloride is listed in the Annex as eye irritant (cat. 2); therefore, the correct classification would be H319: causes serious eye irritation, together with GHS07.

Typical precautionary statements in combination with corrosive substances and mixtures are

- P260: Do not breath dust/fume/gas/mist/vapors/spray.
- P264: Wash . . . thoroughly after handling. (p. e. hands).
- P280: wear protective gloves/protective clothing/eye protection/face protection.
- P301 + P330 + P331: IF SWALLOWED: Rinse mouth. Do not induce vomiting.
- P303 + P361 + P353: IF ON SKIN (or hair): Take off immediately all contaminated clothing. Rinse skin with water [or shower].
- P363: Wash contaminated clothing before reuse.
- P304 + P340: IF INHALED: Remove person to fresh air and keep comfortable for breathing.
- P310: Immediately call a POISON CENTER/doctor/ . . .. (p. e. the supplier).
- P305 + P351 + P338: IF IN EYES: Rinse cautiously with water for several minutes. Remove contact lenses, if present and easy to do. Continue rinsing.

P351 recommends to remove contact lenses, if present and easy to do so. However, it is recommended not to wear at all contact lenses while working with corrosive or irritating materials.

If animal tests have demonstrated that the mechanism for inhalation toxicity is corrosion of the respiratory tract, then EUH071 – 'Corrosive to the respiratory tract' has to be added on the label.

If tests for eye irritation/serious eye damage have produced severe systemic toxicity or even death in test animals, the EUH070 – 'Toxic by eye contact' has to be added on the label.

6.8.2.1 Acidic and alkaline reserve

The acid and alkaline reserve for substances and mixtures having a pH ≤ 2 or pH ≥ 11.5 is elaborated in the publication of Young et al. [52]. The method is applicable for "classical" corrosives, only, that is such substances which act via the pH value of aqueous solutions. The reserve is defined as follows:

- acidic reserve: amount of NaOH (g) needed per 100 g substance or mixture, to arrive at a pH-value of 4.0.
- alkaline reserve: amount of NaOH (g), that is the molar equivalent to moles acid needed per 100 g substance or mixture, to arrive at a pH-value of 10.0.

The substance is titrated with either 2 N NaOH or 2 N H_2SO_4.
Decision logic for classification:
- for an acid: pH – 1/12 acidic reserve ≤ –0.5: classify as corrosive.
- for a base: pH + 1/12 alkaline reserve ≥ 14.5: classify as corrosive.

The procedure shall be explained by theoretical examples, 5% HNO_3 and 20% HNO_3. For simplicity we assume that the density of the solutions is 1 g/cm^3. For HNO_3, the dissociation constant is 22.44, and the molar mass is 63 g/mol. 100 g 5% HNO_3 contains 5 g = 0.079 mol, and the concentration is 0.79 mol/L. In equilibrium, x mol of HNO_3 dissociate, and the equilibrium can be calculated as

$$K = \frac{[H^+] \times [NO_3^-]}{[HNO_3]} \Leftrightarrow 22.44 = \frac{x}{0.79 - x} \Leftrightarrow x = [H^+] = 0.765\frac{mol}{L}$$

The calculated pH is 0.12. Now, by titration the pH has to be brought to 4.0. As HNO_3 is a strong acid, this would mean that approximately 0.765 mol/L NaOH would be needed, which is 0.0765 mol/100 mL (~ 100 g), and this is 3.06 g NaOH. The acidic reserve, therefore, is 3.06. The calculation yields:

$$0.12 - 1/12 \times 3.06 = -0.13$$

As this value is greater than –0.5, the 5% HNO_3 is probably not corrosive, although the pH is below 2. In vitro tests for corrosion should be run for clarification. If we start the calculation with 20% HNO_3, the calculated initial pH is –0.84 and below –0.5, already. Therefore, 20% HNO_3 is to be classified as corrosive (which is the official classification).

For sodium hydroxide, a 1% solution contains 0.25 mol/L. As NaOH is a strong electrolyte, a concentration of 0.25 mol/L of OH^- is assumed. The initial pH value is 13.4, which would call for a classification as corrosive. To achieve a pH of 10 – that is, the concentration of OH^- is 0.0001 – an amount of (0.25–0.0001) = 0.2499 mol/L of acid has to be added, which is 0.02499 mol per 100 mL (~ 100 g). Molar equivalent of NaOH is 0.996 g, which is the alkaline reserve. The alkaline reserve is calculated as

$$pH + \frac{1}{12} \times \text{alkaline reserve} = 13.4 + \frac{1}{12} \times 0.996 = 13.483$$

This value is below 14.5, and 1% NaOH may not be corrosive. Indeed, the official classification for NaOH solutions has the specific concentration limits: skin corr. 1A, H314: $C \geq 5\%$; skin corr. 1B, H314: $2\% \leq C < 5\%$; skin irr. 2, H315: $0.5\% \leq C < 2\%$.

In these explanatory examples, the exercises where purely of theoretical nature. This implies that the electrolytes behave as ideal electrolytes. In reality, always perform pH measurements and titrations.

6.8.3 Sensitization

Sensitization describes the process of the acquisition of an over-sensitivity of the organism (allergy). Organisms need to have a defense system against smaller, parasitic organisms. The immune system can attack intruders via antibodies, via macrophages and via cytotoxic cells. Antibodies add to foreign organisms and cause their precipitation or make them more visible for macrophages and/or cytotoxic cells. Macrophages ingest and digest foreign cells, cytotoxic cells attack them, but they also attack infected own cells, destroying the multiplication basis for the germs. An overshooting immune response is called an allergy.

6.8.3.1 Skin sensitization
In the past, skin sensitizers where either identified by observation at workplaces or case records from consumers, or a guinea pig test was run which was rated positive. The guinea pig tests used either topical administration (Buehler-Tests) or intradermal application of the substance (maximization test). Under current EU regulations, the local lymph node assay in the mouse is the preferred test method, looking after that test substance concentration that induces an activation of the draining lymph node of the painted skin. The classification depends on the concentration of the substance required to have a defined minimum stimulation index (SI). If radioactive thymidine is used as a marker for stimulation, a 3-fold increase in radioactivity in the draining lymph node cells is the benchmark to identify skin-sensitizers.

Skin sensitizers are allocated to one single category: category 1. However, depending on available results, a sub-categorization may be possible into 1A (extreme sensitizers) and 1B (none-extreme sensitizers) (Tab. 48). Category 1A comprises sensitizers which show a high sensitization frequency in human beings, or animal experiments indicate there is potential for a high sensitization frequency; or a small amount of substance is sufficient to induce sensitization in tests with volunteers. Alternatively, if at least 0.2% of the general population, or at least 0.4% of randomly selected, exposed workers are sensitized or there are at least 100 reported and confirmed positive cases, the substance is a category 1A sensitizer. Category 1B comprises

Tab. 48: Categories of skin sensitizers based on animal test results.

Test	Endpoint	Sensitizer category	
		Extreme/strong (1A)	**Moderate (1B)**
Local lymph node assay	% concentration that triggers an SI of 3 (EC 3), detected as incorporated radioactive thymidine.	≤ 2%	> 2%
Buehler test	Number of positive animals	≥ 15% responding at ≤ 0.2% induction dose; or ≥ 30% responding at induction dose > 0.2% and ≤ 20%	≥ 30% and < 60% responding at > 0.2% and ≤ 20% induction dose or ≥ 30% responding at induction dose > 0.2% and ≤ 20% or ≥ 15% responding at > 20% induction dose.
Maximization test	Number of positive animals	≥ 30% response at ≤ 0.1% induction dose, or ≥ 60% response at ≥ 0.1 and < 1.0% induction dose.	≥ 30%, < 60% response at ≤ 0.1–≤ 1.0% induction dose, or > 30% response at ≥ 1.0% induction dose.

sensitizers with moderate to low frequency of sensitization in human beings (or animal experiments indicate a low to moderate potential).

The mouse local lymph node assay (LLNA) results are easier quantifiable than the guinea pig test results. Based on the concentrations of the substance in vehicle required to gain an SI of 3, a categorization as extreme, strong or moderate is possible (Tab. 49).

Tab. 49: Potency of skin sensitizers based on LLNA results, and specific or generic concentration limits for mixtures.

Concentration required for an SI ≥ 3	Potency	Category
> 2%	Moderate	1B
> 0.2%, ≤ 2%	Strong	1A
≤ 0.2%	Extreme	1A

Labelling for skin sensitization is the symbol GHS07 with signal word "WARNING" and hazard statement H317: may cause an allergic skin reaction. For the classification of mixtures, different skin sensitizers are not added up. The mixture is classified as a skin sensitizer if the specific or the generic concentration limit of an ingredient is reached.

The criterion "SI ≥ 3" is used for the incorporation of radiolabeled thymine in the draining lymph node. Other criteria can be chosen, for example lymph node weight, and then other SI-cutoffs may apply, as defined in the respective guidelines.

To calculate the SI cut-off, two different algorithms are applied, depending on the outcome of the test. If the test incorporates concentrations where the SI is below as well as above 3, the EC3 is calculated as

$$EC3 = \frac{3 - [d]}{[b] - [d]} \times ([a] - [c])$$

and if the lowest dosage triggers an SI > 3, already, algortihm to be used is,

$$EC3 = 2^{\left\{ \log_2[C] + \frac{3 - [d]}{[b] - [d]} + (\log_2[a] - \log_2[c]) \right\}}$$

In these equations, [a] = lowest concentration with a response SI > 3, [b] = SI at concentration [a], [c] = highest concentration that creates a response SI < 3, and [d] = SI at concentration [c].

Concerning mixtures, the following concentration limits apply, whereby different sensitizers are not summed up (Tab. 50).

Tab. 50: Classification of mixtures for skin sensitization, generic concentration limits (GCL).

Component	Generic concentration limit	Mixture
Category 1A	0.1%	Category 1A
Category 1B	1.0%	Category 1B
Category 1	1.0%	Category 1

For extreme sensitizers, a specific classification limit as low as at least 0.001% is recommended. If the concentration of the sensitizer is below the classification limit, but still as high as 10% of the classification limit, the identity of the sensitizer has to be disclosed on the label and in the safety data sheet. That is, if the mixture contains 0.08% of a skin sensitizer category 1A and the generic classification limit applies, the mixture itself is not classified as skin sensitizer; the name and identity of this sensitizer, however, has to be put on the label, and the sensitizer has to be mentioned in the safety data sheet. This shall ensure that people already sensitized can avoid contact. Chemical structures of some skin-sensitizers are shon in Fig. 99. Typical precautionary statements used for skin sensitizers are

- P261: Avoid breathing dust/fumes/gas/mist/vapors/spray.
- P272: Contaminated work clothing should not be allowed out of the workplace.
- P280: Wear protective gloves/protective clothing/eye protection/face protection.

- P302 + P352: IF ON SKIN: Wash with plenty of water/ . . .
- P333 + P313: If skin irritation or a rash occurs: Get medical advice/attention.
- P321: Do NOT induce vomiting.
- P362 + P364: Take off contaminated clothing and wash it before reuse.

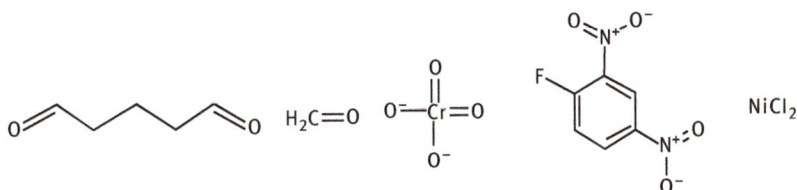

Fig. 99: Examples for skin sensitizers: glutaraldehyde, formaldehyde, chromate, fluoro-2,4-dinitrobenzene, nickel compounds.

6.8.3.2 Respiratory sensitization

Different to skin sensitization, there is currently no generally accepted standard test method for respiratory sensitization classification and labeling. Many respiratory sensitizers were identified by occupational medicine. As for skin sensitizers, there is only one category – 1 – which is subdivided in category 1A – strong sensitizers – and category 1B – other sensitizers. In case of high frequency of occurrence in human beings, the substance is allocated to category 1A; this allocation may also be based on animal experience that indicates a high frequency might be possible in human beings. The severity of effects may be considered. Low or moderate frequency in human beings (actual or prospective) results in allocation into category 1 B. There are in vivo test methods available, p. e. making use of guinea pigs (respiratory response), mice or rats (inflammatory response, IgE antibodies). As for skin tests, there is an induction phase and an elicitation phase. Today, substances are classified as respiratory sensitizers due to experience at workplaces. Typical symptoms (together or isolated) can be: coughing; sneeze attacks; labored breathing; running nose; chest tightness; fever.

Tab. 51: Generic concentration limits for respiratory sensitizers.

Component	Mixture	
	Solid/liquid	Gas
Category 1	≥1.0% (category 1)	≥ 0.2% (category 1)
Category 1A	≥0.1% (category 1A)	≥ 0.1% (Category 1A)
Category 1B	≥1.0% (category 1B)	≥ 0.2% (category 1B)

Classification as respiratory sensitizer is done with the GHS08 symbol, signal word "DANGER" and hazard statement H334: May cause allergy or asthma symptoms or breathing difficulties if inhaled. As for skin sensitizers, a mixture is classified as respiratory sensitizer if the concentration of a respiratory sensitizing component meets or exceeds its specific or the generaic concentration limit (Tab. 51). If a substance or mixture is classified as respiratory sensitizer and as skin sensitizer, the label GHS07 (exclamation mark) has to be omitted. However, if that material is – for example – in addition harmful in contact with skin, the GHS07 symbol has to be added on the label.

If the concentration of the respiratory sensitizer is below the classification limit, but at least 10% of the classification limit, the identity of the sensitizer has to be disclosed on the label and in the safety data sheet. That is, if the mixture contains 0.5% of a respiratory sensitizer category 1, and the generic concentration limit applies (1.0%), the mixture itself is not classified as respiratory sensitizer; the name and identity of this sensitizer, however, have to be put on the label, and the sensitizer has to be mentioned in the safety data sheet, because it is present above one tenth of the respective classification limit. This shall ensure that people already sensitized can avoid contact. Molecular structures for some respiratory sensitizers are shown in Fig. 100. Precautionary statements used for respiratory sensitizers are

– P261: Avoid breathing dust/fumes/gas/mist/vapors/spray.
– P284: [In case of inadequate ventilation] wear respiratory protection.
– P304 + P340: IF INHALED: Remove person to fresh air and keep comfortable for breathing.
– P342 + P311: If experiencing respiratory symptoms: Call a POISON CENTER/doctor/ . . .

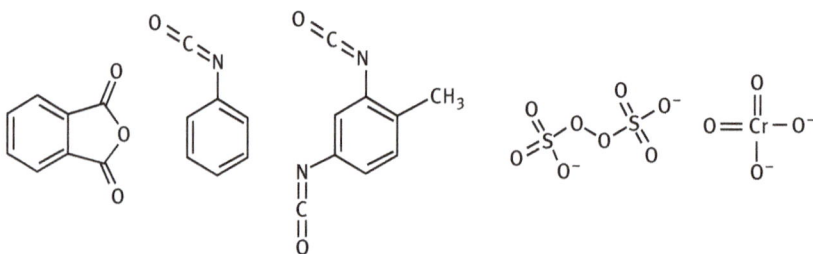

Fig. 100: Examples of respiratory sensitizers: phthalic acid anhydride, phenyl isocyanate, toluenediisocyanate, persulfates, chromates.

6.8.4 Repeated dose toxicity: organ toxicity

Repeated dose toxicity tests aim at identification of sub-lethal, nevertheless critical effects substances can induce after repeated exposure. Severe or even irreversible tissue damage and/or organ failure in subchronic studies (90 d exposure) can result in classification for STOT RE. If the lowest observed adverse effect level (LOAEL) or the lowest observed adverse effect concentration (LOAEC) is below a certain level, the substance is classified for STOT RE category 1 or 2.

For studies of shorter or longer duration, Haber's Rule may be applied (dose × time = constant), but tests of shorter duration than 90 d not showing any adverse outcome up to the top dose (typically 1,000 mg/kg b.w. for oral exposure) are not a proof for the absence of STOT RE; due to the limited exposure time and fewer animals compared to the subchronic study, the subacute study might be too insensitive. However, if adverse effects are produced in a subacute study, the LOAEL or LOAEC may be used for classification (Tab. 52).

Tab. 52: Generic values (GV) for classification of a substance as STOT RE category 1 or 2.

Route of exposure	Unit	Subchronic study (90 d)		Subacute study (28 d)	
		Category 1, GV_1	Category 2, GV_2	Category 1	Category 2
Oral	mg/kg b.w.	LOAEL ≤ 10	10 < LOAEL ≤ 100	LOAEL ≤ 30	30 < LOAEL ≤ 300
Dermal	mg/kg b.w.	LOAEL ≤ 20	20 < LOAEL ≤ 200	LOAEL ≤ 60	60 < LOAEL ≤ 600
Inhalation, gas	ppm (V)/6 h	LOAEC ≤ 50	50 < LOAEC ≤ 250	LOAEC ≤ 150	150 < LOAEC ≤ 750
Inhalation, vapor	mg/L/6 h	LOAEC ≤ 0.2	0.2 < LOAEC ≤ 1.0	LOAEC ≤ 0.6	0.6 < LOAEC ≤ 3.0
Inhalation, dust/ mist	mg/L/6 h	LOAEC ≤ 0.02	0.02 < LOAEC ≤ 0.2	LOAEC ≤ 0.06	0.06 < LOAEC ≤ 0.6

That is, if, for example, in a subacute inhalation study with rats exposed against a solid material dust the LOAEC is 0.5 mg/L/6 h, causing oliguria (impaired kidney function), the substance to be classified as STOT RE category 2. Classification and labelling for STOT RE is listet in Tab. 53.

Tab. 53: Labelling for STOT RE categories 1 and 2.

	Category 1	Category 2
Pictogram	GHS08	GHS08
Signal word	Danger	Warning
Hazard statement	H372: Causes damage to organs *(name organ and exposure route, if only one exposure route is relevant)*	H373: May cause damage to organs *(name organ and exposure route, if only one exposure route is relevant)*

Typical precautionary statements used for STOT RE substances are
- P260: Do not breathe dust/fumes/gas/mist/vapors/spray.
- P264: Wash . . . thoroughly after handling. (*hands, body,* . . . :*specify*)
- P270: Do not eat, drink or smoke when using this product.
- P314: Get medical advice/attention if you feel unwell.
- P501: Dispose of contents/container to . . . (*specify*)

Whether or not a result in a toxicological test is adverse, and whether it is relevant for human beings may be a contentious issue which needs to be left to experts for clarification. One example is impairment of kidney function; if this occurs in male rats only, and the mechanism is the accumulation of a certain protein in tubulus cells of the male rats, this has no relevance for human beings.

Experiences from occupational medicine and cases of intoxication of people may also result in a classification of a substance as STOT RE. This decision has to be left to medical experts.

Mixtures containing a STOT RE substance are classified if the substance exceeds the generic or the specific concentration limit (Tab. 54). Different STOT RE substances in the mixture are not added up as they are assumed to act individually. 10% or more of a STOT RE category 1 substance makes the mixture STOT RE category 1; 1%, but less than 10% of the category 1 substance makes the mixture STOT RE category 2. Ingredients of STOT RE category 2 render the mixture category 2 if they are present at least at 10%.

Tab. 54: STOT RE generic classification limits (GCL) for mixtures.

Component	Mixture	
	Category 1	Category 2
Category 1	$C \geq 10\%$	$1\% \leq C < 10\%$
Category 2		$C \geq 10\%$

Whenever possible, specific classification limit shall be derived for a substance and be used for the classification, *provided the SCL is lower than the GCL.* To do so, the LOAEL is divided by the GV_1 and GV_2 values of Tab. 52. The result is reduced to the next integer 5, 2, 1 or 0. Example A: substance A has a dermal LOAEL of 1.0 mg/kg/d in a subchronic study. Therefore, the result is

$$SCL_1 = \frac{LOAEL}{GV_1} = \frac{1}{20} = 0.05 = 5\%;\ SCL_2 = \frac{LOAEL}{GV_2} = \frac{1}{200} = 0.005 = 0.5\%$$

Example B: In a subacute oral study, the LOAEL is 2.0 mg/kg b.w./day. The substance, therefore, is STOT RE category 1.

$$SCL_1 = \frac{LOAEL}{GV_1} = \frac{2}{30} = 0.067 \approx 6.6\% \rightarrow SCL_1 = 5.0\ \%;$$

$$SCL_2 = \frac{2}{300} = \frac{LOAEL}{GV_2} = 0.0067 \approx 0.66\% \rightarrow SCL_2 = 0.5\%$$

Examples for STOT RE substances are methanol (impaired vision or blindness), dibutyltin compounds (depressed immune system), carbon monoxide (impaired memory), divanadium pentoxide (depressed immune system).

6.8.5 Genotoxicity, mutagenicity

Genotoxicity is the overarching property of any damage – including cell death – that can be induced to and by the carrier of genetic information, the DNA. Mutation means a change in the DNA in such a way that the cell is still viable, and the cell undergoes changes in either transformation towards a malignant cell (cancer), or showing altered functions conveyed to germ cells which cause a changed phenotype (appearance) of the organism.

C&L of mutagens does not mirror the potency of the mutagen in terms of dose; the classification resembles how likely the substance is to act mutagenic in human beings. There are three categories under regulation 1272/2008/EC (Tab. 55).

Tab. 55: Classification of mutagenic substances.

Classification	Category 1A	Category 1B	Category 2
Pictogram	GHS08	GHS08	GHS08
Signals word	Danger	Danger	Warning
Hazard Statement	H340: May cause genetic defects (*State route of exposure if it is demonstrated that other routes of exposure do not cause the hazard*).	H340: May cause genetic defects (*State route of exposure if it is demonstrated that other routes of exposure do not cause the hazard*).	H341: Suspected of causing genetic defects (*State route of exposure if it is demonstrated that other routes of exposure do not cause the hazard*).

- Category 1
    - Substances known to induce heritable mutations or to be regarded as if they induce heritable mutations in the germ cells of humans.
    - 1A: Substances known to induce heritable mutations in the germ cells of humans (epidemiological evidence).
    - 1B: *positive result(s) from in vivo heritable germ cell mutagenicity tests in mammals; or* positive results in somatic cells, but there is evidence that

germ cells in humans can be affected, or positive results in human germ cells without evidence for transmission to the progeny.
- Category 2
  - Substances which cause concern for humans owing to the possibility that they may induce heritable mutations in the germ cells of humans.
  - somatic cell mutation in vivo in mammals
  - other positive in-vivo tests, together with positive in vitro tests, or structural alerts

For the C&L of mixtures, individual mutagens are not added up. The generic classification limits are 0.1% for category 1A and 1B mutagens, and 1% for category 2 mutagens. Specific concentration limits have preference (Tab. 56).

Tab. 56: Generic concentration limits for mutagenic substances.

Component	Mixture		
	Category 1A	Category 1B	Category 2
1A	≥ 0.1%		
1B		≥ 0.1%	
2			≥ 1.0%

If a mixture contains less than 1.0%, but at least 0.1% of a category 2 mutagen, a safety data sheet must be made available on request.

Precautionary statements used for mutagenic substances category 1 are
- P201: Obtain special instructions before use.
- P202: Do not handle until all safety precautions have been read and understood.
- P280: Wear protective gloves/protective clothing/eye protection/face protection.
- P308 + P313: If exposed: Call a POISON CENTER or doctor/physician.
- P405: Store locked up.
- P501: Dispose of contents/container to . . . (*specify*)

Many mutagens are also classified as carcinogens. Fig. 101 shows a few substances classified as mutagens but not carcinogens.

6.8.6 Carcinogenicity

Epidemiologic studies can identify increased cancer risk in specific exposed human populations. A group of people exposed to a certain agent is compared to a group without exposure. All other properties of both groups should be as equivalent to each other as possible. If the substance under investigation is a carcinogen, the

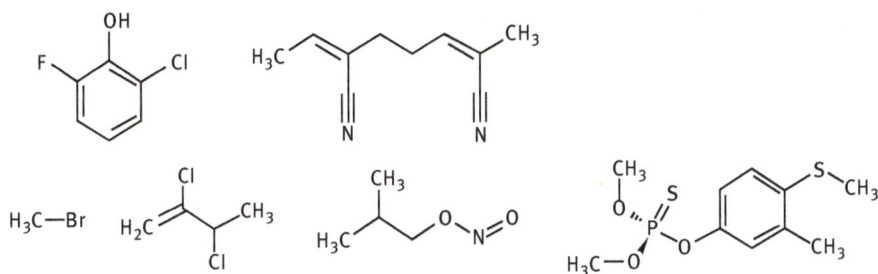

Fig. 101: Examples of substances currently classified as mutagens category 1B (top row) and category 2 (bottom row), but not classified as carcinogens.

exposed group shows more cases of cancer after a certain time (typically several years or decades). Critical is the appreciation of confounders and the proper identification of exposure. Confounders can be, for example, age and gender, personal habits (smoking, drinking), unknown exposures. By definition, statistical significance is achieved if the error for a false-positive association is below 5%; therefore, in theory in every 20th investigation it may occur that an association is seen which has no biological background but is a chance-finding. Positive findings in epidemiological studies need to be checked on biological plausibility.

Today's cancer tests in animals are run typically with rats and mice. 50 animals per gender dose (control, low, mid and high dose) are exposed for 18 (mice) or 24 months (rats). After this time, the animals of the control and high dose are investigated for tissue changes indicating cancer. If an increased tumor incidence is observed, the other two exposure groups are evaluated for this specific tumor type as well.

A carcinogen is identified if, due to statistical tests, the substance causes a significant
– increase in cancer cases at the end of the observation period, or
– increase in cancer sites in the organism, or
– reduction in onset-time of cancer.

As for mutagens, carcinogens are classified in three categories mirroring the strength of an association of exposure and cancer (Tab. 57)
– Category 1: Known or presumed carcinogen.
  – Category 1A: substance known to have carcinogenic potential for human beings, typically based on human evidence (epidemiological studies).
  – Category 1B: substance is presumed to have carcinogenic potential in human beings; either there is human evidence (but not strong enough for category 1A), or data from animal experiments raise concern: that is, two independent, positive studies at exposures which are relevant for human beings, or there is one positive animal study and convincing mutagenic potential.

– Category 2: Suspected human carcinogen. There is some evidence from human experience or from animal studies, but not sufficiently convincing to place the substance in category 1. This may happen because the methodologies of the investigations are questionable, increase in cancer incidence is borderline, absence of mutagenicity, increased cancer incidence at very high exposures, only etc.

Tab. 57: C&L of carcinogenic substances.

Classification	Category 1A	Category 1B	Category 2
Pictogram	GHS08	GHS08	GHS08
Signals word	Danger	Danger	Warning
Hazard Statement	H350: May cause cancer (*State route of exposure if it is demonstrated that other routes of exposure do not cause the hazard*).	H350: May cause cancer (*State route of exposure if it is demonstrated that other routes of exposure do not cause the hazard*).	H341: Suspected of causing cancer (*State route of exposure if it is demonstrated that other routes of exposure do not cause the hazard*).

An example for highlighting the route of exposure is the case of cobalt and its compounds. In ANNEX VI of Regulation (EU) No. 1272/2008, several cobalt compounds are listed with H350i: May cause cancer if inhaled. Certain cobalt compounds are not classified as carcinogen. Cobalt is an essential element and the central atom in vitamin B_{12}. Chronic inhalation studies with $CoSO_4$ as aerosol in rats and mice caused increased tumor incidences.

Typical Precautionary Statements in combination with a classification as carcinogen category 1 are
– P201: Obtain special instructions before use.
– P202: Do not handle until all safety precautions have been read and understood.
– P280: Wear protective gloves/protective clothing/eye protection/face protection.
– P308 + P313: If exposed: Call a POISON CENTER or doctor/physician.
– P405: Store locked up.
– P501: Dispose of contents/container to . . . (*specify*).

A mixture has to be classified as carcinogenic category 1 if it contains 0.1% of an individual carcinogen category 1, and category 2 if it contains at least 1% of an individual carcinogen category 2 (Tab. 58). If SCLs are available, they overwrite the GCLs.

Examples for *category 1A* carcinogens are asbestos, benzene, vinyl chloride, smoking, passive smoking, ethanol, 2-amino-naphthalene, 4-amino-biphenyl and 4,4′-diaminobiphenyl.

Tab. 58: Generic concentration limits for carcinogens in mixtures.

Component	Mixture		
	Category 1A	Category 1B	Category 2
1A	≥ 0.1%		
1B		≥ 0.1%	
2			≥ 1.0%

Examples for *category 1B* carcinogens are 2-amino-toluene, buta-1,3-diene, chromium (VI), ethylene oxide.

Examples for *category 2* carcinogens are aniline and thio-urea.

6.8.7 Toxicity to reproduction

Interest in toxicity to reproduction has gained much impetus with the thalidomide-catastrophe in the 1960's, when an analgesic given to pregnant women caused crippled legs and arms in their children. Toxicity to reproduction has two main branches:
- Developmental toxicity.
- Disturbed fertility

An additional effect that may cause a classification is a potential hazard for the breastfed child, if a substance can be transduced from the mother to the child via milk.

Concerning developmental toxicity, agents can affect the development of the young organism. The embryo may die in the uterus, an abortion may occur. The young organism may have lacking or delayed development on a physical basis (bones, tissues, body size) and/or on a mental/neurological basis. Retarded development may ameliorate over time, or it may be persistent. The malformation of limbs, or lack of limb formation is called teratogenicity.

Agents can disturb male and/or female fertility. This may happen by interference with sexual hormones (so the maturation of the germ cells is affected), or by damage to the relevant tissue or germ cells over their different stages. To detect effects on fertility, male and female animals receive the substance before mating, so a whole estrous cycle or the time span for complete sperm maturation is covered.

The preferred animal for reproduction toxicity testing is the rat. Mice or rabbits are the typical "second" species. Tests in a second species may be required by the agency if results from the first test are not clearly adverse, or if the exposure situation for the general population calls for a high reliability that adverse effects in human beings can be excluded.

A screening study is run with 10 male and female rats which receive the test substance prior to mating. After mating, the females are dosed until day 4 after birth. All animals are investigated histopathological, and litters are evaluated according to gender distribution, number of pups, birth weight and any malformation.

A teratogenicity study starts with pregnant dams which are exposed over the time from just after implantation to just before birth. Pups are excised by caesarian section and investigated for skeletal and visceral malformation.

In a one-generation study, animals are exposed prior to mating, during gestation and after birth until the pups are juvenile. In a two-generation study, the progeny from the parent generation is mated and allowed to litter, and exposure is maintained across the generations.

Epidemiological studies can investigate the sperm quality or the reproduction success of exposed workers.

Classification is done according to the likelihood of the substance expected to cause reproduction toxicity effects in human beings (Tab. 59).

- Category 1: Known or presume human reproductive toxicant.
    - Category 1A: known human reproductive toxicant, typically based on human evidence.
    - Category 1B: presumed human reproductive toxicant, mainly based on data gained with experimental animals. The quality of effects and the exposure conditions raise concern that similar effects may occur in exposed human beings.
- Category 2: Suspected human reproductive toxicant. Substances are classified as category 2 if there is some evidence for toxicity to reproduction, but the data are not sufficient for category 1. This may occur if it is uncertain whether the effect observed is owed to a specific sensitivity of the species and was not observed in other species, and/or high dosages cause the effect in combination with maternal toxicity.

If an effect in offspring is very likely attributable to maternal toxicity, the substance may not be classified as toxic for development. For example, if significant reduced body weight in offspring occurred in dams with severely reduced body weight gain, only, the substance is most likely not toxic to reproduction. As is typical for CMR toxicity, the debate on questionable results has to be left to experts in that field.

Tab. 59: C&L for reproduction toxicity.

	Category 1A	Category 1B	Category 2	Effects via lactation
Pictogram	GHS08	GHS08	GHS08	None
Signal word	Danger	Danger	Warning	None
Hazard Statement	H360 (F, D): May damage fertility or the unborn child.	H360 (F, D): May damage fertility or the unborn child.	H361 (f, d): Suspected of damaging fertility or the unborn child.	H362: May cause harm to breastfed children.

Precautionary Statements used for substances toxic to reproduction are
- P201: Obtain special instructions before use.
- P202: Do not handle until all safety precautions have been read and understood.
- P280: Wear protective gloves/protective clothing/eye protection/face protection.
- P308 + P313: If exposed: Call a POISON CENTER or doctor/physician.
- P405: Store locked up.
- P501: Dispose of contents/container to . . . (*specify*).

The two different branches of reproduction toxicity – damage to fertility and damage to development – can result in different categories for fertility against developmental toxicity. For example, a substance may be category 1 for developmental effects and category 2 for fertility; in that case, H360 is used solely, but the capital "D" is added for category 1 concerning development, and "f" is added for category 2 concerning effects on fertility (and vice versa). The result is
- H360 (D, f): May damage the unborn child; suspected of damaging fertility,
- H360 (F, d): May damage fertility; suspected of damaging the unborn child.

Examples for substances in
- category 1A: lead and lead-compounds, mercury and mercury compounds, carbon monoxide, carotenoids;
- category 1B: chromium (VI), boric acid, nickel compounds, perfluorinated alkane sulfonates;
- category 2: branched nonylphenol, toluene, ethylene glycol monomethyl ether;
- H362: hexabromo-cyclododecane, perfluorinated alkane sulfonates, chlorinated paraffins (C_{14} – C_{17});

Generic concentration limits for reproduction toxicants in mixtures are listed in the following table (Tab. 60).

Tab. 60: Generic concentration limits for reproduction toxicants in mixtures.

Component	Mixture			
	Category 1A	Category 1B	Category 2	Effects via lactation
Category 1A	≥ 0.3%			
Category 1B		≥ 0.3%		
Category 2			≥ 3.0%	
Effects via lactation				≥ 0.3%

Specific concentration limits shall be allocated to substances toxic for reproduction (fertility and development), if possible. For doing so, the 10% effective dose (ED_{10}) has to be identified (dose creating 10% increase in adverse effects against the background). Preferably, the ED_{10} should be calculated by Benchmark Dose software. Alternatively, it may be estimated by linear interpolation, provided a NOAEL and a LOAEL could be identified in the study. In case of absence of a NOAEL, an ED_{10} can be derived by linear interpolation between LOAEL and the control group; SCLs based on this latter procedure can only be used if they are below the GCL.

The exercise of linear interpolation shall be explained by examples (additional examples see [47]). Assume a substance decreases the testicular weights of male rats in a one-generation study, and it also causes malformations in offspring. The hypothetical data are summarized in the following table (Tab. 61).

Tab. 61: Data from a fictive one-generation study.

Dose [mg/kg b.w./day]	Offspring with malformations [%]	Mean testicular weight of males [µg]
0	2	750
50	2	730 (NOAEL)
200	3 (NOAEL)	600 (LOAEL)
800	15 (LOAEL)	520

The substance causes developmental and fertility effects, so two ED_{10} have to be derived. A 10% increase in malformations above background is 12%, and a 10% decrease in mean testicular weight is 675 µg:
- Effect ($ED_{10,\ D}$) = 12% offspring with malformations;
- Effect ($ED_{10,\ F}$) = 675 µg mean testicular weight.

Data are sketched on the following graphs (Fig. 102):

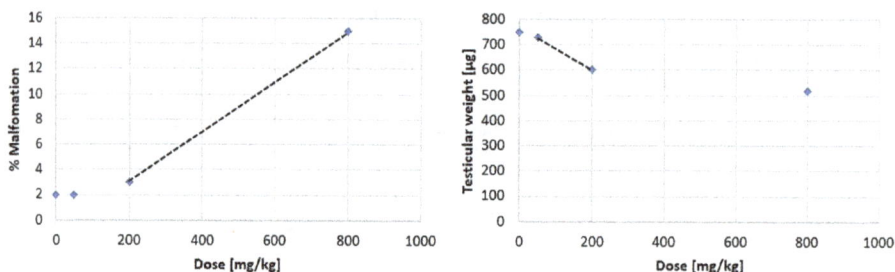

Fig. 102: Graphical presentation of the dose-response data given in Tab. 61.

Linear interpolation means creation of a straight line equation, $Y = A + B \times X$; we will do so for both endpoints. For developmental effects between NOAEL and LOAEL, we have the slope

$$\frac{15-3}{800-200} = B = 0.02$$

At the NOAEL, we can write $3\ [\%] = A + 0.02 \times 200 \qquad A = -1$; now we can calculate the ED_{10}:

$$12\% = -1 + 0.02 \times ED_{10,\ D} \Leftrightarrow ED_{10,\ D} = 650\ \text{mg/kg}$$

For fertility effects, the slope is

$$\frac{600-730}{200-50} = B = -0.867$$

At the NOAEL, we have $730 = A - 0.867 \times 50 \qquad A = 773.35$; now we can calculate the $ED_{10,\ F}$:

$$675 = 773.35 - 0.867 \times ED_{10,\ F} \Leftrightarrow ED_{10,F} = 113\ \text{mg/kg}$$

As a result, concerning developmental effects the substance is in the low potency group, and for fertility effects it is in the low potency group (Tab. 62).

Tab. 62: C&L of mixtures containing components toxic to reproduction, specific (group-) concentration limits.

Component	Mixture	
	Category 1	Category 2
High potency, $ED_{10} < 4$ mg/kg b.w./day	0.03%	0.3%
Medium potency, $4 \le ED_{10} \le 400$ mg/kg b.w./day	0.3%	3.0%
Low potency, $ED_{10} > 400$ mg/kg b.w./day	3.0%	3.0% (. . . 10%)

The 10% limit may be considered for substances with low potency and an ED_{10} above 1,000 mg/kg.

6.9 Annex I, part IV: environmental hazards

As explained in Chapter 4, ecotoxicity deals with the question whether or not a substance can negatively influence ecosystems. Investigations of the substance aim at the identification of the Predicted No Effect Concentration (PNEC). The PNEC has to be found for every relevant environmental compartment that may be contaminated.

In praxis, these compartments are the water, the sediment and the soil. For water and sediments, tests are typically run with freshwater species. Tests should be run with marine species as well if exposure of these compartments can occur. In absence of marine species data, it is assumed that marine species are 10 times more sensitive than freshwater species.

For C&L, the GHS has defined rules concerning the aquatic environment. Guidance how to classify according to effects in the sediment or soil compartment is expected in the future.

6.9.1 Acute and chronic aquatic toxicity

Tests for acute aquatic toxicity are run with comparatively short exposure times. They are aiming at incidences, and they provide a first insight into the aquatic toxicity of the test substance and help to define concentrations for chronic tests. The need for testing several species has been explained in Chapter 4.

6.9.2 C&L for aquatic toxicity

Currently only data on aquatic toxicity can lead to a C&L in ecotoxicity. Due to the tonnage bands defined in Regulation (EC) No. 1907/2006, chronic data will be available only for substances that achieve annual tonnage bands at or above 10 t/a. In absence of chronic data, a substance may, nevertheless, be allocated to a chronic aquatic toxicity category, if it is not rapidly biodegradable and/or is bioaccumulative; a substance is rated as bioaccumlative if the experimental bioconcentration factor (BCF) is 500 or more, or – if the BCF is not available – the octanol-water partition coefficient is 10,000 or more (log $K_{OW} \geq 4$). Note that the degradability of the substance is called *rapid*, not *readily* as in the OECD test guidelines 301-series.

A substance is regarded as rapidly degradable if one of the following conditions is fulfilled:

- in test on ready biodegradability, 60% O_2 – consumption or 60% CO_2 – production or 70% DOC removal (other than adsorption) is achieved within 28 days.
  - if these endpoints are reached within a 10-d window after adaptation, the substance is rated as readily biodegradable.
- in surface water simulation tests or in hydrolysis tests, the half-life of the substance is below 16 days, *and* it can be demonstrated that the degradation products are not hazardous to the environment.
- in case there are only data on biological oxygen demand (BOD) and chemical oxygen demand (COD), a ratio of BOD/COD > 0.5 indicates the substance is rapidly degradable.

For the environmental properties, hazard phrases of the 400-series are used. For acute aquatic toxicity, only category 1 is defined. Due to nonbioaccumulation and rapid degradation, the substance is not expected to be a critical burden on the long term. For chronic toxicity, there are in total four categories.

The minimum data set required for aquatic toxicity C&L are acute data on algae, daphnia (crustacean) and fish. The lowest value achieved in one of these species defines the category of aquatic toxicity (Tab. 63).

Tab. 63: Acute aquatic toxicity data (mg/L), allocation to categories.

Test-system	Category 1, acute or chronic[a]	Category 2, chronic[a]	Category 3, chronic[a]
Fish 96-h LC_{50}	$C \leq 1$	$1 < C \leq 10$	$10 < C \leq 100$
Daphnia EC_{50}	$C \leq 1$	$1 < C \leq 10$	$10 < C \leq 100$
Algae 72-h IC_{50}	$C \leq 1$	$1 < C \leq 10$	$10 < C \leq 100$

[a]Substance either not rapidly degradable or bioaccumulative.

For classification of a substance as chronic aquatic toxic, data from chronic tests replace data from acute tests if the most sensitive species from the acute test series is covered, or if the chronic data end up in a lower (= more severe) category than the acute data. For chronic data, a distinction is made between rapidly degradable and non-bioaccumulative substances on the one hand, and either nonrapidly degradable or bioaccumulative substances on the other hand. If a substance is either nonrapidly degradable or bioaccumulative, there is concern that higher concentrations may be reached in the aquatic system compared to rapidly degradable, non-bioaccumulative substances. As a consequence, the cutoff values for the aquatic categories are a factor of 10 higher (Tab. 64 and 65).

Tab. 64: Chronic aquatic categories based on chronic data (mg/L) for nonrapidly degradable or bioaccumulative substances.

Test system	Category 1, chronic	Category 2, chronic	Category 3, chronic
Fish NOAEC or EC_{10}	$C \leq 0.1$	$0.1 < C \leq 1.0$	$1.0 < C \leq 10$
Daphnia NOAEC or EC_{10}	$C \leq 0.1$	$0.1 < C \leq 1.0$	$1.0 < C \leq 10$
Algae NOAEC or EC_{10}	$C \leq 0.1$	$0.1 < C \leq 1.0$	$1.0 < C \leq 10$

EXAMPLE: A substance has been tested for acute aquatic toxicity. The LOG K_{OW} is 4.5, biodegradation is 70% O_2 consumption after 28 days. The acute LC_{50}/EC_{50} data are 12 mg/L for fish, 0.35 mg/L for daphnia and 3 mg/L for algae. What is the appropriate C&L?

Tab. 65: Chronic aquatic categories based on chronic data (mg/L) for substances which are rapidly degradable and nonbioaccumulative.

Test system	Category 1, chronic	Category 2, chronic	Category 3, chronic
Fish NOAEC or EC_{10}	$C \leq 0.01$	$0.01 < C \leq 0.1$	$0.1 < C \leq 1.0$
Daphnia NOAEC or EC_{10}	$C \leq 0.01$	$0.01 < C \leq 0.1$	$0.1 < C \leq 1.0$
Algae NOAEC or EC_{10}	$C \leq 0.01$	$0.01 < C \leq 0.1$	$0.1 < C \leq 1.0$

Note: a substance which has – for example – the lowest acute toxicity testing value at 0.2 mg/L, and which is not rapidly degradable, is *classified* as acute aquatic toxic category 1 and chronic aquatic toxic category 1. It is *labeled* as chronic category 1, only, as this labelling has the same category, but it is stronger in warning than acute category 1.

The lowest acute value is 0.35 mg/L, and that leads to acute aquatic toxicity category 1. As the substance is bioaccumulative, it is also classified as chronic aquatic toxicity category 1. The labeling would be GHS09, signal word "WARNING" and H410: Very toxic to aquatic life with long-lasting effects.

Now assume that for the algae toxicity in our example, the EC_{10} shall be 1.5 mg/L; can the substance be allocated to chronic category 3 (see Tab. 64)? No, the chronic data point is not available for the most sensitive species in acute tests, the daphnia. Only if a chronic daphnia test is performed and the EC_{10} or NOAEC would be higher than 0.1 mg/L, the substance can be allocated to a higher (less severe) chronic category.

What is the appropriate C&L for our example if the NOAEC in the chronic daphnia test was 0.15 mg/L? In that case, chronic data are available for the species that was most sensitive in acute tests, and the substance is classified as acute aquatic toxic category 1 and chronic aquatic toxic category 2. The labeling, then, is GHS09, wording "WARNING," H400: Very toxic to aquatic life, and H411: Toxic to aquatic life with long-lasting effects.

The chronic category 4 is a safeguard for such substances which show no aquatic toxicity up to their water solubility, but the water solubility is below 100 mg/L, and the substance is nonrapidly degradable and bioaccumulative. The classification and labelling in accordance to the categories of aquatic toxicity is summarized in Tab. 66.

Tab. 66: C&L for aquatic toxicity.

	Category 1, acute	Category 1, chronic	Category 2, chronic	Category 3, chronic	Category 4, chronic
Pictogram	GHS09	GHS09	GHS09	–	–
Signal word	Warning	Warning	–	–	–

Tab. 66 (continued)

	Category 1, acute	Category 1, chronic	Category 2, chronic	Category 3, chronic	Category 4, chronic
Hazard Statement	H400: Very toxic to aquatic life.	H410: Very toxic to aquatic life with long-lasting effects.	H411: Toxic to aquatic life with long-lasting effects.	H412: Harmful to aquatic life with long-lasting effects.	H413: May cause long-lasting harmful effects to aquatic life.

Concerning mixtures with substances showing aquatic toxicity it may occur that category 1 substances may have much lower LC_{50} or EC_{50} or NOAECs than the cutoff concentration limit for category 1. Assume that a substance has been classified as acute toxic to aquatic organism category 1 due to an LC_{50} of 0.08 mg/L. If you produce a 10% solution of this substance in water, that mixture will have a LC_{50} of 0.8 mg/L and, therefore, has to be classified as category 1! If the exact LC_{50}, EC_{50} or NOAECs are not known, and only the category 1 is mentioned in the labeling, M-factors have to be included into the classification. That is, for the substance in our example, the M-factor is 10, because a 10-time dilution still leaves the mixture in category 1.

- If the LC_{50}, EC_{50} or NOAEC is 10 times below the cutoff level of category 1, the M-factor is 10.
- If the LC_{50}, EC_{50} or NOAEC is 100 times below the cutoff level of category 1, the M-factor is 100.
- If the . . ..

For the classification of mixtures, existing ecotoxic data of the mixture should be given preference. If these are not available but the ingredients have been tested, it is assumed that substances act additive in their toxicity to aquatic organisms (Tab. 67).

Tab. 67: Classification of mixtures based on aquatic toxicity categories of the ingredients.

Components	Mixture
$\Sigma(M \times C)_{i, \text{ acute1}} \geq 25\%$:	Acute 1
$\Sigma(M \times C)_{i, \text{ chronic 1}} \geq 25\%$	Chronic 1
$\Sigma(\text{chronic2}) + 10 \times \Sigma(M \times C)_{i, \text{chronic1}} \geq 25\%$	Chronic 2
$\Sigma(\text{chronic3}) + 10 \times \Sigma(\text{chronic2}) + 100 \times \Sigma(M \times C)_{i, \text{chronic1}} \geq 25\%$	Chronic 3
$\Sigma(\text{chronic4}) + \Sigma(\text{chronic3}) + \Sigma(\text{chronic2}) + \Sigma(\text{chronic1}) \geq 25\%$	Chronic 4

The content of Tab. 67 shall be illustrated by an example. Assume a mixture of 2% substance A (chronic category 1, M = 10) and 6% of substance B (chronic category 2) in water. What is the C&L for this mixture?

- Check for category 1: $10 \times 2\% = 20\% < 25\%$; not chronic category 1.
- Check for category 2: $10 \times \{10 \times 2\%\} + 6\% = 206\% \geq 25\%$; category chronic 2.

For the calculation of a mixture, take all substances category 1 into account present at $\geq 0.1\%$ (or $0.1\%/(M\text{-factor})$), and substances of all other categories if present at $\geq 1\%$.

If LC_{50}, EC_{50} and NOAEC data are available, they shall be taken into consideration. The toxicity of the mixture is calculated as

$$\text{Acute toxicity:} \frac{\sum C_i}{EC_{50,\,\text{mix}}} = \sum \frac{C_i}{EC_{50,\,i}}$$

If $EC_{50,\,\text{mix}}$ is 1 mg/L or below, the mixture is acute aquatic toxic category 1.

For chronic aquatic toxicity, the calculation is

$$\text{Chronic toxicity:} \frac{\sum C_i + \sum C_j}{NOEC_{\text{mix}}} = \sum \frac{C_i}{NOEC_i} + \sum \frac{C_j}{0,1 * NOEC_j}$$

Here, the suffix "i" stands for rapidly, nonbioaccumulative substances and "j" stands for substances that are not rapidly degradable or bioaccumulative. These calculations have to be run for every aquatic species separately! The calculated $NOEC_{\text{mix}}$ is to be matched against Tab. 65, as the nondegradable or bioaccumulative components where sufficiently covered by the "penalty factor" 0.1 in the equation shown above, as this factor makes the nonrapidly degradable and/or bioaccumulative substances 10 times more potent than those which are rapidly degradable and nonbioaccumulative (see Tab. 64 compared to Tab. 65).

In case the mixture has some components with known NOAEC values and other components where there is only the category available, those components with NOAEC are treated as a sub-mixture, for which the category is calculated. With this sub-mixture category result, the complete mixture is classified. This procedure is demonstrated with a hypothetical mixture as given in Tab. 68.

Tab. 68: Mixture where some components have specific data.

Component	Percentage	NOAEC [mg/L]			LOG K_{ow}	Degradability	Chronic Category
		Fish	Daphnia	Algae			
A	5	0.02	0.15	4.0	3	Rapidly	2
B	15	0.25	0.05	0.20	3	Not rapidly	2
C	10	–			5	Rapidly	1
Water	70	–			–		–

For A, there is no concern with respect to biodegradation and bioaccumulation. Due to the NOAEC of 0.02 mg/L. it belongs to chronic category 2. B has a higher NOAEC, but it is nonrapidly degradable, so it is also category 2. C is classified as chronic category 1, and NOAEC data are not available. First, A and B are added together. For fish, we have

$$\frac{\sum C_i + \sum C_j}{\text{NOEC}_{\text{mix}}} = \sum \frac{C_i}{\text{NOEC}_i} + \sum \frac{C_j}{0,1*\text{NOEC}_j} \leftrightarrow \frac{20}{\text{NOAEC}_{\text{mix}}}$$

$$= \frac{5}{0.02} + \frac{15}{0.1 \times 0.25} \Leftrightarrow \text{NOEC}_{\text{mix, fish}} = 0.0235$$

For daphnia, the calculation results in

$$\frac{\sum C_i + \sum C_j}{\text{NOEC}_{\text{mix}}} = \sum \frac{C_i}{\text{NOEC}_i} + \sum \frac{C_j}{0,1*\text{NOEC}_j} \leftrightarrow \frac{20}{\text{NOAEC}_{\text{mix}}}$$

$$= \frac{5}{0.15} + \frac{15}{0.1 \times 0.05} \Leftrightarrow \text{NOEC}_{\text{mix, daphnia}} = 0.006$$

For algae, the NOEC$_{\text{mix, algae}}$ is

$$\frac{\sum C_i + \sum C_j}{\text{NOEC}_{\text{mix}}} = \sum \frac{C_i}{\text{NOEC}_i} + \sum \frac{C_j}{0,1*\text{NOEC}_j} \leftrightarrow \frac{20}{\text{NOAEC}_{\text{mix}}}$$

$$= \frac{5}{4} + \frac{15}{0.1 \times 0.2} \Leftrightarrow \text{NOEC}_{\text{mix, algae}} = 0.027$$

For the combination substance A + substance B, the lowest NOAEC is 0.006 mg/L. According to Tab. 65, the combination is to be classified as chronic category 1. Now, 20% (A + B) and 10% C in water makes the aqueous mixture chronic category 1 (category 1 substances are $10\% + 20\% \geq 25\%$). The appropriate labeling is GHS09, "WARNING," H410: Very toxic to aquatic life with long-lasting effects.

Typical precautionary statements to be used for a substance with chronic aquatic toxicity category 1 are:

- P273: Avoid release to the environment.
- P391: Collect spillage.
- P501: Dispose of contents/container to . . . (*specify*).

6.10 Annex I, part V: additional hazards

This chapter of the regulation provides a possibility to request a C&L for hazards which are not unanimously allocated to either physical, health or environmental hazards. Currently, there is only one endpoint listed, that is "Hazardous to the ozone layer."

6.10.1 Background

"Ozone depletion" addresses substances that are sufficiently stable in the atmosphere to reach the stratosphere in a critical amount, but there they can interfere with the ozone formation (Fig. 193). "Classics" for ozone depleting substances are the chlorofluorocarbons (CFCs), which were used as refrigerants in refrigerators and air conditioners, propellants and blowing agents. Their benefit was beneath to their physical-technical properties their inertness to fire and explosion and their harmlessness with respect to toxicity. However, in the stratosphere, they are cleaved by hard UV radiation, releasing chlorine-radicals which then interfere with the ozone formation.

UV radiation in the stratosphere cleaves di-oxygen molecules. Oxygen radicals formed react with di-oxygen to ozone; ozone itself is cleaved by UV-radiation, and ozone will achieve a steady-state equilibrium (Fig. 103).

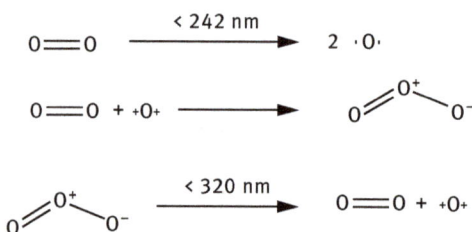

Fig. 103: Stratospheric ozone balance reactions.

The "benchmark" for ozone depletion potency is the fluoro-trichloromethane (trichlorofluoromethane), CFC-11 or R-11, where "R" stands for refrigerant, and 11 stands for 1 carbon atom and 1 fluorine atom (rest chlorine atoms). As can be seen in Fig. 104, the halogen radicals are acting like catalysts. Besides chlorine radicals, also bromine radicals act deleterious on stratospheric ozone. Reduced ozone concentration means that UV radiation can more easily arrive earth surface and cause damage in biological molecules of exposed organisms. An increase in skin cancer is associated with a diluted stratospheric ozone layer.

6.10.2 C&L for ozone depletion

An international panel defines a substance as being ozone depleting or not, and they allocate a potency as "R-11-equivalents" (see Regulation (EC) No. 1005/2009). A substance rated as ozone depleting is labeled as listed in Tab. 69. The generic concentration limit for mixtures is 0.1%.

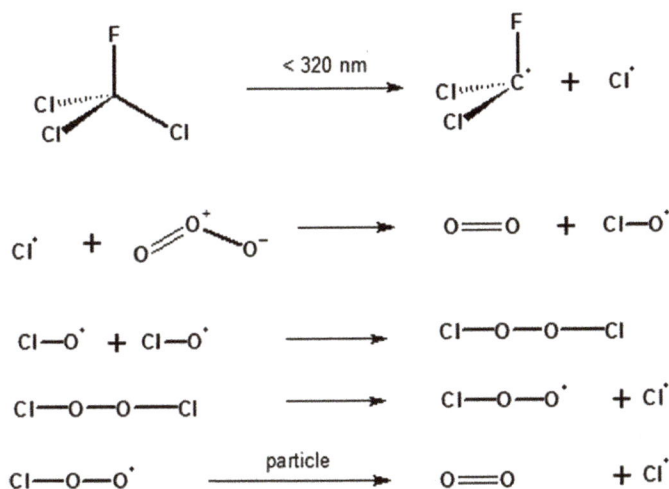

Fig. 104: Ozone decay reaction sequences as caused by CFC-11 [53].

Tab. 69: Ozone depleting substances.

Label	GHS09
Signal word	Warning
Hazard Statement	H420: Harms public health and the environment by destroying ozone in the upper atmosphere
Precautionary Statement	P502: Refers to manufacturer/supplier for information on recovery/recycling.

6.11 Annex II: special rules for labeling and packaging of certain substances and mixtures

Annex II lists additional requirements for the European Union, going beyond the GHS; for example, there are additional hazard phrases, the EUH. These go beyond the GHS Hazard statements and warn again certain risks that may easily encounter, are can easily be generated under normal handling and use.

6.11.1 Part I: supplemental hazard information

- EUH014: Reacts violently with water.
  - Example: elemental sodium.

- EUH018: In use may form flammable/explosive vapor–air mixture.
- EUH019: May form explosive peroxides.
  - Examples: diethyl ether; tetrahydronaphthalene.
- EUH029: Contact with water liberates toxic gas.
  - Example: thionyl chloride.
- EUH031: Contact with acids liberates toxic gas.
  - Example: sodium dithionite.
- EUH032: Contact with acids liberates very toxic gas.
  - Example: zinc phosphide.
- EUH044: Risk of explosion if heated under confinement.
  - Example: ammonium perchlorate.
- EUH059: Hazardous to the ozone layer.
  - Example: bromo methane.
- EUH066: Repeated exposure may cause skin dryness or cracking.
  - Example: n-pentane. Degreasing of the skin can result in skin irritation and fosters contact dermatitis like symptoms.
- EUH070: Toxic by eye contact.
  - Example: 2-nitro-2-phenyl-propane-1,3-diol. This substance is harmful if swallowed and by skin contact.
- EUH071: Corrosive to the respiratory tract.
- EUH201 – Contains lead. Should not be used on surfaces liable to be chewed or sucked by children.
  - Lead is explicitly toxic for the neuronal development of children.
- EUH202 – Cyanoacrylate. Danger. Bonds skin and eyes in seconds. Keep out of the reach of children.
  - Background for this phrase is numerous incidents with children.
- EUH203 – Contains chromium (VI). May produce an allergic reaction.
  - This is meant for cement which has Cr^{6+} concentrations below the C&L limit, but at or above 2 ppm. Cr^{6+} is a strong skin allergen, and people having acquired a skin allergy already need to be warned.
- EUH204 – Contains isocyanates. May produce an allergic reaction.
  - Organic isocyanates are skin sensitizers, but predominantly strong respiratory sensitizers. The symptoms are coughing, chest tightness, shortness of breath and may be life threatening. Those already sensitized need to be warned. From the authors own experience, it can be said that people with respiratory sensitization may be much more sensitive than the best analytical isocyanate detectors on the market.
- EUH205 – Contains epoxy constituents. May produce an allergic reaction.
  - This is for epoxy resins with a mean molecular weight below 700 g/mol. They can induce and trigger an allergic skin reaction, and allergic contact dermatitis is comparatively common in construction workers, where two-component glues based on these resins are used. In case the mean molecular

weight is at least 700 g/mol, the resin can no longer penetrate the skin to a sufficient amount to induce allergy.
- EUH206 – Warning! Do not use together with other products. May release dangerous gases (chlorine).
    - Background of this EUH statement is the use of sanitary cleaners which contain hypochlorite, equivalent to at least 1% active chlorine. These release chlorine if the pH is shifted into the acidic region, which is the case if they are mixed with – for example – cleaners containing citric acid.
- EUH207 – Warning! Contains cadmium. Dangerous fumes are formed during use. See information supplied by the manufacturer. Comply with the safety instructions.
    - Alloys containing cadmium release CdO when heated under soldering or brazing. CdO can be inhaled, causing damage to the mucous membranes, inducing a critical lung inflammation and causing cancer in case of repeated exposure.
- EUH208 – Contains (*name of sensitizing substance*). May produce an allergic reaction.
    - This phrase is to be used if the sensitizer is present between 10% and less than 100% of its SCL or GCL. People who have acquired a sensitization shall be informed, so they can avoid contact.
- EUH209 – Can become highly flammable in use.
- EUH209A – Can become flammable in use.
    - These phrases are used for mixtures which have a flash point between 60 and 93 °C and which contain 5% or more of a flammable of highly flammable substance. The flash point would not require a C&L as flammable. However, during use of the product, components causing an increase in the flash-point may evaporate, leaving a flammable residue.
- EUH210 – Safety Data Sheet available on request.
    - This is for mixtures which have not to be classified as hazardous, but which contain sensitizers at 10% or more of the SCL or GCL or CMR substances category 2 present at 10% or more of their SCL or GCL.

6.12 Annexes III to V

Annex III lists the Hazard Statements and their translation into the official languages in the European Union.

Annex IV lists the Precautionary Statements and their translation into the official languages in the European Union. In that table, cross-reference is provided for the appropriate hazard endpoints to which these Precautionary Statements shall be linked to.

Annex V lists the hazard pictograms and the according hazard classes and hazard categories.

6.13 Annex VI, VII and VIII

The subject of annex VI is the harmonized C&L. Part 1 is an introduction to the list of substances with EU-harmonized C&L, providing some back-ground information and interpretation support to understand the entries in Part 3 (Part 3 is the table of substances with an EU-harmonized C&L.). Part 2 is about the dossier to be submitted if a harmonized C&L is to be proposed. A modified extract of that list is added as APPENDIX 1. Appendix 1 is meant as explanatory example and does not necessarily coincide with current C&L, specific limits etc.!

Annex VII provides some information about the transformation of risk phrases ans sefety recommendations from the previously valid regulation into the standards of Regulation (EU) No. 1272/2008/EC.

Annex VIII addresses information to be forwarded to Poison Centers for mixtures classified as hazardous brought on the EU market. In case of accidents with mixtures, the information disclosed by the label may not be sufficient for the appropriate treatment of contaminated persons. The manufacturer or importer has to generate a Unique Fomula Identifier (UFI), an alphanumeric code which identifies the mixture unanimously, and which has to be submitted to the agency together with the disclosed composition of the formulation. The UFI has to be put on the label, well visible and ineradicable, starting with "UFI:...". In the information submitted, hazardous substances must be identified, and their content needs to be mentioned as concentration range, depending on the hazard class of the individual substance. For more details see Annex VIII of the regulation.

6.14 Addendum: C&L of aerosol cans, directives 75/324/EEC and 94/1/EC

C&L of aerosol cans has to comply with European Union directives 75/324/EEC and 94/1/EC. The maximum permitted content is 1 L. The cans have to be labelled with
- "Pressurized container: protect from sunlight and do not expose to temperatures exceeding 50° C. Do not pierce or burn, even after use."
- "Do not spray on a naked flame or any incandescent material," unless the aerosol dispenser is designed for that purpose.
- *Flammable* or the symbol of a naked flame if the contents include more than 45% by weight, or more than 250 g of flammable components.
- "Keep away from sources of ignition – No smoking."
- "Keep out of the reach of children."

6.15 Exercises

(1) Given is the following mixture which is intended for spray application (= mist, dust) (Tab. 70); what is the appropriate classification concerning acute inhalation toxicity?

Tab. 70: Composition of the mixture.

Component	LC50	Category	ATE
Solid, 6%	1–5 mg/L/4 h	4	1.5
Solid, 11%	0.6 mg/L/4 h	3	= LC50
Solid, 10%	6.0 mg/L/4 h	–	None
Liquid, 40%	1.1 mg/L/4 h	4	= LC50
Water, 35%	–	–	None

(2) Classify a mixture of 10% N,N'-bis(3-aminopropyl)methylamine (H302,H311) and 50% hexamethylendiamine (H302,H312), the other components are not toxic.

(3) Taking the previous example, now assume that 40% of the mixture are unknown or not tested.

(4) Classify a mixture of 12% hexamethylene diamine (H302; H312) and 50% boric acid trimethylester (H312; LD_{50}/dermal = 1,980 mg/kg). The other components are not toxic.

(5) A mixture consists of the following ingredients, and the additivity approach is applicable: 1% of substance A which is corrosive to the skin (cat. 1) and 2% of substance B which is irritating to the eyes (cat. 2). All other components are not irritating or corrosive. What is the appropriate C&L concerning eye effects?

(6) A mixture contains 40% of a base A with the following specific concentration limits given below, and a further 2% of base B which is corrosive cat. 1B, H314. The remaining component of the mixture is water. The bases act additively. What is the appropriate C&L? Specific concentration limits for A:
   - C ≥ 70%: H314, 1A
   - 50% ≤ C < 70%: H314, 1B
   - 35% ≤ C < 50%: H315
   - 8% ≤ C < 50%: H318
   - 5% ≤ C < 8%: H319

(7) Take the previous example and imagine, base A is reduced to 4%, and it is not acting additively. Component B is maintained at 2%, the remaining component is water. What is the result for C&L?

(8) Imagine you have a mixture of the following three tensides (Tab. 71). The table lists the NOAECs for chronic aquatic toxicity, and data concerning bioaccumulation potential and biodegradability. What is the appropriate classification for this mixture for chronic aquatic toxicity?

Tab. 71: Composition of mixture.

Component	Content	NOAEC (mg/L)			Biodeg.	log Kow
		Fish	Daphnia	Algae		
A	20%	0.02	2.0	1.0	Rapidly	2.1
B	40%	0.8	0.6	4.0	Rapidly	5.7
C	40%	0.25	1.2	0.1	Not rapidly	1.3

(9) How has an aqueous solution to be classified and labeled that contains 10% N, N-dimethylaniline and 0.05% cobalt sulfate (see Annex)? Calculate the inhalation toxicity for vapor.

(10) Assume a mixture of 5% chloroacetic acid, 5% 1-chloro-4-nitrobenzene, 81% butanone, 2% 4-tert.-butylphenol, 2% 2-methyl-5-tert-butylthiophenol and 5% water. With the aid of table in the annex, calculate the C&L concerning human health effects and environmental effects. Assume that local effects are additive, and respiratory toxicity shall be evaluated for the vapor phase.

(11) Calculate the appropriate C&L against human health and environmental effects of a mixture that contains 8% phosphoric acid, 22% 2,2-dichlorovinyl-dimethylphosphate, 20% tetrahydrofurane, 40% ethane-1,2-diol, 1% bis(2-chloroethyl) ether and 9% water. Look up the substance data in the annex. What information has to be put on the label? Inhalation toxicity shall be calculated for mist.

(12) In an acute oral toxicity test, the LD50 for rats above 50, but below 300 mg/kg b. w.; cramps preceded death, which was most likely a result of paralysis of muscles so breathing was no longer possible. What is the appropriate classification?

(13) In an acute dermal toxicity study, the LD50 was higher than 1,000, but below 2,000 mg/kg b.w. Decay of red blood cells was the obvious reason of death. Further, deceased and surviving animals showed a degeneration of the retina (the light receiving, sensitive network of nerves in the eye). What is then appropriate C&L?

(14) In a subchronic study where the substance was applied on the skin, the LOAEL is 15 mg/kg b.w. Derive the specific concentration limits for STOT RE.

(15) In a reproduction toxicity study, a substance affects development and was classified as category 1. The LOAEL was 80 mg/kg which caused a reduction of

birth weight of 20% against control. The NOAEL at 40 mg/kg caused a reduction in birth weight of 4% against control. Calculate the ED_{10} by linear interpolation and allocate the substance to a potency group.

(16) A mixture contains the following substances, and for three of them, the most sensitive species in acute aquatic tests was daphnia; for daphnia, chronic NOAECs were reported: Substance A (10%, rapidly degradable, not bioaccumulative, NOAEC = 0.2 mg/L), substance B (30%, not rapidly degradable, not bioaccumulative, NOAEC = 0.9 mg/L), substance C (50%, rapidly degradable, bioaccumulative, NOAEC = 8 mg/L) and substance D (20%, rapidly degradable, not bioaccumulative, category 2). How is the mixture to be classified and labeled against aquatic toxicity?

7 A brief view on the classification and labeling of dangerous goods

Dangerous goods are substances, mixtures or articles that can pose a hazard to the public and environmental health and safety, due to their intrinsic properties and/or due to the mode of transport. Regulations concerning the safe transport of hazardous substances in the EU date back to 1957, when, for example, the agreement concerning the International Carriage of Dangerous Goods by Road was ratified [48]. This regulation covers all aspects of the transport of dangerous goods; this is not only classification and labeling but also testing of the goods, appropriate packaging, requirements for trucks and drivers, driver training and so on. Annex A covers the classification in part 2, and a list of dangerous goods in part 3 [54].

Safety data sheets for hazardous substances and mixtures list the appropriate classification and labeling in section 14, split up according to road, rail, ship (inland and sea) and air transport.

A good overview of the labeling of dangerous goods is provided by ChemSafetyPro [55]. The nine classes the dangerous goods are allocated to are

- Class 1: Explosives
- Class 2: Gases
- Class 3: Flammable liquids
- Class 4: Flammable solids; self-igniting materials in contact with air; materials releasing flammable gas in contact with water
- Class 5: Oxidizers, organic peroxides
- Class 6: Toxic and infectious materials
- Class 7: Radioactive goods
- Class 8: Corrosive goods
- Class 9: Others

7.1 Labels for dangerous goods

Dangerous goods are subdivided into nine classes, and in the following sections the labeling will be presented, focusing on road transport. These labels have parallels in the hazardous substance regulations, but some are unique. One material may be labeled with more than one label.

7.1.1 Class 1: explosive substances and articles

Class 1 covers explosives. As in case for dangerous substances, this class is further split into subclasses 1.1 . . . 1.6, and the globally harmonized system (GHS) for

https://doi.org/10.1515/9783110618952-007

classsification, labelling and packaging of hazardous substances, has copied the allocation of substances to these classes.

Fig. 105: Labels for explosives, dangerous goods classes 1.1–1.6; GHS label on the right.

7.1.2 Class 2: gases

The main hazard of this class is the physical state of the material. However, distinctions are made depending on whether the gas has additional hazardous properties, for example, toxic or flammable gases. The labels can vary accordingly.

Flammable gases are allocated to class 2.1. The symbol is that of flammable goods; only the number "2" in the label hints on the physical state.

Fig. 106: Label for flammable gases; GHS label on the right.

Compressed gases present an inherent hazard as overheating may result in rupture of the confinement, and in case of mechanical damage, released gas may be very cold, causing freezing and expelling oxygen into air.

For toxic gases, the label shows skull and crossbones, and the number "2" in the label indicates the gaseous state.

Fig. 107: Compressed gas, class 2.2; GHS label on the right.

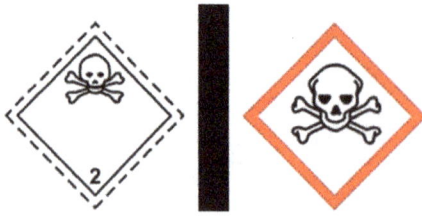

Fig. 108: Toxic gas, class 2.3; GHS label on the right.

7.1.3 Class 3: flammable liquids

Flammable liquids are further subdivided according to their flash point. Goods labeled as flammable liquid are products with a flash point not higher than 60 °C. Classification criteria are the same for flammable hazardous substances, flammable liquids and flammable dangerous goods. Criteria are summarized in the following table [48].

Tab. 72: Classification criteria of flammable dangerous goods.

Packaging group	Flash point (closed cup)	Initial boiling point
I	---	≤35 °C
II	<23 °C	>35 °C
III	≥23 °C, ≤ 60 °C	>35 °C

Fig. 109: Label for class 3, flammable liquids; GHS label on the right.

7.1.4 Other flammable goods

7.1.4.1 Class 4.1: flammable solids, self-reactive substances, polymerizing substances and solid desensitized explosives

Concerning flammability, flammable gases and flammable liquids were covered, so far. Class 4.1 covers those solid substances and articles which are flammable or desensitized explosives, self-reactive liquids and solids and polymerizing substances

Fig. 110: Label for class 4.1.

7.1.4.2 Class 4.2: substances liable to spontaneous combustion

Class 4.2 covers
- Pyrophoric substances and mixtures: substances and mixtures which even in small quantities ignite in contact with air within five minutes.
- Self-heating substances and articles: without external energy supply, they are liable to self-heating in contact with air. Other than pyrophoric substances, they will ignite only in large amounts (kilograms) and after long periods of time (hours or days).

Fig. 111: Label for class 4.2.

7.1.4.3 Class 4.3: substances that on contact with water emit flammable gases

Class 4.3 covers substances and articles that react with water. Due to the reaction, flammable gases are emitted, which may form explosive mixtures with air.

Fig. 112: Label for class 4.3.

7.1.5 Oxidizing substances and organic peroxides

7.1.5.1 Class 5.1: oxidizing substances

Oxidizing goods can transform a flammable good into a highly flammable good. In road transport, they pose a special hazard as trucks/cars transporting these goods are typically driven with gasoline. When spilled, special care has to be taken if gasoline is dripping out; contact to the oxidizer may cause fire or even explosion.

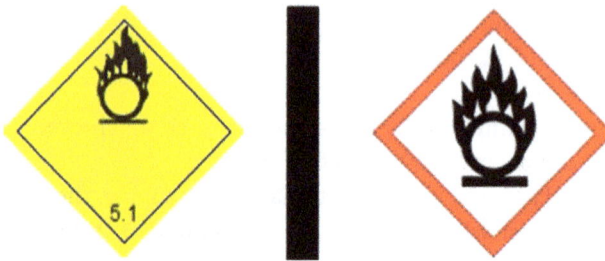

Fig. 113: Label for class 5.1; GHS label on the right.

7.1.5.2 Class 5.2: oxidizing peroxides

Organic peroxides are not necessarily strong oxidizers, but they are frequently heat sensitive; if ignited, they burn vigorously like flammable substances mixed with oxidizers.

Fig. 114: Label for class 5.2.

7.1.6 Class 6: toxic and infectious substances

7.1.6.1 Class 6.1: toxic substances

For the allocation of a material into class 6.1, acute toxicity data are relevant (Tab. 73). The criteria for acute oral and dermal toxicity are the same as for dangerous substances, acute toxicity categories 1–3. Concerning acute inhalation toxicity, substances tested as dusts and mists are covered by this class, if the LC_{50} values for one-hour exposure are at or below the levels listed in the following table. Toxic gases are allocated to class 2.

Tab. 73: Classification criteria for class 6.1 [48].

	Packaging group	Oral toxicity, LD_{50} [mg/kg]	Dermal toxicity, LD_{50} [mg/kg]	Inhalation toxicity, 1 h LC_{50} for dusts and mists [mg/L]
Highly toxic	I	≤5	≤50	≤0.2
Toxic	II	>5, ≤50	>50, ≤200	>0.2, ≤2
Slightly toxic	III	>50, ≤300	>200, ≤ 1,000	>2, ≤4

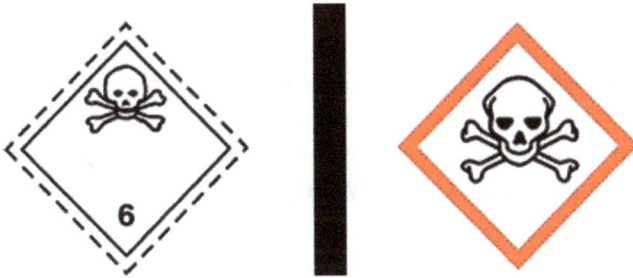

Fig. 115: Label for dangerous goods class 6.1; the GHS label (right) is applicable for toxic substances.

7.1.6.2 Class 6.2: infectious substances

Infectious substances are not subject to Regulation (EU) No. 1272/2008 for the classification, labeling and packaging.

Fig. 116: Label for dangerous goods class 6.2.

7.1.7 Class 7: radioactive materials

Radioactive goods are covered by class 7 and not subject to Regulation (EU) No. 1272/2008 for the classification, labeling and packaging. Class 7 materials are subdivided according to the activity in combination with the respective radionuclide. Higher class means lower activity.

Fig. 117: Labels for radioactive goods classes 7.1, 7.2 and 7.3 and fissile material.

7.1.8 Class 8: Corrosive substances

It was the dangerous goods regulation that first extended corrosivity from biological effects to physical effects, that is, corrosion of material. Material corrosion has a consequence for the appropriate packaging. For dangerous goods and hazardous substances, the classification criteria concerning corrosivity classes 1A ... 1C are the same (Tab. 74).

Tab. 74: Classification criteria for corrosive substances [48].

Packaging group	Exposure time	Observation period	Effect
I	≤3 min	≤60 min	Irreversible damage to intact skin
II	>3 min, ≤1 h	≤14 days	Irreversible damage to intact skin
III	>1 h, ≤4 h	≤14 days	Irreversible damage to intact skin
III	–	–	Corrosion rate on either steel or aluminum surfaces exceeding 6.25 mm a year at a test temperature of 55 °C when tested on both materials

Fig. 118: Label for class 8 dangerous goods; GHS label on the right.

7.1.9 Class 9: miscellaneous dangerous substances and articles

Class 9 covers dangerous goods that cannot be allocated to other classes, but presents a certain hazard that calls for specific attention in case of accidents. Originally, this class was for material that can present a hazard because of its mere bulk appearance. Meanwhile, this class is extended – to name examples – to substances that are hazardous to the environment and genetically modified organisms. Further labels give hints on conditions to be avoided.

Substances and articles in class 9 are subdivided as follows:
- M1: Substances that, on inhalation as fine dust, may threaten health.
- M2: Substances and articles that, in the event of fire, may form dioxins.
- M3: Substances evolving to flammable vapor.
- M4: Lithium batteries.
- M5: Live-saving appliances.
- M6–M8: Environmentally hazardous substances:
    - M6: Pollutant to the aquatic environment, liquid.
    - M7: Pollutant to the aquatic environment, solid.
    - M8: Genetically modified microorganism or organism.
- M9–M10: Elevated temperature substances:
    - M9: Liquid.
    - M10: Solid.
- M11: Other substances and articles presenting a danger during carriage, but not meeting the definitions of other classes.

The classification criteria for M6 and M7 are the same as for substances of acute aquatic toxicity category 1 or chronic aquatic toxicity categories 1 and 2 according to Regulation (EU) No. 1272/2008.

Fig. 119: Label for class 9 (left) and labels pointing on environmental hazard, upright position and maximum temperature/heat sensitivity.

7.2 Orange plate

In road and rail transport of dangerous goods, trucks and railway wagons need to carry an orange warning plate if certain amounts of dangerous goods are exceeded. These orange plates (Fig. 120) shall warn other participants and emergency staff in case of accidents, so they can take care for self-protection. Transport units carrying dangerous goods shall display the rectangular orange-colored plates, set in a vertical plane. They shall be affixed at the front and the rear of the transport unit.

Fig. 120: Neutral orange warning plate: the carriage is any kind of dangerous good.

When a specific dangerous good is transported, the plates have a hazard identification number (HIN; Kemmler code) in the upper field and a product identifier, the UN-number in the lower field.

Emergency call: Mention the presence of the orange warning plate. Name numbers in the upper and lower fields of the plate.

The UN number is an identifier for a specific product or product class. The HIN or Kemmler code is based on the nine classes of dangerous goods. It has at least two digits, starting with the main hazard, followed by numbers indicating either no further hazard (number is zero), strengthening the hazard (same number again) or an additional hazard (other number, 1–9). For example, 30 represents flammable liquid and 33 stands for highly flammable liquid. A black "X" at the beginning of the HIN warns for high reactivity in contact with water.

Tab. 75: Hazard Identification Number (Kemmler code).

Number	Meaning first digit	Meaning additional digit
0	Not applicable	No additional warning
1	Explosive product	Explosion hazard
2	Gas	Gas formation
3	Flammable liquid	Flammability
4	Flammable solid or self-heating product	(Only as intensifier)
5	Oxidizer or organic peroxide	Oxidizer
6	Toxic or infectious	Toxicity or infection
7	Radioactive	Radioactive
8	Corrosive	Corrosion
9	Other	Other

As an example, see the following orange plates which may be seen in everyday life (Fig. 121).

Fig. 121: Orange plate for gas oil (left) and gasoline (right).

1202 is the UN code for gas oil; the HIN is 30, meaning "flammable liquid" and no further hazard. For gasoline, the UN number is 1203, and the HIN is 33, now meaning "very flammable liquid," as the first three is boosted by the second three.

In Fig. 122 you see the orange plate for elemental potassium – as the UN number says. The HIN is interpreted as such: the material that reacts with water ("X") is a flammable solid, self-ignitable or water-reactive material ("4"), whereby a gas is released ("2") which is flammable ("3").

Fig. 123 shows examples for road- and railay transport vessels carrying the orange plate.

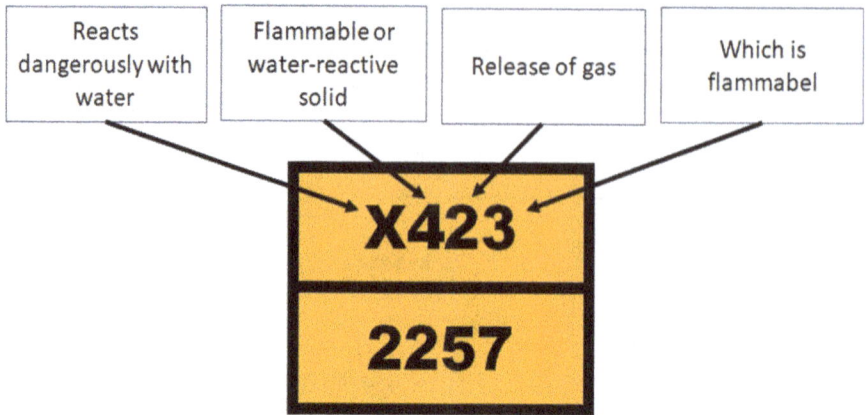

Fig. 122: Orange plate for elemental potassium.

Fig. 123: Transport of dangerous goods: orange plates at a truck and a railway wagon. The orange stripe around the railway waggon tells us the transported good is a gas.

7.3 Classification and labeling between dangerous goods and hazardous substances

It may be asked why there are two different systems for the classification and labeling of products that present a physical–chemical, toxicological or environmental hazard. This section provides some background information, and the case of incidents will be addressed. As explained in this chapter, Title III of Regulation (EC) No. 1272/2008 permits that in case of multiple packaging for a hazardous substance which are also dangerous goods, the outer package (transport package) may carry the classification and labeling as dangerous good, only.

7.3.1 Differences in the classification and labeling

As demonstrated in the previous sections, classification and labeling for dangerous goods has many features in common with the regulations 1272/2008/EC concerning classification and labeling of hazardous substances with respect to acute effects. However, there are some differences, and these address the hazardous goods category 4, which provides much more detailed labels for hazardous goods than for hazardous substances. As an illustrative example, see the following orange plate (Fig. 124).

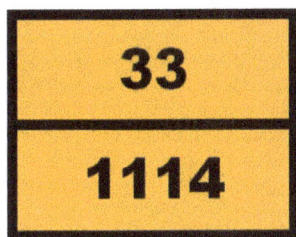

Fig. 124: Which hazard is indicated?

The HIN tells us that this is a highly flammable liquid. There are obviously no other dangers. When looking up the UN number, it is the number for benzene! Benzene is a category 1A carcinogen; why is there no indication for the benzene toxicity? The reason is that dangerous goods regulations require that the packaging does not allow any release of the contained material. Therefore, during normal transport and handling of containers, there is no exposure. In this respect, cancer is not an issue because it is the iterative, repeated exposure against benzene that finally causes cancer. That means that in case of an incidence, there may be high exposure, but this exposure occurs once. But the spilled benzene poses a fire hazard and, therefore, labeling it as highly flammable liquid is most indicated.

Acute inhalation toxicity is another example for differences between transport regulation and the regulation for handling and use; for transport classification, the 1 h-LC_{50} is used as a criterion.

As mentioned in Chapter 6, substances and mixtures classified as hazardous according to Regulation (EU) No. 1272/2008 must have labels on every package. If the hazardous product is a dangerous good as well, the transport packaging may carry the labels for hazardous goods, only, provided the inner packaging is labeled according to the hazardous substance regulation. If there is one packaging only for a product that is a hazardous substance/mixture and a hazardous good, labels fulfilling the requirements of both regulations have to be added (Fig. 125).

Fig. 125: Labels for classification as dangerous good (left) and hazardous substance (right) placed on a bucket. UN 1759: corrosive solids, n.o.s. (not other specified).

7.3.2 The International Chemical Environment (ICE) scheme for traffic incidents with dangerous goods

The example benzene mentioned earlier has shown that the dangerous goods classification and labeling scheme is valuable for self-protection, emergency calls and emergency actions, but it leaves the reader with a conflict: as long as the substance is flammable only, I could draw the conclusion that I will try to provide first aid to injured people; I "just" need to avoid open flames and sparks. But what if I find out that the material is a carcinogen? Not only the first responder but also fire fighters and paramedics will find themselves buried under piling up questions concerning self-protection, precautionary measures and appropriate actions to limit damage; expert advice is highly desirable. The European Chemical

Industry Council has installed an EU-wide support system, called "ICE" [56]. ICE can be approached by companies, but also by emergency institutions like fire brigades, police and so on in case of accidents with dangerous goods. Support is organized on three levels.

On level I, emergency responders on the scene of incident are connected to chemical experts, preferably speaking the same language (Fig. 126). These chemical experts are members of the chemical industry who have every day experience with the product under consideration. The substance experts can provide recommendations to the emergency staff by phone on what to do (and what not to do). In most cases, this is sufficient to handle the accident in an appropriate way. However, sometimes more complicated situations occur, and conversation by phone or skype is not sufficient to bring the expert into the full picture of the scene. At level II, experts knowing the chemical are brought to the scene of incidence and consult local emergency staff. Level III, finally, means that not only experts but also special equipment probably together with the works fire brigade of a chemical company provides support at the scene of incidence.

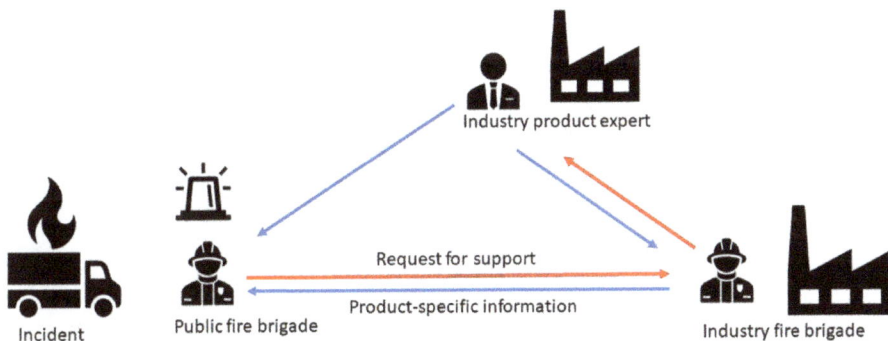

Industry product expert

Request for support

Product-specific information

Incident   Public fire brigade   Industry fire brigade

Fig. 126: ICE scheme, level 1: consultation via phone.

As there are many chemicals transported in the EU, every fire brigade and every police station should have the emergency number of the ICE scheme readily available, and they are encouraged to make use of it in any case.

7.4 Exercises

(1) If a package is labeled as carrying a hazardous substance, does this transport label substitute the label(s) for dangerous goods?
(2) If a packaging carries the labels according to dangerous goods regulations, does this substitute the label for classification as hazardous substance?

(3) In case of accidents with dangerous goods, what should be done to acquire sub-stance-specific information? What specific aid can be provided?
(4) Is the dangerous goods class 6 classification and labeling merely identical with the hazardous substance classification and labeling for toxicological hazards or what are the differences?
(5) What hazard is indicated by the following HIN? 40, 23, 366, 268, 48, 56, 255, 336 and X44.

8 Regulations concerning notification and marketing of substances in the European Union

8.1 Regulation (EC) No. 1907/2006 on the Registration, Evaluation and Authorisation of Chemicals

For the Registration, Evaluation and Authorisation of Chemicals (REACh) regulation, the helpdesk of the European Chemicals Agency (ECHA) provides a lot of supporting materials and guidance documents. See, for example: https://echa.europa.eu/de/regulations/reach/understanding-reach.

The regulation is subdivided into different titles and annexes:
- Title I: General issues
- Title II: Registration
- Title III: Data sharing and avoiding unnecessary testing
- Title IV: Information in the supply chain
- Title V: Downstream users
- Title VI: Evaluation
- Title VII: Authorization
- Title VIII: Restriction
- Titles IX–XV: Fees, the agency, information, competent authorities, enforcement, transitional measures
- Annexes I–XVII, for example,
    - Annexes IV and V: Exemptions from the obligation to register.
    - Annex VII–X: standard information requirements for substances manufactured or imported in dependence on annual tonnage.
    - Annex XIV: List of substances subject to authorization.
    - Annex XVII: Restrictions on the manufacturing, placing on the market and use of certain dangerous substances, mixtures and articles.

8.1.1 Title I: General issues

Title I describes the scope of the regulation and some general aspects. The following products are *exempted from the regulation 1907/2006*, most of them due to specific legislations:
- Radioactive substances
    - If they are subject to directive 2013/59/EC
- Products under custom supervision
    - They are not yet on the EU market

https://doi.org/10.1515/9783110618952-008

- Nonisolated intermediates
- Waste as defined in directive 2006/12/EC
- Substances in scientific research and development, provided strictly controlled handling is assured
- Labeling and packaging of dangerous goods for transport
  - These are classified and labeled according to dangerous goods regulations (see Chapter 7).
- Exempt from Titles II (registration), V (downstream users), VI (evaluation) and VII (authorization) are:
  - Pharmaceuticals in their finished form, ready for consumer use
    - They fall under directives 2001/82/EC and 2001/83/EC
  - Medical devices
    - Directives 93/42/EC, 98/78/EC and Regulation (EC) No. 2017/745 are applicable
- Exempt from title IV (information in the supply chain) are the following mixtures in their ready to use form:
  - Cosmetic products in their finished form, ready for consumer use
    - Cosmetics are subject to Regulation (EC) No. 1223/2009
  - Medicinal products
  - Certain medical devices
  - Foodstuff and feedstuff
- Exempted from titles II (registration), V (downstream communication) and VI (evaluation) are
  - Substances in Annexes IV and V
  - Food- and feedstuff
    - These products are subject to regulation 178/2002/EC
    - These are also exempted from title VII (authorization)
  - Reimported substances and mixtures, provided the composition does not deviate from that in the registration dossier
  - Recycling of registered substances, if the recycling process is covered by the registration dossier of the substance
  - Polymers are exempted from titles II and VI, provided the content of non-registered substances is below 2%
- Exempted from title IV (information in the supply chain):
  - Medicinal products for humans or veterinary use
  - Cosmetic products as defined in directive 76/768/EEC (European regulation 1223/2009)
  - Medical devices
    - Directives 98/78/EC and regulation (EC) No. 2017/745 are applicable
  - Foodstuff and feedstuff
    - These products are subject to regulation 178/2002/EC

Title I lists up several definitions (chapter 2, article 3):

A *substance* is a chemical, element or molecule that has a main component with a CAS-No., including all by-products, impurities and additives.

A *mixture* is a mix of substances, where the single substances maintain their chemical identity.

An *article* is defined by its shape and design and not primarily by its chemical composition. For example, a glove made from PVC and plasticizer is not a *mixture of PVC and plasticizer*, but the *article* "PVC-glove."

A *polymer* is a substance consisting of molecules which differ mainly by the number of repeating building blocks in a monomer unit. A molecule consisting of at least three monomers in one chain plus an initiator (or another monomer) is the most abundant molecule by weight, but it is less than 50% by weight of the substance. The substance shows a molecular weight distribution as a consequence of the polymer reaction and differences in the number of monomer units.

Scientific research and development is any experimentation, analysis or chemical research under controlled conditions. Although there is no volume limitation, volumes are typically below 1 t/year.

8.1.2 Title II: registration of substances

Generally, substance intended for the EU market as such, in mixtures or in articles at an annual volume of 1 to or more per legal entity of a company, need to be registered prior to marketing. The content of the data set to be submitted depends on the annual market volume. In higher tonnage bands, more data points need to be filled up. For animal welfare reasons, the registrant first has to propose a testing program to the agency. The agency can comment on the program, considering intended uses and markets. The agency will allocate a registration number, and with this number marketing can start. For annual volumes below 1 to, it is sufficeint to inform the agency, and a registration number is not required. In this notification, the identification of the substance, the full address of the company, a classification and labelling proposal, all known data concerning health and environment and endorsed uses are submitted.

In the registration process, the following data need to be submitted (see ANNEX VI of Regulation (EU) No. 1907/2006):

- Name and address of the registrant
- Identity of the substance
  - For *monoconstituent substances*, where a defined molecule makes up at least 80% of the composition, and for *multiconstituent substances*, where no component achieves a content of 80%, all components contributing 0.1% or more need to be identified (including by-products, impurities, additives);
  - If the substance belongs to the class of *UVCB substances* (substances of *unknown* or *variable* composition, complex reaction products or *biological*

materials [57]), at least those components should be identified that contribute 10% or more to the composition
- Description of the production process
- Intended uses
- Not supported uses (uses advised against)
    - These are uses where the registrant knows or has reason to assume that they pose an undue risk to human health and/or environment
- The annual tonnage bands
- Proposal for classification and labeling (C&L)
- All known data, plus those requested by Annexes VII and VIII, depending on the tonnage band,
    - Physical and chemical properties
    - Physical chemical safety
    - Environmental behavior
    - Ecotoxicity
    - Toxicity
    - Human experience, first aid, exposure information
    - Testing proposals, if Annexes IX and X apply
    - Exposure information if the annual market volume is below 10 t
    - A chemical safety report (CSR) if the annual market volume exceeds 10 t

This information has to be submitted electronically by use of the free available IUCLID software (International Uniform Chemical Information Database). Some information like market volume and detailed composition are confidential (international antitrust regulations!). Health, safety and environment relevant data will be made publicly available. Compared to the preceding chemicals regulation in the European Union, a substance is now registered *in combination with endorsed uses*.

Polymers are exempted from REACh provided that the monomers and other ingredients have a full registration. Only up to 2% of the polymer may consists of nonregistered substances (monomers or otherwise bound in the polymer chain), or these nonregistered substances do not arrive at and exceed an annual market volume of 1 ton. Although a full data set is required, monomers are regarded as intermediates. Note: Additives (stabilizers, flame retardants, etc.) need to have their own registration.

Substances in articles need a registration if they are intended to be released from the article and if the annual EU market volume is 1 ton or more. If the substance is not intended for release, there is no need for a registration. However, if the not intendedly released substance in the article is present at 0.1% or more, and if it is a substance of very high concern (see below), the manufacturer or importer has to inform the agency. The agency than may request a registration.

Manufacturers and traders outside the European Union can engage an "*only representative*" (OR) in the European Union for fulfilling the registration demands.

The OR is taking over the role of an EU importer. Any trade and communication have to be organized via the OR (Fig. 127).

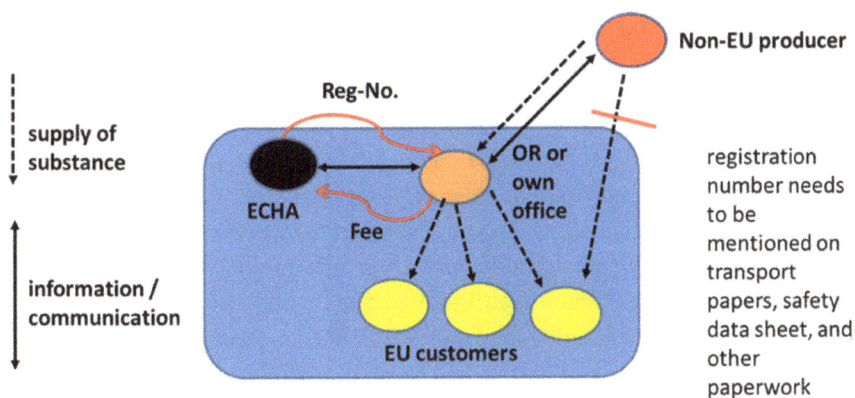

Fig. 127: The "only representative" (OR).

Substances brought on the market for *product and process-oriented research and development (PPORD)* need not a registration, but the agency has to be notified. In this notification, the producer/importer informs the agency about the tonnage, the research program, involved downstream users, measures taken to protect health and environment and all known properties of the substance. The agency assigns a notification number that needs to be communicated in the supply chain. PPORD is endorsed for a maximum of five years, but this can be extended for another five years on mutual agreement.

If the annual market volume of a substance is 10 t or more, a chemical safety assessment (CSA) is mandatory. This CSA requires to model the potential exposure of workers, consumers and the environment against endorsed uses of the substance, and to match the exposure with the "derived no effect levels" (DNELs) for human health effects and predicted environmental concentrations (PEC) with predicted no effect concentrations for ecotoxicity (PNECs). Based on these evaluations, the CSA needs to define safe uses and has to identify areas of concern and to propose countermeasures.

For intermediates handled and used within the chemical industry, a somewhat simpler registration can be run. For *on-site isolated intermediates*, the registration needs to comprise only the identity of the substance, identity of the manufacturer, available substance data, a classification proposal, a short general description of the use(s) and detailed information concerning risk managing measures.

For *isolated and transported intermediates,* the same information requirements are sufficient, if involved parties guarantee handling under strictly controlled conditions (rigorously contained over whole life cycle, means and measures are installed to minimize exposure/emission; only trained staff has access to the substance . . .). However, if the annual market volume is 1,000 t or more, a data set according to Annex VII has to be generated.

After registration is complete, it has to be kept up-to-date by the registrant: changes in tonnage, areas of use, substance data and so on call for an update of the registration dossier.

8.1.3 Title III: data sharing and avoidance of unnecessary animal testing

Data generated for the registration must meet certain quality criteria which are set out in Annex XI. For *animal welfare* reasons, tests on vertebrate animals shall be performed only if unavoidable. In-silico methods and in-vitro methods is given preference (see section classification and labeling, skin effects and eye effects as an example). Tests concerning human health and environmental effects have to be run according to accepted guidelines, explicitly according to regulation 440/2008/EC and against "Good Laboratory Praxis" (GLP). GLP is a sophisticated quality assurance system that shall guarantee to retrieve all relevant information on the performance of the test over several years (who did what and why; all observations are recorded). Vertebrate animal tests need an endorsement by animal welfare experts/committees; registrants of the same substance are obliged to share vertebrate animal test data. For that reason, a new registrant first has to file an inquiry to the agency, whether a registration of the substance he wants to bring on the market exists already. Animal welfare is also the reason why tests according to annexes IX and X have to be proposed to the agency. These annexes list tests which need to use a large number of vertebrate animals.

8.1.4 Title IV: information in the supply chain

A supplier of a substance/mixture has to submit a safety data sheet (SDS) to the customer, if
 – the product is classified and labeled as hazardous, or
 – the product fulfils criteria of PBT and vPvB, or
 – the product fulfils the criteria of "substances of very high concern" (SVHC) for other reasons:
     – for example, if the substance is interfering with the hormone system.

If the product is not classified, but contains 1% (solids or liquids) or 0.2% (gases) of a substance hazardous to health or the environment in general, or 0.1% of a carcinogen category 2, or of a substance toxic to reproduction category 1A, 1B or 2, or a sensitizer category 1, or has effects via lactation, or is – according to criteria laid out in Annex XIII – persistent, bioaccumulative and toxic (PBT), or it is very persistent and very bioaccumulative (vPvB), or it contains a substance with a community occupational exposure limit: a SDS has to be made available on request. Annex II describes the contents and further requirements of the SDS. The SDS has to be submitted not later than the first delivery of the substance. An updated SDS has to be submitted whenever new information relevant for the safety has become available, and if the authorization or restriction status has changed. Updated SDS has to be sent to all customers who received the product within the last 12 months.

For classified substances with an annual market volume of at least 10 t, a CSR has to be created. For these substances, an annex to the safety data sheet has to be added where endorsed uses are listed and safe uses are described. To harmonize and guide the description of endorsed and safe uses, a list of descriptors like process categories (PROC), product categories (PC), article categories (AC) and environmental categories (ERC) is provided as separate guidance document by the ECHA [58].

If *articles* contain 0.1% or more substances of very high concern (SVHC; see below), the supplier shall submit information on safe use and disposal of the articles.

Workers must always have free access to safety relevant information. In that respect, the safety data sheet (SDS) is an open document.

Manufacturers and importers must keep the safety relevant information available for up to 10 years following the last delivery of the product.

8.1.5 Title V: downstream users

A downstream user purchasing a product form a supplier in the European Union (manufacturer or importer) can assume all REACh obligations are fulfilled for products which are classified and labeled and have an annual market volume of at least 10 to, provided he finds his uses covered in the extended safety data sheet (eSDS). If his uses are not covered, he may ask the supplier to update his CSA and CSR. In that case, he has to submit information necessary to cover the specific use in the CSA/CSR.

If the downstream user wants to keep his uses confidential, he needs to run his own "downstream user registration process!" He has to run his own CSA covering his confidential use, and a notification to ECHA is required. A downstream user registration is not necessary if the product does not require a CSA, or if the annual volume of the substance used in the respective process is less than 1 ton.

8.1.6 Title VI: evaluation

This title addresses mainly the agency and competent authorities who need to check the quality and completeness of submitted data and proposals. Each European national competent authority could bring substances on the list for evaluation based on the hazardous properties which are set by ECHA as requirement for evaluation (community rolling action plan = CoRAP); the list is updated every year.

8.1.7 Title VII: authorization

Authorization means a ban of substances having properties which trigger high concern for health or environment. All market participants are requested to seek for substitutes. Until substitutes are found, an authorization may be granted to use a SVHC substance for certain purposes and for a limited period of time. The SVHC substances are listed in annex XIV, which is a living document and updated continuously. Authorization is not required for
– Registered plant protection products (PPPs)
    – They are authorized under the umbrella of a specific regulation (see Section 8.2.1)
– Registered biocides
    – They are authorized under the umbrella of a specific regulation (see Section 8.2.1)
– Substances used as fuels in motors or in mobile or fixed combustion plants
– Intermediates

Criteria for authorization are
– Carcinogenic, mutagenic or toxic to reproduction category 1A or 1B,
– Substances that are PBT,
– Substances that are vPvB,
– Substances with properties that give reason for equivalent concern:
    – These may be, for example, substances having deleterious effects to certain species in the environment.

Authorization may be granted, if
– risks for health and environment are adequately controlled (not possible for CMR substances), or
– socioeconomic outweigh risks to health and environment,
– the applicant submits a study plan to identify alternative substances,
– search for technical alternatives was not successful, so far, and
– the authorization is not in contradiction to restrictions as laid down in Annex XVII.

A granted authorization is typically limited to 5 years, and it is specific for the substance, the process and use and the parties involved. If the applicant cannot find technical viable alternatives, he may apply for another authorization period. An existing authorization and its sunset date have to be communicated in the supply chain. The sunset date defines the latest date after which any further use is possible only if an authorization is granted. Every stakeholder in the supply chain has to file an authorization for his specific applications.

8.1.8 Title VIII: restriction on the manufacturing, placing on the market and use of certain dangerous substances, mixtures and articles

Other than authorization, substances, mixtures or articles may be restricted. In contrary to authorization, restriction means all uses are in general permitted but restricted in detail. Restricted substances, mixtures and articles are included in annex XVII; the entry in annex XVII identifies the restricted products and lists precisely the prohibited uses and the conditions of permitted uses. To give an example: entry number 5 is benzene. Benzene must not be present in toys at levels of 5 ppm and higher. It shall not be placed on the market as such and in mixtures at 0.1% and above, with the exemption of motor fuels compliant with the respective EU directive on motor fuels, and substances and mixtures used in closed industrial processes are exempted.

The reader should look up the Annex XVII of Regulation (EU) No. 1907/2006 for more details. However, entries 28–30 shall be mentioned at this place. These entries say that substances which are carcinogens, mutagens or which are toxic to reproduction (CMR) category 1A or 1B must not be placed on the market for the general public; this is valid for the substances as such or in mixtures where the general or specific concentration limits are arrived or exceeded. The label must cite the sentence "restricted to professional users." By way of derogation, certain medicinal or veterinarian products, cosmetics, fuels and oil products and artist paints may be exempted.

Annex XVII is a living document. Experts responsible for chemical regulations are requested to check this annex on a regular basis.

8.1.9 The guidance documents

Regulation (EC) No. 1907/2006 names many obligations to be followed by those who want to bring substances, mixtures and articles on the market. More detailed descriptions how these requirements can be fulfilled are laid down in a series of guidance documents. As these guidance documents are not directly binding legal texts, they define best practice and provide the opportunity to focus on practicability.

As these guidance documents in total comprise many pages, and because they are updated continuously and are open source documents, the reader is encouraged to look up the ECHA website: https://echa.europa.eu/support.

Here, one part shall be mentioned, nevertheless, as it provides great help in the estimation of exposure. The Guidance on Information Requirements and Chemical Safety Assessment, Chapter R.12 [58], describes the exposure assessment and risk characterization for hazardous chemicals in their identified uses. The use description includes any use of the substance as such and in mixtures and any subsequent service life in articles resulting from a use. The use description is allocated to the life cycle stage of the substance. Key elements for describing a use are life cycle stage (LCS), sectors of use (SU), product categories (PC), process categories (PROC), environmental release categories (ERC) and article categories (AC). The latter were called descriptors and are used already in the annex to the safety data sheet for description of permitted uses.

The *SU* covers the market description and is meant to provide information on the sector of economy or market where the use takes place: industrial use (SU 3), professional use (SU 22) and consumer use. Further examples are SU 1 (agriculture, forestry, fishery), SU 9 (manufacture of fine chemicals), SU 12 (manufacture of plastic products) and SU 24 (scientific research and development).

Examples for *PCs* are adhesives and sealants (PC 1), antifreeze and deicing products (PC 4) or PPPs (PC 27).

PROCs describe the tasks and process types for products from the occupation perspective, like PROC 3 (closed batch process; little danger of exposure because the whole process happens predominantly in a contained way), PROC 4 (use in batch manufacture of a chemical where significant opportunity for exposure arises, e.g., during charging) or PROC 7 (industrial spraying which requires specific equipment or training for the personnel).

ERCs summarize activities which share a certain likelihood and pathway of releases into the environment, for example manufacture of substances in dedicated plants (ERC 1), industrial use resulting in manufacture of another substance (ERC 6a – intermediate use) or low release of substances included into or onto articles and materials during their service life in outdoor use, such as metal, wooden and plastic construction and building materials (ERC 10a).

Examples for *ACs* are vehicles (AC 1), leather articles (AC 6) or plastic articles (AC 13).

IT tools to estimate exposure and release are made available by ECHA. In most of these tools, besides the physical–chemical properties the user needs to tick appropriate SU, PC, PROC, ERC and AC. The programs then forecast exposures and releases. Where a calculated risk characterization ration (RCR) is lower than 1 – which means exposure is lower than a derived no effect level (DNEL), or if a predicted environmental concentration (PEC) is lower than the predicted no-effect concentration (PNEC) – a threat to health and environment is not expected.

With regards to measured data, the REACh guidance document chapter R.14 addresses occupational exposure. Measured data are accepted, if
- they are generated with a validated method;
- they are personal sampling data or
  - typically taken in the breathing zone of the employees;
- area monitoring data must allow a robust estimate of personal exposure;
- they are representative in terms of
  - numbers: sufficient data points over several different,
  - working process covered, including rare, but predictable events,
    - see European Standard EN 689;
- data – sampling and analysis – were generated by a competent (certified) lab;
- data allow to calculate the mean and the standard deviation as well as 75th and 90th percentiles,

- the 90th percentile (90% of data do not exceed this value) is taken forward as reasonable worst case,
  - this would require at least 10 valid data points for a specific exposure scenario.

Concerning validated methods and frequency and locations of measurements, there is room for definition by member states. Please always observe national guidance on this issue. The REACh guidance document chapter R.16 deals with environmental exposure. In this document, the estimate of losses to the environment against tonnages and processes/uses is described. Concerning measured data, the same quality standards have to be met as for workers exposure data.

8.2 Further regulations addressing the marketing of hazardous substances

8.2.1 Pesticides, plant protection products and biocides

Pesticide is the overarching terminus for agents that act on living organisms with the purpose to control them. Pesticides are subdivided in plant protection products (PPP) and biocides. These products have typically a dispersive use, the general population is on purpose or by chance likely to be exposed, and these products are designed to provoke an effect in living organisms. Because of this, a more stringent regulation than REACh is required. The EU regulations for these products are:
- Pesticides: Directive 2009/128/EC
- Biocides: Regulation 528/2012/EC
- PPPs: Regulation 1107/2009/EC

The use of these products needs close control and authorization, and benefits must at least level off the risks. Less risky alternatives shall be promoted. Allocation to different classes is due to the target organisms: insecticides, herbicides, bactericides and so on.

EU member states are obliged to care for sustainable use under protection of environmental goods like ecosystems, groundwater and nontarget organisms (including the general population) and appropriate control and design of equipment used to apply pesticides, and adequate training of those using pesticides.

PPPs are not necessarily only substances or mixtures thereof, but these may also be other organisms – including genetically modified ones – to protect plants, support or inhibit plant growth or to preserve plant products. In Germany, PPP can be sold by certificated competent experts only (§11 Chemikalienverbotsverordnung). Distinction is made between PPP for professional users only, and those for the general population.

The authorization process covers methods of use: against/for which organisms, effectiveness, at what time in the year, latest use before harvest. Data have to be provided concerning analytical methods, including metabolites in plants and wildlife and the analysis of these metabolites, effects on target organisms and nontarget organisms (like lady-beetles and honey-bees).

PPP which are accessible for the general public and which might be mixed up with food must contain repellents; the repellents may either provoke a pungent smell or disgusted taste. (*Author experience: in a poison center, a mixture was presented which was used for an unsuccessful suicide; the odor of that liquid was able to kill any appetite*).

Biocides are typically formulations made of the active substance (may be an organism) and additives; if these additives have hazardous properties, they are named "substance of concern." Biocides may further contain production aids. These are substances present in the biocide not because they need to be present for the biocidal effect, but because they were used somewhere in the production process and cannot (completely) be extracted. The active substance needs an authorization, and every ready to use biocide product needs a separate authorization as well.

Authorizations are granted for a certain time period and cover conditions of use (target organisms, permitted areas of use, form of application, application time, competence of users).

Biocides are allocated to several classes:
- general biocidal products and disinfectants;
- disinfection, general hygiene, drinking water treatment;
- preservatives;
- wood, leather, textiles, rubber;
- pest control agents;
- rodenticides, insecticides and so on;
- others;
- conservation of feed and food, antifouling-agents and so on.

Biocides must not be sold to the private consumer if they are classified as
- acute toxic categories 1–3;
- carcinogenic or mutagenic or toxic to reproduction category 1;
- PBT, vPvB;
- endocrine modulators;
- developmental neurotoxicants or immunotoxins.

8.2.2 Waste

Directive 2008/98/EC defines waste and the hierarchy of waste treatment (avoidance > material recycling > raw material recycling > energetic recycling > disposal).

Annex I defines 16 classes of waste to support appropriate waste handling and waste treatment. Annex III defines properties to identify hazardous waste.

Hazardous waste is subject to directive 91/689/EC. It is either the properties of the waste which classify it as hazardous waste (properties as laid down in 1272/2008/EC), or waste stemming from certain processes, p. e. tar in distillation residues of refineries or from pyrolysis processes.

Hazardous waste must not be diluted to render it nonhazardous waste. Hazardous waste can be treated by certified converters, only, and hazardous waste must be retrievable.

8.2.3 Ozone-depleting substances

Regulation (EC) No. 291/2011/EC is the EU implementation of the Montreal protocol to control substances that can cause stratospheric ozone depletion. Substances falling under this regulation are listed in the annexes. The classics are the chlorinated fluoro-carbons. Due to their nonflammability, nontoxicity and heat capacity they had ample applications as blowing agents, propellants and liquids in climate conditioning plants (cool-houses, refrigerators, air-conditioning). Halons were used as very effective fire-fighting agents. Methyl bromide is an industrial chemical and was used as a fumigant (on application, authorities may grant specified and limited uses of this fumigant).

Use of these compounds is restricted if not prohibited. The regulation and its restrictions cover

- production,
- marketing,
- use,
- recycling and
- destruction

of listed ozone-depleting substances. The purpose is to avoid release into the atmosphere as efficient as achievable. Therefore, phase-out programs depending on the identity of the substance and the use are installed. Maintenance work for installations and destruction can be done only by certified companies.

Mass balances have to be calculated and recorded to ensure there are no losses by any company involved in trading, use and handling. Records have to be sent to national authorities every year.

8.2.4 Dual-use products and "prior informed consent"

Regulation 428/2009/EC is the EU implementation of the Chemical Weapons Convention. It not only covers chemicals, but also machinery and equipment that can be used for mass-destructive weapons. Annex I lists regulated items, annex II ff. list nations having signed the treaty (or not), special items and so on. In case of international shipment, you have to check whether substances or components of mixtures are in annex I, and if yes, whether and how the target country is listed in the following annexes. In case of doubt, consult the national agency. Thiodiglycol, phosphor oxychloride, 2-chloroethanol and dimethylamine are examples of substances listed.

Regulation 111/2005/EC addresses the monitoring of drug precursors, and 1277/2005/EC lists drug precursors and substances that can be used for the production of illicit (illegal) drugs. The latter describes obligations to be fulfilled if you trade/ship listed substances.

Regulation 649/2012/EC is the "Prior Informed Consent" regulation and shall ensure, that very dangerous products are not "lost" somewhere in the world. If certain products are to be exported outside the European Union, depending on the target country an official paper may be required where an authority of the receiving country endorses the shipment and import.

8.2.5 Greenhouse gases

Gases absorbing in the infrared (IR) region of the spectrum can act as greenhouse gases by increasing the radiative force. If these gases let light in the near-UV and visible region pass, but absorb the radiation of heat (IR region), they act similar to window glass used in greenhouses; colloquially these gases are called greenhouse gases [59]. The absorption of IR radiation and re-emission in all directions contributes to the radiative forcing activity; this contribution is stronger the higher the absorption coefficient of the substance and the less likely it absorbs in a region where other greenhouse gases are active already, namely water and carbon dioxide. The lifetime in atmosphere determinates for how long the parent molecules are active. With increasing concentration, the absorption of radiation at the specific wavelength becomes saturated so the contribution to radiative forcing declines. However, the "thickness" of the intransparent atmospheric layer is increasing as well.

The concentration, wavelength of absorption, the absorption coefficient and the lifetime of a substance contribute to the total radiation forcing. The benchmark is carbon dioxide, and IR-active substances are matched against an amount of carbon dioxide that would result in the same radiative forcing over a defined lifetime (typically calculated for 20, 100 or 500 years). The contribution to radiative forcing is expressed as mass equivalents of carbon dioxide.

As greenhouse gases have most likely an influence on the climate, their release into the environment is undesired and regulated. In the European Union, this is done in Regulation (EU) No. 517/2014. Annexes I and II of the regulation list substances which are regulated due their high CO_2 equivalents which are based on the high IR absorption in combination with a relative long lifetime in the atmosphere.

All listed substances are fluorinated (Fig. 128). The fluorination is reason for the IR absorption and environmental stability of the fluorocarbons (Fig. 129). Many of these substances or their primary degradation products are persistent (although that is not a direct subject of this regulation).

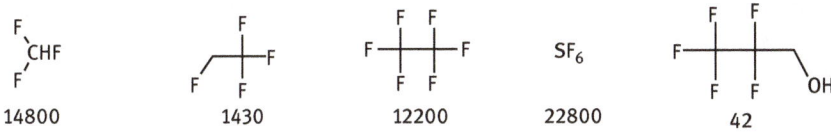

Fig. 128: Some fluorinated greenhouses gases listed in Regulation (EU) No. 517/2014 and their CO_2 – equivalents.

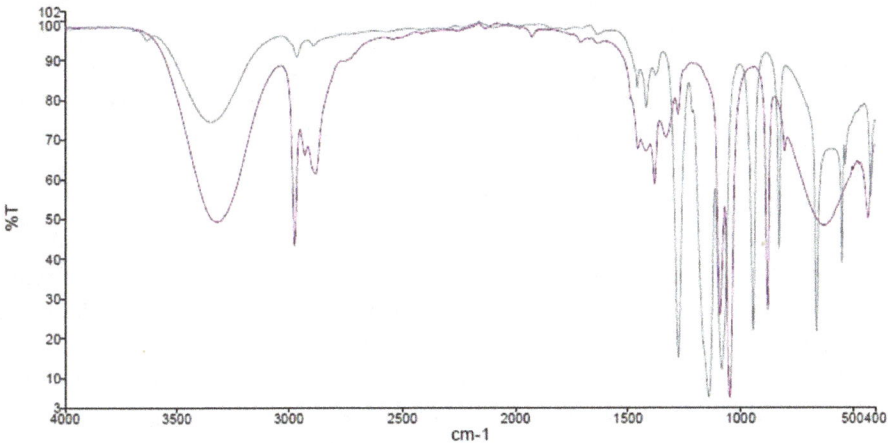

Fig. 129: FT-IR spectra (ATR) of ethanol (pink) and trifluoroethanol (blue). Trifluoroethanol has additional, strong C-F absorption at about 1,100 and 1,300 cm^{-1}.

The prime targets of the regulation are to
- Avoid emission of these gases in the atmosphere
    - refilling and dismantling activities concerning installations that use these substances (air conditioning, large heat-pumps for cool-houses and cool-trucks, blowing agents for polymer foams) can be done by authorized companies, only;
    - depending on the size of the installment, leak-checks and leak controls have to be performed on a regular basis; above 500 t CO_2-equivalents, leakage detection systems are required;
    - purchases, losses, recoveries have to be balanced and listed in an inventory.
- Phase out of the greenhouse gases

- Marketing is restricted by
  - type of use;
  - type of equipment; and
  - sunset-dates for certain uses and equipment.

The listed greenhouse gases can be marketed and used by permission only, and equipment containing these gases needs to be labeled accordingly to ensure appropriate dismantling at the end of life.

8.2.6 Persistent organic pollutants

Persistent organic pollutants (POPs) are covered by Regulation (EU) No. 2019/1021 which replaces earlier regulations. It is the EU implementation of the so-called Stockholm convention where signing states commit themselves to reduce (and finally prohibit) the manufacture, placing on the market and use of substances which are very persistent in the environment and which may undergo long-range transport. Originally, this convention addressed the "dirty dozen" (Fig. 129), chlorinated pesticides and chemicals and impurities/by-products which turned out to be very persistent in the environment.

Due to a sufficiently high vapor pressure and long half-lives in the atmosphere, these substances are also subject to long-range transport. For example, polybrominated biphenyls – used as flame-retardants – were detected in the fat of polar bears although release happened only in moderate climates.

According to the regulation, the manufacture, marketing and use of substances in Annex I of the regulation as such, in mixtures and in articles is prohibited. Use as reference material for analytical purposes is not restricted, and also articles containing less material than the cut-off mentioned in Annex I are not restricted. For example, tetra- and higher brominated diphenyl ethers, perfluoro octane sulfonic acid and its salts and derivatives, thereof, are listed in Annex I, Part A, with restriction limits of 10 ppm. Hexachlorocyclohexane with all isomers, hexachloro benzene and pentachloro phenol, its salts and esters are listed without cut-off. For polychlorinated bi-phenyls (PCBs), member states are requested to take articles actively out of the market by 31 December 2025, the latest if they contain more than a certain amount of PCBs.

Production and use of POPs in closed systems as on-site intermediate is possible if this exemption is mentioned in the annex, and a kind of a notification and authorization process has been passed. Note this procedure is stricter than the REACh regulation, which does not request an authorization for intermediates.

Annex II is currently empty; it is intended for POPs whose manufacturing and use is permitted provided that specific restrictions are met.

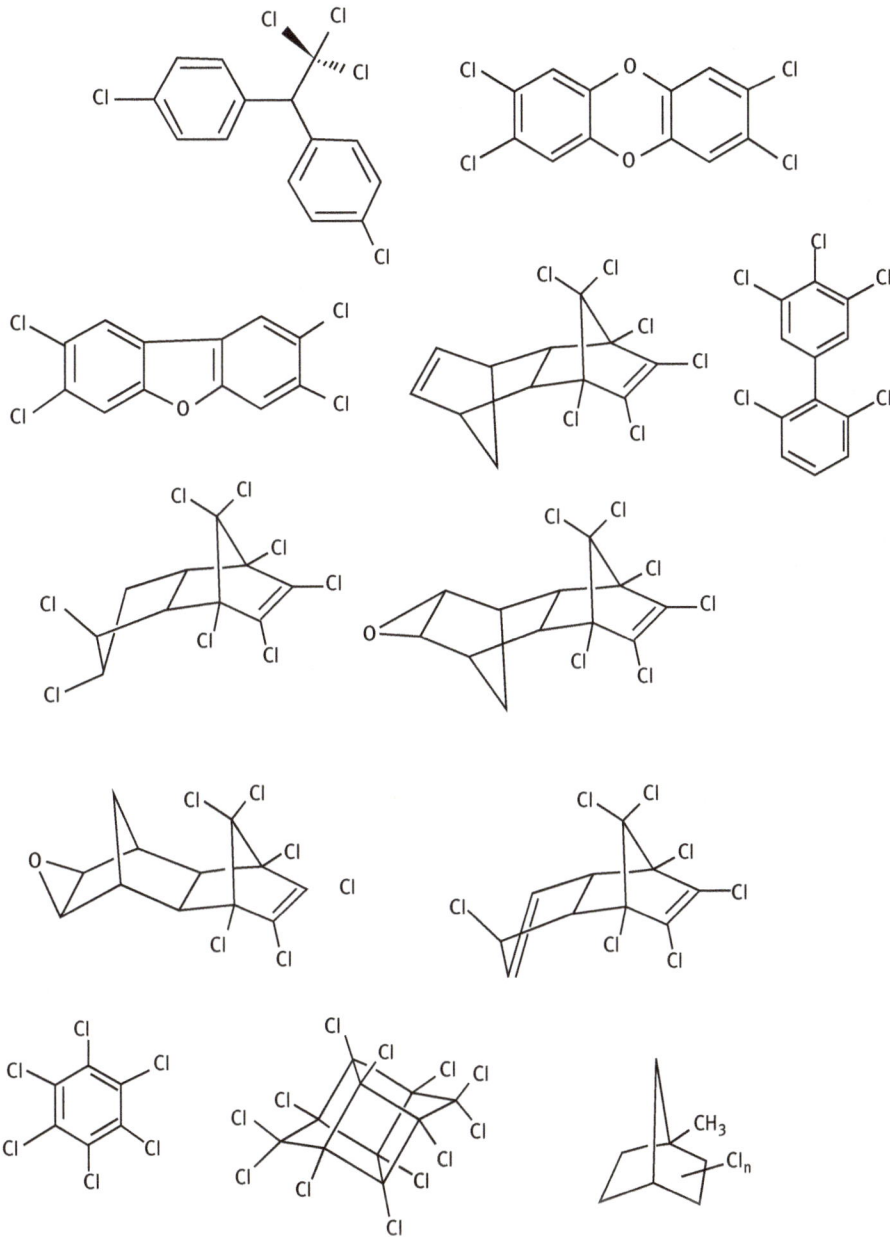

Fig. 130: Original "Dirty Dozen": Dichloro-diphenyl-trichloro ethane (DDT), 2,3,7,8-tetrachloro dibenzo-p-dioxin (a PCDD), 2,3,7,8-tetrachloro dibenzo-furane (a PCDF), aldrin, 3,4,6,2',6'-pentachloro biphenyl (a PCB), chlordane, dieldrin, endrin, heptachlor, hexachlorobenzene, mirex and toxaphen (isomers).

Annex III lists substances where member states are obliged to draw up a release inventory covering releases to air, water and land. Part A lists substances where environmental monitoring is mandatory. Currently polychlorinated dibenzo-*p*-dioxins (PCDD), -dibenzofuranes (PCDF) and PCBs are listed. Part B mentions substances where the need for mandatory monitoring shall be reevaluated on a regular basis. Examples for substances listed in part B are hexachlorobenzene, polycyclic aromatic hydrocarbons (PAHs) and polychlorinated naphtalenes.

Annex IV lists all those POPs which require a special treatment of waste if they are present at or above a certain limit in the waste. For example, if tetra- and higher brominated diphenyl ethers exceed 1,000 ppm, perfluoro octane sulfonate 50 ppm, or PCDD/PCDF 15 ppb, the waste has to be subject to special treatment as defined in Annex V to assure that the risk posed by the POPs is annihilated.

8.3 Exercises

(1) If I am going to produce a substance from ethanol and ethylene oxide, both purchased from a supplier who has registered these substances, do I need to register my product if I place 10 t/year on the EU market? The substance is not new, and others have it on the market since decades.

(2) A company produces a product **A** from ethylene glycol (glycol, $M = 61$ g/mol) with ethylene oxide ($M = 44$ g/mol), a product **B** from the same starting materials and a product **C** from 1,2,3-propanetriol (glycerol, $M = 92$ g/mol) with ethylene oxide. Glycol, ethylene oxide and glycerol have been registered. Size exclusion chromatograms reveal that for **A** the most abundant molecule is present at 40% (by weight) and has a molecular weight of 238 g/mol; for **B,** the most abundant molecule is present at 56% with a molecular weight of 150 g/mol; for **C**, the most abundant molecule is present at 45% and has a weight of 224 g/mol. Is there an obligation to register these products if they are sold on the EU market?

(3) If a company imports articles that contain substances of very high concern, are they exempted from registration because the substances of very high concern are not intended for release?

(4) A company produces a catalyst which undergoes several, subsequent steps in the market: (a) The catalyst is produced and consumed onsite. (b) A selected customer receives 20 t for a mutual research and development program; other customers purchase in total about 500 kg of the catalyst per year; 2 t/year are exported to the USA. (c) A customer in the EU purchases 80 t/year as intermediate. d) The catalyst is sold as intermediate at amounts of 1,500 t/year. What are the obligations under REACh, if there are any?

(5) If I am importing printer cartridges into the European Union containing a certain, nonregistered solvent, am I obliged to register that solvent? If yes, can my non-EU supplier run the registration?

(6) If I purchase chemicals only from EU suppliers who have registration numbers and I am producing articles, am I exempt from registration obligations?

(7) Imagine you are working in the "chemical regulations group" of a company. Your company wants to participate in the market margin a certain chemical has. You have not yet registered the substance. For not attracting the notice of competitors, the management asks you to run all investigations in-house for the tier-one registration (1 to <10 t/year). What is your view on that?

(8) A mixture which I bought on the market is not classified. Why and when I am obliged to submit a safety data sheet?

(9) What is the meaning of authorization and how can a company use substances under authorization?

(10) Can all chemical substances become subject to authorization, provided they show SVHC properties?

(11) A company produces annually 1,500 t of ester E. From this ester, 700 t is supplied directly for the production of pharmaceuticals, and 800 t is used as a solvent in industry. What has to be registered?

(12) Are there other possibilities to restrict the use of chemicals than authorization?

(13) Ozone depleting substances, POPs and greenhouse gases could be regulated in the scope of Annex XVII of REACh. What is the benefit of separate regulations, what are differences in the specific regulations against REACh?

(14) Ethanol may not only be used as a solvent but also as a biocide. Why is there a specific legislation for biocides?

(15) In what respect goes the notification of PPPs beyond the REACh regulation?

(16) Can a national agency prohibit the marketing of a substance although there is a valid authorization?

9 Workplace safety

9.1 Workplace safety: Council Directive 89/391/EEC

The *Council Directive of 12 June 1989 on the introduction of measures to encourage improvements in the safety of workers at work* aims at health and safety for workers in general (in Germany, the implementation is named "Arbeitsschutzgesetz"). The employer is responsible for workplace safety, including prevention of occupational risks, and he has to provide information and training to minimize risks.

The first priority is avoidance of risks. If risks cannot be avoided beyond a certain level, risks have to be limited at their source. That includes that workplaces, equipment and processes have to be optimized in terms of safety and health protection. This is an iterative task as from time to time adaptation to technical progress is necessary.

In general, the "dangerous" shall be replaced by the "nondangerous" or the "less dangerous" in all aspects. For example, if the floor is slippery, it has to be made up so that danger of gliding is avoided. If poor illumination can cause accidents, appropriate illumination has to be installed. Collective/area measures (p.e., illumination of the work area) should be given preference over individual measures (p.e., equip workers with a pocket light).

Workers should be given the appropriate instructions to be able to avoid risks. Before uptake of work, the employer needs to evaluate the workplace for any kind of risks (risk assessment). In this directive, chemicals are mentioned explicitly. Identified risks need to be reduced and managed. He/she has to consider that workers cannot be charged with tasks they are not fit for or not educated for. Before new technologies are introduced, the employer has to evaluate them against health and safety, together with the workers or their representatives.

If there are working areas with specific and considerable risks, only workers having received adequate instruction shall have access. This is typically the case for substances classified as carcinogens, mutagens, toxic to reproduction or respiratory sensitizers.

Workers must receive safety training specific for their workplace at the beginning of their employment and on a regular basis thereafter and whenever processes at the workplace are to be changed. Such training has to be given during worktime without any disadvantages on the workers income. Workers must be informed about all potential risks at the workplace and protective measures taken. This includes external workers working on the site. If workers from different companies work together on a site, all workers shall enjoy the same level of protection, and employers need to work together on that issue. You are not allowed to "outsource the risk"; employees of a subcontractor working at your site must be given the same level of training, protective equipment and safety supervision as own employees.

https://doi.org/10.1515/9783110618952-009

Cost inferred by health, safety and hygiene must not be imposed onto workers. That is, the employer has to cover the costs for the purchase of safety equipment like lab coats, safety glasses, safety boots, but also for first-aid equipment, medical examination of employees as this is required by legislation and services by experts (air monitoring, workplace risk assessment, etc.).

The employer shall identify one or more people of the workforce who shall take care for all aspects of workplace safety. If these workers agree to become charged with specific tasks, they must be granted an adequate education and equipment, and they must not have any disadvantage because of their special tasks. Workers delegated for health and safety must have full access to all safety relevant information, including the workplace risk assessment, reports from external experts, incident and accident records and monitoring results. The employer has to consult them concerning all existing or future situations and activities that may have an impact on safety. These workers must be given the chance to communicate with inspection agents. If deemed necessary, the employer has to hire support from external experts for health and safety. Irrespective of any delegation, the employer stays in charge for compliance with the directive.

The employer is responsible for procurement of first aid, firefighting and emergency installations and arranges necessary contacts with external specialized services. An adequate number of workers shall receive training for first aid and firefighting.

In case of imminent danger, measures have to be in place to inform and warn the workforce immediately. Workers must not be asked to resume work in an area with imminent danger (unless in exceptional and duly substantiated cases). The design of the workplaces and organization of work must ensure that workers can avoid any imminent danger (p.e., escape routes, emergency exits, alarm systems, etc.).

The employer must keep records of incidents that resulted in an absence of a worker for three days or more.

Workers must accept safety training and abstain from any activities that could infringe safety measures. If they detect irregularities that can affect safety, the must inform their superiors without delay.

Depending on the kind of work to be performed, workers shall be given the chance of medical surveillance and regular health controls.

9.2 Hazardous agents Directive 98/24/EC

The *Council Directive 98/24/EC of 7 April 1998 on the protection of the health and safety of workers from the risks related to chemical agents at work* is a special follow-up legislation extension of the workplace safety Directive 89/391/EEC, as handling of hazardous agents calls for additional measures. In Germany, this directive is implemented as "Gesetz zum Schutz vor Gefährlichen Stoffen (Chemikaliengesetz)" and the "Gefahrstoffverordnung."

9.2.1 Definitions

The directive provides some definitions, for example
- Hazardous chemical agent: any chemical agent that is classified as hazardous or may become hazardous under conditions of use or which has an occupational exposure limit (OEL).
- Activity involving chemical agents: handling, filling, storage, on-site transport, disposal and destruction of chemical agents.
- OEL: the maximum concentration of a substance in the air at the workplace that will not induce a disease if the workers are exposed for 8 h/day and 5 days/week until retirement.
- Biological limit: highest concentration of a substance or their degradation products in body specimens (blood, urine, etc.) or change in biochemical or physiological markers (p.e., hormone level in blood, methemoglobin (MetHb) level, blood pressure, etc.) that will not result in impaired health.

9.2.2 Employers obligation

The workplace risk assessment must address hazardous agents. In line with the general workplace safety directive, hazardous agents have to be replaced by the less hazardous agents. A very first indication is the labeling of the products; however, care has to be taken. A less stringent label does not always indicate lower risk. For example, in a factory, the solvent dimethyl formamide was used to clean machine parts due to its dissolution power for cured plastics. When dimethyl formamide became labeled as toxic to reproduction category 1B, the company switched to the technically acceptable and at that time unlabeled dimethyl acetamide. However, after the toxicological tests for effects on development were finalized for dimethyl acetamide, it turned out to be toxic to reproduction 1B as well, so the company switched to another technical alternative N-methyl pyrrolidone. When test results became available for N-methyl pyrrolidone leading to a classification as toxic to the development category 1B, there was a move to the technically feasible and still nonlabeled N-ethyl pyrrolidone. Finally, it turned out that this solvent is toxic to reproduction category 1B (development) as well. As this example shows, substituting a substance due to a certain hazard by another substance which was not (yet) tested against this specific hazard can end up in a hare race.

Actually, the risk has to be reduced which can be described as product of hazard times exposure:

$$Risk = Hazard \times Exposure$$

Exposure can be reduced by either reducing the height of exposure or time of exposure. The employer has to document the investigation into substitution potentials

and the result. If technically feasible, the less risky material must be given preference. Health and safety professionals and the company occupational medical expert need to be involved in investigations for substitution.

9.2.2.1 Risk assessment

In the workplace risk assessment, besides search for substitution the employer has
to collect information about the hazardous properties of the substances, potential
risks and how measures can be taken to limit risks and how these measures can be
installed. Exposure has to be minimized and to be controlled and in that scope also
rare activities have to be covered like maintenance work and sampling. Gathering
information and updating the risk assessment has to be done on regular intervals.

9.2.2.2 Preventive measures

Measures to prevent incidents are mentioned in a more general way in the directive.
Existing OELs need to be controlled, and area ventilation, or better point of source ventilation can help to match the OELs. Amounts of hazardous agents at the workplace
shall be kept to a minimum, and containments should be kept closed. Containments
and pipes must always be identifiable: what porducts does it contain, what are
the hazards associated with these products? Concerning pipes, what is the direc-
tion of flow?

Ignition sources should be avoided or kept to a minimum (open flames, sparks
and hot surfaces). Processes should be run as closed processes. Number of exposed
workers, exposure levels and exposure time should be kept low. Working procedures should be automated/mechanized in such a way that they are intrinsically
safe; for example, remotely controlled pumps allow the refilling of drums in an exhausted cabinet and result in smaller exposure than mechanical manual pumps.
Sanitation, social rooms and so on help to improve personal hygiene.

9.2.2.3 Information, training and incidents

Workers must have free access to all safety relevant information. Safety training has
to be provided before uptake of work, then on repeated intervals (in Germany at least
annually) or whenever processes or hazardous agents change. In this training the employees need to be informed about the hazards posed by the chemicals and machines,
appropriate protection and how to use this. Examples are given in Figs. 131 and 132.

An action plan has to be designed for emergencies, and emergency response has
to be exercised in regular drills, involving external experts as necessary. Provisions
for first aid and firefighting need to cover special risks posed by the chemicals. For
example, if substances are handled that would react violently with water, other extinguishing media than water have to be kept readily available. Decontamination
procedures can be performed by specially instructed and equipped workers only or

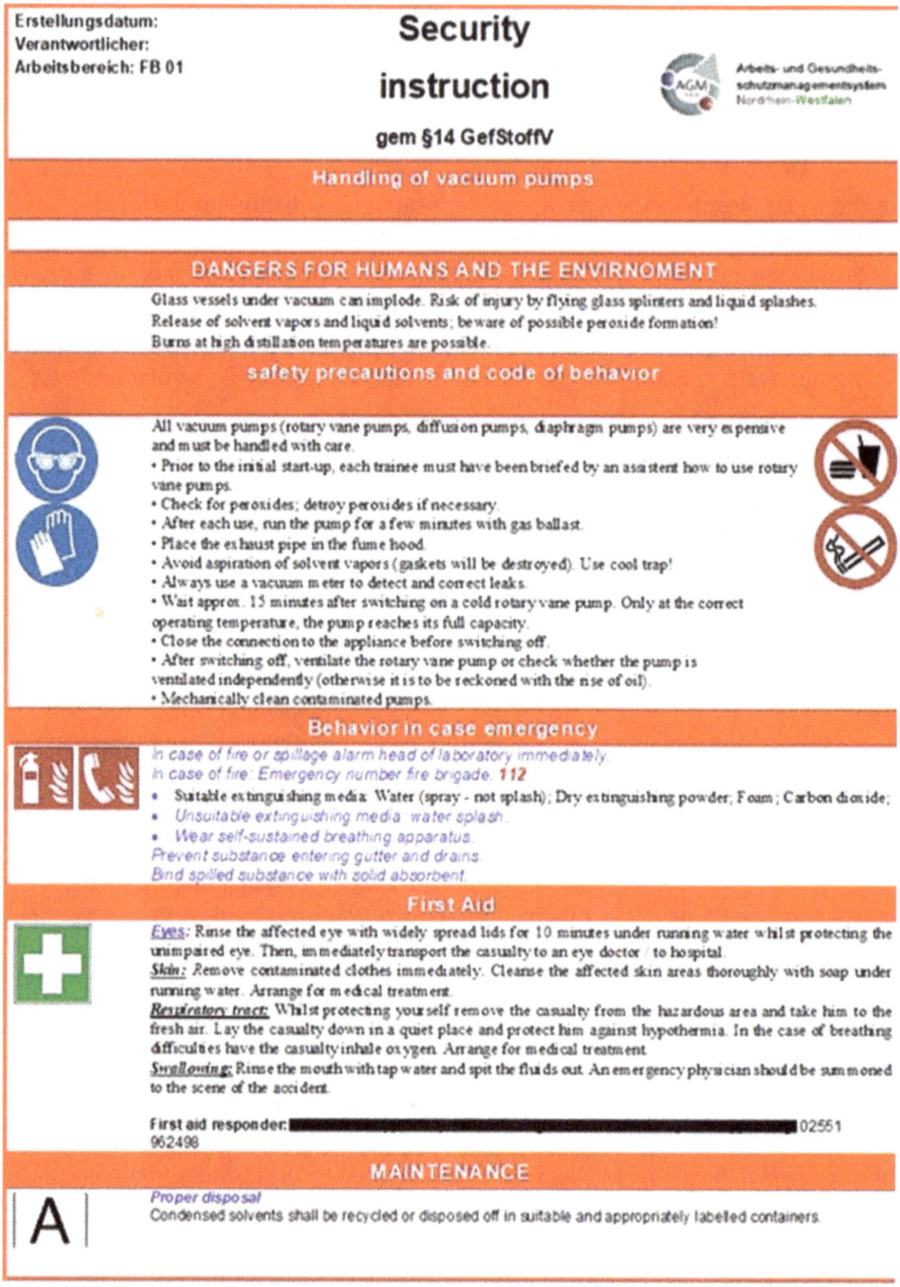

Erstellungsdatum: Verantwortlicher: Arbeitsbereich: FB 01	**Security** **instruction** **gem §14 GefStoffV**	AGM · Arbeits- und Gesundheits- schutzmanagementsystem Nordrhein-Westfalen

Handling of vacuum pumps

DANGERS FOR HUMANS AND THE ENVIRNOMENT

Glass vessels under vacuum can implode. Risk of injury by flying glass splinters and liquid splashes.
Release of solvent vapors and liquid solvents; beware of possible peroxide formation!
Burns at high distillation temperatures are possible.

safety precautions and code of behavior

All vacuum pumps (rotary vane pumps, diffusion pumps, diaphragm pumps) are very expensive
and must be handled with care.
• Prior to the initial start-up, each trainee must have been briefed by an assistent how to use rotary
vane pumps.
• Check for peroxides; detroy peroxides if necessary.
• After each use, run the pump for a few minutes with gas ballast.
• Place the exhaust pipe in the fume hood.
• Avoid aspiration of solvent vapors (gaskets will be destroyed). Use cool trap!
• Always use a vacuum meter to detect and correct leaks.
• Wait approx. 15 minutes after switching on a cold rotary vane pump. Only at the correct
operating temperature, the pump reaches its full capacity.
• Close the connection to the appliance before switching off.
• After switching off, ventilate the rotary vane pump or check whether the pump is
ventilated independently (otherwise it is to be reckoned with the nse of oil).
• Mechanically clean contaminated pumps.

Behavior in case emergency

In case of fire or spillage alarm head of laboratory immediately.
In case of fire: Emergency number fire brigade. **112**
• Suitable extinguishing media: Water (spray - not splash); Dry extinguishing powder; Foam; Carbon dioxide;
• Unsuitable extinguishing media: water splash.
• Wear self-sustained breathing apparatus.
Prevent substance entering gutter and drains.
Bind spilled substance with solid absorbent.

First Aid

Eyes: Rinse the affected eye with widely spread lids for 10 minutes under running water whilst protecting the
unimpaired eye. Then, immediately transport the casualty to an eye doctor / to hospital.
Skin: Remove contaminated clothes immediately. Cleanse the affected skin areas thoroughly with soap under
running water. Arrange for medical treatment.
Respiratory tract: Whilst protecting yourself remove the casualty from the hazardous area and take him to the
fresh air. Lay the casualty down in a quiet place and protect him against hypothermia. In the case of breathing
difficulties have the casualty inhale oxygen. Arrange for medical treatment.
Swallowing: Rinse the mouth with tap water and spit the fluids out. An emergency physician should be summoned
to the scene of the accident.

First aid responder: ▆▆▆▆▆▆▆▆▆▆▆▆▆▆▆▆▆▆▆▆▆▆▆ 02551
962498

MAINTENANCE

A

Proper disposal
Condensed solvents shall be recycled or disposed off in suitable and appropriately labelled containers.

Fig. 131: Example for a safety instruction for running machinery and equipment (no warranty).

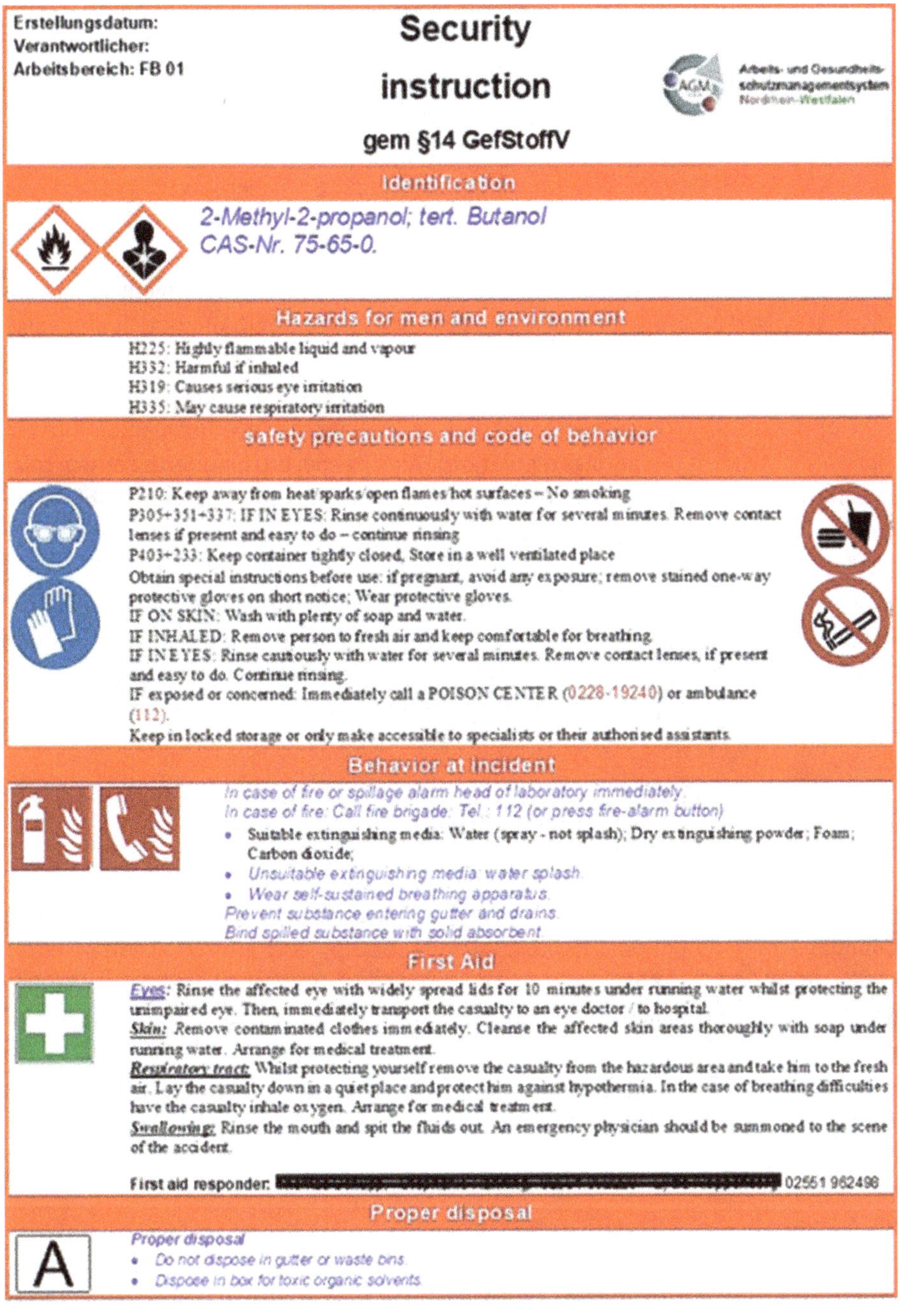

Erstellungsdatum: Verantwortlicher: Arbeitsbereich: FB 01	**Security** **instruction** **gem §14 GefStoffV**	AGM	Arbeits- und Gesundheits- schutzmanagementsystem Nordrhein-Westfalen

Identification

2-Methyl-2-propanol; tert. Butanol
CAS-Nr. 75-65-0.

Hazards for men and environment

H225: Highly flammable liquid and vapour
H332: Harmful if inhaled
H319: Causes serious eye irritation
H335: May cause respiratory irritation

safety precautions and code of behavior

P210: Keep away from heat/sparks/open flames/hot surfaces – No smoking
P305+351+337: IF IN EYES: Rinse continuously with water for several minutes. Remove contact lenses if present and easy to do – continue rinsing
P403+233: Keep container tightly closed, Store in a well ventilated place
Obtain special instructions before use: if pregnant, avoid any exposure; remove stained one-way protective gloves on short notice; Wear protective gloves.
IF ON SKIN: Wash with plenty of soap and water.
IF INHALED: Remove person to fresh air and keep comfortable for breathing.
IF IN EYES: Rinse cautiously with water for several minutes. Remove contact lenses, if present and easy to do. Continue rinsing.
IF exposed or concerned: Immediately call a POISON CENTER (0228-19240) or ambulance (112).
Keep in locked storage or only make accessible to specialists or their authorised assistants.

Behavior at incident

In case of fire or spillage alarm head of laboratory immediately.
In case of fire: Call fire brigade: Tel.: 112 (or press fire-alarm button)
- Suitable extinguishing media: Water (spray - not splash); Dry extinguishing powder; Foam; Carbon dioxide;
- Unsuitable extinguishing media: water splash.
- Wear self-sustained breathing apparatus.
Prevent substance entering gutter and drains.
Bind spilled substance with solid absorbent.

First Aid

Eyes: Rinse the affected eye with widely spread lids for 10 minutes under running water whilst protecting the unimpaired eye. Then, immediately transport the casualty to an eye doctor / to hospital.
Skin: Remove contaminated clothes immediately. Cleanse the affected skin areas thoroughly with soap under running water. Arrange for medical treatment.
Respiratory tract: Whilst protecting yourself remove the casualty from the hazardous area and take him to the fresh air. Lay the casualty down in a quiet place and protect him against hypothermia. In the case of breathing difficulties have the casualty inhale oxygen. Arrange for medical treatment.
Swallowing: Rinse the mouth and spit the fluids out. An emergency physician should be summoned to the scene of the accident.

First aid responder: ▓▓▓▓▓▓▓▓▓▓▓▓▓▓▓▓▓▓▓▓▓▓▓▓▓▓▓▓▓▓▓▓ 02551 962498

Proper disposal

A

Proper disposal
- Do not dispose in gutter or waste bins.
- Dispose in box for toxic organic solvents.

Fig. 132: Safety instruction for a hazardous substance (no warranty).

by external professional emergency staff. Emergency information covering special risks (p.e., large amounts of flammable substances, gas cylinders and confined space), properties of substances and a hazardous substances registry (amount, C&L and safety data sheets) has to be kept available.

9.2.3 Medical surveillance

Exposure to certain hazardous substances may trigger compulsory medical surveillance; this is the case – between others – where binding biological limits exist. Workers have to be informed before they are asked to take up work that falls into this area. For other substances, participation in voluntary medical surveillance has to be offered to the workers. Activities and/or substances triggering medical surveillance are laid down in national regulations. As for exposure monitoring at workplaces, results from medical surveillance have to be stored in a retrievable way. Workers can request getting insight into their monitoring and medical surveillance results at any time. When a disease is allotted to an exposure or when a biological value was exceeded, the worker has to be informed immediately by the company doctor or other adequate medical staff familiar with the workplace. If a worker diseased by workplace exposure cannot return to his workplace (p.e., in case of respiratory allergy), the employer has to offer an alternative job.

9.2.4 Annex I

Annex I lists the EU harmonized, binding OEL. Currently, only lead (Pb) and its inorganic compounds are listed with an OEL of 0.15 mg/m^3. In addition, the Scientific Committee on Occupational Exposure Limits (SCOEL; see: https://ec.europa.eu/social/main.jsp?catId=148&langId=en&intPageId=684;) published recommendations on OELs. These recommendations are not legally binding, but member state agencies may adopt them. For several hundred substances argumentations for OELs have been published. From 2019 onward, the Risk Assessment Committee (RAC) of the European Chemicals Agency (EChA) has taken over the role of the SCOEL [60].

9.2.5 Annex II

Annex II lists the binding biological limit values (BLV). Currently, only lead in blood is listed. However, the SCOEL had published recommendations for BLVs for many substances (https://ec.europa.eu/social/main.jsp?catId=148&langId=en&intPageId=684;), and biological guidance values (BGV) where a "safe" value cannot be defined, p.e., 1-hydroxypyrene in urine as a marker for exposure against polycyclic aromatic hydrocarbons (PAH). This task has been forwarded to the EChA RAC by 2019.

9.2.6 Annex III

Annex III lists prohibited substances and activities. Currently some substances known as strong carcinogens in human beings are listed, for example, benzidine, 4-aminobiphenyl, 4-nitrobiphenyl and 2-aminonaphthalene.

9.3 Some more aspects of monitoring and exposure

Methods for air and biological monitoring must be reliable, robust and reproducible. In the SCOEL documents, an overview of methods together with references is provided (https://ec.europa.eu/social/main.jsp?catId=148&langId=en&intPageId=684). Typically, the primary literature cited there specifies what to measure, where, when and over what time period. At workplaces, there is typically exposure against more than just one substance having an OEL. In that case, a general approach is to assume additive activity of the substances and to require

$$\sum_i \frac{C_i}{OEL_i} \leq 1$$

where C_i is the measured value of compound i.

9.3.1 Air monitoring

For air monitoring methods, in general, there is a distinction between area monitoring and personal monitoring.

Area monitoring is done by a stationary device that is placed somewhere at the working area (Fig. 133, 134). Of course, it should be placed at locations where exposure to workers can occur. This should also cover reasonable worst-case situations, p. e., maintenance and cleaning activities. The advantage of area monitoring is that hotspots can be identified by appropriate selection of sampling stations. An indirect benefit is that no worker is handicapped by carrying the sampling device, and the "good behavior" bias is less likely (when people realize their activities are monitored, they may act more carefully and deliberative than usual). The downside of area monitoring is that the concentration measured is not automatically identical to the concentration that is present at the breathing zone of the worker.

Personal monitoring is measuring the concentration in the breathing area of the person so a more precise picture of personal exposure can be generated. The

Fig. 133: 2Area monitor (© Drägerwerk AG & Co. KGaA, Lübeck. Alle Rechte vorbehalten. All rights reserved).

downside is that due to awareness of monitoring, the respective person may act much more carefully than in routine tasks. Further, some monitors require the use of battery driven pumps which is a handicap at work.

Member states may edit precise descriptions concerning workplace measurements. The German Technical Guide for Hazardous Substances No. 402 (TRGS 402) is an example. See https://www.baua.de/EN/Service/Legislative-texts-and-technical-rules/Rules/TRGS/TRGS-402.html for the English version and https://www.baua.de/DE/Angebote/Rechtstexte-und-Technische-Regeln/Regelwerk/TRGS/TRGS-402.html for the German version; on that website, there is also a link to a list of certified laboratories which are entitled to perform officially workplace measurements. According to that guidance, exposures may be estimated on physical basis, p.e., saturated vapor concentration at relevant temperature. However, beware of hot materials, hot surfaces and liquids under pressure! If aerosols can occur, calculation results for gases may underestimate the real exposure to a large extend!

If peak exposures can occur, for example, opening a drum and filling day tanks, short-term- or peak-exposure limits are the benchmark. Sampling should mirror the short-term exposure situation; having one peak event of, p.e., 2 min, and then going on with sampling for a further 58 min, equilibrates the peak exposure, making the results looking fine although the peak exposure limit was exceeded. Only direct reading instruments (i. e., integrating concentrations over a few seconds, only) are reliable and may raise immediate alarm (Fig. 134). Nondirect methods, where the sampling device is to be analyzed in a lab consequently, blur the peak exposure if they sample more than just the peak exposure time. Such behavior is non-professional (not to say criminal).

Fig. 134: Classics in air monitoring: the Draeger tubes and manual pump (© Drägerwerk AG & Co. KGaA, Lübeck. Alle Rechte vorbehalten. All rights reserved).

9.3.2 Estimation of respiratory exposure

The uptake of substances by inhalation can be estimated, if the concentration in air is known. As a first, cursory approach, the concentration in air can be calculated on the basis of the ideal gas law:

$$\frac{n}{V} = \frac{p}{R \times T}$$

with n/V in mol/m^3 with p as vapor pressure (Pa) at the given temperature (K), and R being the gas constant, 8.314 J / (mol × K). Over the day, an adult person inhales 20 m^3 air; a worker with light activity inhales about 10 m^3 air during an 8 h workshift. For a person working at ambient temperature (298 K) with 2-pentanone (vapor pressure of 5,260 Pa at 298 K; M = 86 g/mol) can take up the substance by inhalation:

$$\text{Uptake} = 10\,\text{m}^3 \times \frac{n}{V} = 10\,\text{m}^3 \times \frac{5260\,\text{Pa} \times \text{mol} \times \text{K}}{8.314\,\text{J} \times 298\,\text{K}} = 2.12\,\text{mol} = 182\,\text{g}$$

As worst case, it was assumed that the inhaled pentanone is completely absorbed in the lungs.

Free IT-tools are available to calculate the exposure of workers. Examples are the Advanced REACh Tool (ART [61, 62]), the Chemicals Safety Assessment and Reporting Tool of the ECHA (CHESAR, [63]) and the Targeting Risk Assessment Tool of the European Centre for Ecotoxicology and Toxicology of Chemicals (ECETOC TRA, [64, 65]).

CHESAR and ECETOC TRA have many program parts in common as these were developed mutually; actually, CHESARs human exposure calculation is based upon ECETOC TRA, and environmental exposure is based upon EUSES (European Union System for the Evaluation of Substances) [31]. Other than ART, beyond worker exposure, these programs also allow to calculate exposure of consumers and the environment, and they address dermal uptake. The user is requested to define the process categories (PROCs), operational conditions (OCs), risk management measures (e.g., local exhaust ventilation yes/no), environmental release categories, as defined in the annex.

All these programs request some physical–chemical properties to be put in, like molecular weight, boiling and melting points and vapor pressure (see Chapter 2). Excellent online training tools are available, and these are updated in a regular manner, so a detailed description of the programs in a textbook like this is not necessary (and might soon be outdated). However, some features of the ART shall be described here to provide an understanding of the principles of such programs.

Over the last decades, many workplace exposure data have been collected in the EU, and ART makes use of this experience. An important physical factor for exposure is either the dustiness of powders or the vapor pressure of liquids. Concerning the vapor pressure of mixtures, ART has imbedded a XLUNIFAC calculation program for activity coefficients which than are used to estimate the real vapor pressure of the compound at the temperature under consideration. ART then requires a description of the workplace situation:
- Is the emission source within or outside a 1 m breathing zone?
- Class of working activity, for example, brushing, pouring and spraying
- Surface area covered per hour
- Kind of local exposure control, if any (glove box, local exhaust ventilation, etc.)
- Categorical description in how far the emission source is enclosed
- General hygiene and tidiness at workplace
- Application outdoors or indoors
- For indoors:
  - Volume of room
  - Ventilation rate

- – Ventilation system
- – Is there near-field carryover (e.g., another worker doing the same exercise nearby)?

With these data the program is run and delivers a result naming a certain percentile (the default is the 75-percentile) and a confidence interval for the calculated value (the default is interquartile distance). This percentile is based on the existing data base of measured exposures for given exposure scenarios. 75-Percentile means that the calculated value covers 75% of all the measured data that were taken to derive parameters for the multiple linear regression for the model. It may be asked whether or not a higher percentile should be chosen "for safety reasons"; a higher percentile means a poorer robustness of the algorithm (higher scatter), and increased likelihood that "unusual" measured data due to – for example – poor OCs like failure of ventilation are incorporated. The confidence interval can be set to 80%, 90% or 95%, which causes an increasing interval. In case of substances which are classified as carcinogen, mutagen, toxic to reproduction, sensitizers or substances with specific target organ toxicity (STOT), the user may desire to introduce higher percentiles and higher confidence limits.

Whether or not such calculations may replace workplace measurements is up to national or regional regulations. For example, in Germany the control of respiratory exposure is described in the TRGS 402 [66] (see earlier). According to this guidance, calculation methods shall be used preferably if the data base covers the substance and comparable workplaces. In case of substances classified as carcinogen, mutagen or toxic to reproduction, or if the calculated values are narrow to exposure limits, performing respiratory exposure measurements is recommended. Measured data shall be cross-checked with calculated data. For measurements, validated methods have to be used by appropriately trained personal.

9.3.3 Biomonitoring

For biomonitoring, the compound under investigation itself or its transformation products may be analyzed in body specimens, mostly urine and sometimes blood. In addition, alterations in physiological behavior may be investigated such as lung capacity, blood pressure, reaction time and memory performance – to name a few. Data/performance of exposed people can be investigated and matched against non-exposed controls or against earlier data generated for the same people. The benefit of biomonitoring is that a picture is generated concerning how much material was biologically available; for example, at 1 mg/m^3 aniline in air, the heavy working technician inhales more air per time than his computer-screen watching colleague. Further, differences in the reaction toward the aniline and its metabolism can be investigated: people may be fast acetylators or slow acetylators (transform aniline to acetyl-aniline); the fast acetylators are less prone to develop methemoglobinemia than the slow acetylators. Downside of biomonitoring may be nonacceptance by workers (fears the employer is looking for more than just the chemical), concerns raised by nonprofessional communication and the difficulty to detect peak exposures. In many countries, biomonitoring results are strictly confidential, and only the occupational physician and the worker himself are allowed to know the specific personal results.

9.4 Carcinogens and mutagens: Directive 2004/37/EC

The *Directive 2004/37/EC of the European Parliament and of the Council of 29 April 2004 on the protection of workers from the risks related to exposure to carcinogens or mutagens at work* acknowledges the specific risk posed by carcinogens and mutagens. Carcinogens and mutagens in the scope of this directive are substances classified as carcinogens and mutagens category 1A or 1B. For these substances, in most cases it is impossible to define a safe exposure level. Therefore, exposure has to be reduced "as low as reasonably achievable."

The employer needs to demonstrate on a regular basis why these substances have to be used and what are the results of the search for substitutes. Processes have to be as enclosed as technically achievable.

Carcinogens and mutagens are to be handled only in designated working areas with restricted access, and the number of workers in that area shall be kept as low as possible.

National implementations of the EU directives like the German "Gefahrstoffverordnung" extend this requirement also to acute toxic substances category 1–3 and respiratory sensitizers.

Ventilated air of the working area has to be filtered before it is emitted. The employer has to take care of best available technology and equipment and keep it up to date. Organizational and technical safety measures shall have preference before personal protective measures. The employer is responsible for the appropriate personal protective equipment (PPE), its laundry and maintenance and timely replacement. Personal hygiene rooms must be made available for workers, and separate storage of working suits and personal clothing must be ensured. Further, the employer is responsible to keep records on monitoring data concerning exposure, biological values in combination with identities of exposed workers, work procedures, tasks, identity of carcinogens and mutagens, their amount handled and exposure times. These records have to be kept available for up to 40 years after the worker has ceased that specific task. Health surveillance is extended beyond employment.

The annexes of that directive list carcinogenic substances and procedures, recommendations for the health surveillances and carcinogenic and/or mutagenic substances with EU limit values, p.e. vinyl chloride and benzene.

9.5 Protection of vulnerable subpopulations: Directive 92/85/EC (protection of pregnant and breastfeeding women) and Directive 94/33/EC (protection of young people at work)

Pregnant and breastfeeding women, unborn children and young people can be more vulnerable than the healthy adult person. The directives cover all aspects of protection, not only hazardous chemicals but also physical stress, risk of infection and so on. In this section, however, the focus is on hazardous chemicals.

The workplace risk assessment and protective measures need to cover vulnerable subpopulations. If risks cannot be reduced sufficiently, alternative jobs have to be offered or the employees shall be given free vacation for the critical time period. Free time has to be granted for prenatal examinations and maternal leave. During maternity leave, demission is prohibited. The annex I of Directive 92/85/EC lists compounds dangerous for the unborn child; annex II lists substances where any exposure of pregnant women is prohibited, and currently lead compounds as a general class are mentioned. Directive 94/33/EC covers all aspects concerning the protection of young people at work. This includes working time (total hours per day and per week, prohibition of night-shift work), physical burden and so on, but certainly also working with hazardous agents. In general, young people must not work with agents labeled as

- Acute toxicity categories 1, 2 and 3
- Skin corrosion category 1
- Flammable gases and liquids category 1 and 2
- Flammable aerosols category 1
- Explosive categories 1, 1.1, . . ., 1.5 (1.6 exempt)
- Self-reactive substance classes A–D (E, F exempt)
- Organic peroxide classes A, B (C–F exempt)
- STOT SE categories 1 and 2 (3 exempt)
- STOT RE categories 1 and 2
- Respiratory sensitizers, skin sensitizers
- Carcinogens, mutagens, categories 1A, 1B and 2
- Substances toxic to reproduction categories 1A and 1B
- Lead and its compounds
- Asbestos

Concerning toxicological hazards, only substances labeled not more critical than GHS07 (exclamation mark) with the wording WARNING are permitted, but not skin sensitizers. If for the education success working with critical compounds is required, this is permitted provided that exposure is minimized and certainly below OELs, and work is performed under close supervision by an experienced and competent worker.

9.6 Personal protective equipment

While proper ventilation and segregation of emission sources shall limit the respiratory exposure, PPE primarily aims at limiting dermal exposure to hazardous substances.

All clothing used in a workplace where hazardous substances are handled shall not be easily ignitable. This can be achieved by a minimum content of cotton or other poly-carbohydrates of at least 67% in the textile. In case of handling highly or extremely flammable substances, special fire-resistant clothing is recommended.

The general lab coat protects against splashes. It can quickly be taken off without to be torn over the head.

Goggles, or face shields protect the eyes and the face.

Gloves can be worn to protect the hands against hazardous substances. Certain gloves are designed to protect against mechanical or thermal injury. "Chemical" gloves shall protect the dermal absorption of substances and against corrosion and irritation. Depending on the material thickness and composition of polymers, different qualities are available for different classes of hazardous substances.

There is no universal material fit for protection against all kinds of chemicals!

Different international standards address protective gloves, between others EN 420. Depending on the hazardous substances handled, different kinds of polymers may be more or less suited for skin protection. Materials used for chemical gloves are, to name a few examples:
- Natural rubber (latex)
- Butyl-nitrile-rubber
- Nitrile rubber
- Polyvinylchloride

Being based on the same chemical identity of the polymer material and thickness, the quality may differ as there are several different ways to synthesize polymers. Of course, the thickness has an influence on the breakthrough time of substances. This time is to be measured in standardized tests, and suppliers of protective gloves should be able to name breakthrough times for common solvents, acids and bases.

Breakthrough time: the time period a glove can be worn before significant amounts of a specific chemical can reach the skin.

After work, the rubber material may have absorbed a certain amount of the chemical, and the potential for decontamination and reuse has to be evaluated by experts for occupational hygiene. In laboratories, thin disposable gloves are common. The breakthrough time is in general very short. Being aware of this, in case of splashes these one-way gloves have to be disposed off without delay, but there is no need for panic (at least the disposable gloves should be selected on this basis). While dropping off disposable gloves, contamination by touching the outside of the glove shall be avoided (Fig. 135).

Gloves should be worn only if needed. Important: don't touch items while wearing gloves which will be touched by others with bare hands (Fig. 136)! The polymer materials of gloves do not allow for the exchange of humidity, so wearing gloves is softening the stratum corneum and a burden to the skin. Skin care programs are to be part of occupational hygiene and training of employees.

Wearing respiratory protection should not be the regular case at the workplace. Half-masks may be sufficient for very limited concentrations of aerosols and gases in the atmosphere.

Fresh-air masks with external air supply according to the standard EN 138 belong to a kind of respiratory equipment with the least burden to the worker, but even in these cases the vision is necessarily handicapped because full-masks are worn.

For maintenance work and in case of emergencies, the use of respiratory protection may become mandatory. In case of incidents, workers might need respiratory protection to escape from the area or to handle a spillage. Typically, employees need to be checked for appropriate fitness, and such nonregular situations need to be trained in emergency exercises.

Filter masks present resistance to breathing, they are fit for filtering only those gases and aerosols they are defined for, and the ambient air must contain sufficient oxygen (at least 17%, and 19% for CO filters). The standard EN 141 addresses gas filter cartridges, and EN 143 addresses particle filters. Besides the specification of gas types, filter classes in combination with full-face masks define a factor by which the OEL-factor may be exceeded [67] (Tab. 76, 77).

Concerning the filtration of gases, neither $30 \times$ of the OEL nor the maximum concentration as given by the filter class must be exceeded. For example, the current OEL for butanone is 200 ppm [68]. If a worker wears a filter mask suitable for

Tab. 76: Full-face mask with particle filter classes [67]. The OEL factor is a multiplayer that is applied to the occupational exposure limit of the relevant particulate material and sets the concentration in air that must not be exceeded, although these filters are used.

Filter class	OEL factor	Remark
P1	4	Not for CMR or radioactive substances and not for infectious material.
P2	15	
P3	400	

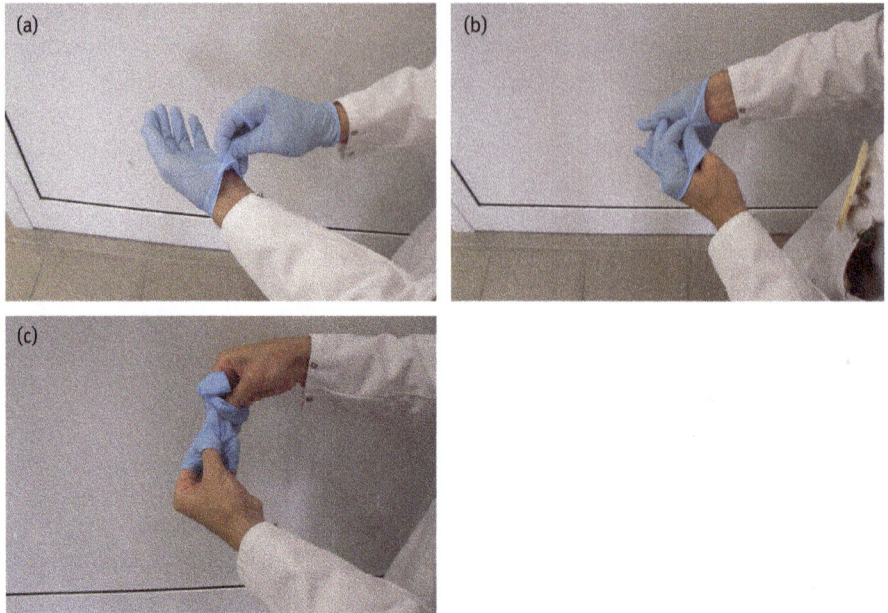

Fig. 135: How to put off contaminated gloves.

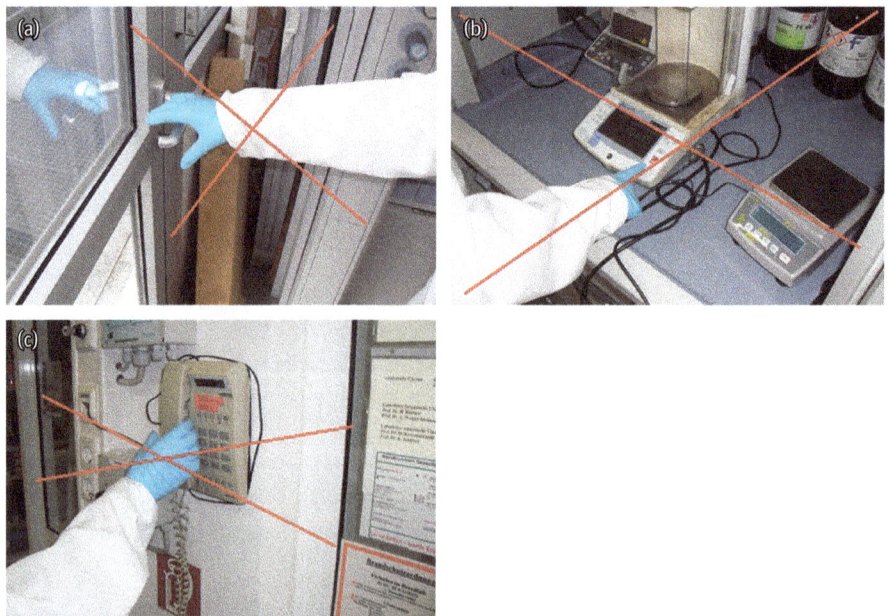

Fig. 136: Do not touch items with gloves that will be touched by other people with bare hands!

butanone, the concentration in air may reach 500 ppm for a filter class 2 and 6,000 ppm for a filter class 3. According to Tab. 77, the lower value of OEL × 30 (here: 200 ppm × 30 = 6,000 ppm) and maximum permitted concentration defines the tolerable level in air.

Self-contained respiratory protective equipment is independent on ambient air and available as compressed air breathing apparatus (EN 137) (Fig. 137) or as regenerative apparatus; in the latter, exhaled air is "recycled," and CO_2 is converted to O_2:

$$CO_2 + 2KO_2 = K_2CO_3 + 1.5O_2$$

Tab. 77: Full-face or partial mask with specified gas filters [67].

Filter class	OEL factor	Remarks
1	30	Maximum permitted concentration 1,000 ppm
2	30	Maximum permitted concentration 5,000 ppm
3	30	Maximum permitted concentration 10,000 ppm

Fig. 137: Full protective suit and pressurized gas respiratory protection (© Drägerwerk AG & Co. KGaA, Lübeck. Alle Rechte vorbehalten. All rights reserved).

As the self-contained respirators are independent on the composition of ambient air, they are the preferred respiratory protective equipment for emergency staff. The factor by which the OEL may be exceeded depends on details of the equipment, the mask and whether a full protective suit is worn or not.

9.7 Brief view on dealing with spillages, fire and first aid

Regulations dealing with workplace safety request employers to take care of first aid, fire safety and firefighting capabilities and prevention of product release into the environment, depending on product properties. "What if"-scenarios need to be exercised regularly to make sure that damage to life, environment and properties is minimized in case of incidents.

9.7.1 General measures in case of incidents

In case of incidents, the most important point is to secure the area, warn others and prevent healthy people to venture naively into the area and become injured. It may happen that you need to escape as quickly as possible. Rescue injured colleagues if you can do so without taking unacceptable personal risk.

Rescuing people must always be given preference over saving equipment.

In terms of rescuing other persons, what is an acceptable risk and what is not sometimes has to be decided on short notice. The risk depends on the properties of the products released, the amount and temperature of the product and additional circumstances like fire, ventilation or available escape routes, to name a few. Regular emergency exercises with varying exercise scenarios will certainly help to draw the right conclusions in case of real emergencies. The following initial steps can be taken under many different scenarios (Fig. 138):

 – raise alarm: warn others
 – quickly decide whether the area must be evacuated without delay
   – if yes: rescue injured persons, if possible
 – call internal/external help as appropriate
   – first-aid responders, paramedics, ambulance, fire brigade
 – prevent others from running naively into affected area
 – fight small fires, if this is possible without risk
 – provide first aid
   – take care so as to not become contaminated

The general process is illustrated in Fig. 138. The next chapters will address predominantly firefighting and first aid.

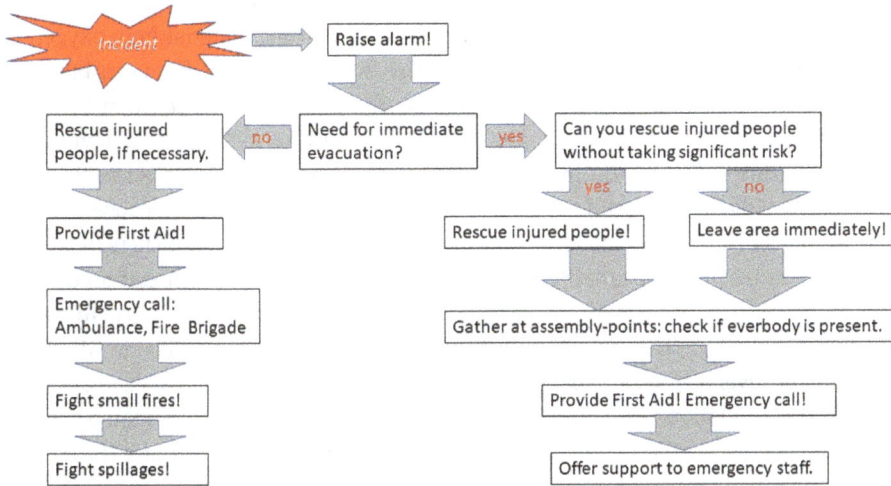

Fig. 138: Proposal for logical steps to be followed in case of an incident. First aid, emergency call, fighting fire and spillages can be done in parallel, if enough appropriately trained staff is available.

9.7.2 Firefighting

Firefighting needs to be educated and trained in special courses. At this place, only a few general remarks concerning firefighting shall be made.

Fire fighters need to protect themselves against thermal damage and against inhalation of toxic gases. Self-sustained breathing apparatus is compulsory for staff at the fire front, irrespective of the material that is burning. Although synthetic polymers were frequently discussed as a cause for release of explicitly toxic fire gases like hydrogen cyanide (HCN), the conditions of the fire have the greatest influence on the smoke gas toxicity, mainly temperature and oxygen supply [69]; natural materials are not necessarily better than synthetic materials [70]. About 50% of casualties caused by fires are attributable to fire gas intoxication [71]. Carbon monoxide (CO) is the most prominent toxic gas in fires, sometimes supported by HCN. Unconscious victims rescued from burning areas have a carbon monoxide intoxication, and hydrogen cyanide may be involved on top. Irritating gases may trigger inflammatory processes in the airways that may end up in edema of the glottis and/or the lung. Especially in case of limited air supply (smoldering fires and indoor fires), materials are more likely decomposed under release of toxic compounds like CO and HCN; in case of sufficient air supply, CO and HCN are oxidized and appear as trace compounds, only.

Fire-fighting material needs to be suitable for the burning items. Water is the most accessible extinguishing media but not always useful or even dangerous in case of fires of nonwater-miscible solvents, light metal fires, organometallics and other compounds or other materials that react with water heavily (see Fig. 14). Extinguisher foam is an option for many solvent fires as it separates the burning material from oxygen in air. Extinguishing powders as well can separate burning material from air, but they also trap radicals and by this way interrupt the radical chain reaction in fires. Brominated halogenated solvents, the halons, are very powerful fire extinguishing agents, but due to their stratospheric ozone depleting potency their use is strictly regulated. Carbon dioxide is useful for fighting fires of solvents and gases. Because it leaves the equipment tidy and intact, it is frequently held ready in fire extinguishers in laboratories and in electrical installations. However, where large amounts of carbon dioxide can be released by stationary firefighting equipment, alarm systems need to be installed ensuring a rapid evacuation of the area; the sprayed carbon dioxide will expel air from low-level regions and people are in danger to be suffocated in a "lake of CO_2."

Solid combustion residuals and ashes may be contaminated by many hazardous compounds, so the carcinogenic polycyclic aromatic hydrocarbons (PAH) and polychlorinated dibenzodioxins and polychlorinated dibenzofuranes.

Extinguishing water can carry many different dissolved toxins. It should be prevented from entering gutters and drains. Even in absence of specific hazardous toxins, the content of dissolved organic carbon with high biochemical oxygen demand can cause shortness of dissolved oxygen in receiving water bodies.

9.7.3 First aid: high-priority measures

High-priority measures in case of first aid for people being contaminated:
- First aid responders must first put on personal protective equipment (PPE) to avoid becoming contaminated
- Take off contaminated clothing
- Ensure circulation and breathing
- Stop dangerous bleeding
- Take measures to prevent shock
- Wash affected skin with water for local toxins
    - Soap and water or special decontaminant for systemic toxins
- Emergency call: poison center and ambulance, if deemed necessary
- If the intoxicated person vomits or loses urine or feces, save samples for analysis

The first aid responder must make sure he cannot become contaminated himself while providing support to the injured person. This can mean putting on protective gloves and providing artificial breathing via instruments only if the toxin is fatal by inhalation or fatal by skin contact. In such cases, poison centers can provide extremely important information via the phone; at poison centers, you can consult experts in medicine and toxicology who may provide information on easy and highly effective measures (Do you have the emergency numbers for ambulance, fire brigade and poison center written down next to your phone?)

9.7.4 After skin contact

In case the injured person (in the following named "patient") has contaminated skin and/or clothing, dropping off the contaminated clothing and storing it securely has a high priority; it may be needed later for analysis of the toxin. Showers in the lab may be used but beware of undercooling the patient!

Liquid acids or bases on the skin or in eyes shall be washed off with plenty of tap water. In case of corrosive solid particles, visible particles should first be tried to be gently removed "dry" with a piece of towel before intensive washing starts. Eyes must be kept wide open and washing water should have only short passage over the rest of the body. Take care so as the washing water does not enter the other eye. Never try to neutralize an acid (a base) by washing the skin with diluted base (acid)! Only for hydrogen fluoride (HF), a special solution may be used (see the following, antidotes).

In case of acid or base on skin or in eye: wash with tap water, only. Never try to neutralize!

Keep on washing until
- the patient asks for a break, or
- paramedics take over.

Concerning systemic toxicants on the skin, wash the skin with water, or with liquid soap and water to stop further dermal absorption. Hydrophobic compounds may be washed off with polyethylene glycol or polypropylene glycol of molecular weight >4,000 g/mol, if these materials were successfully tested against the relevant toxic substance; an example is polymeric methylene diphenyl di-isocyanate, which is not water soluble but cures in the presence of water, and where polypropylene glycol was demonstrated to be an effective skin decontaminant [72].

N-Chloro-*para*-toluene-sulfonamide (chloramine T) is used as a skin decontaminant in military and civil defense forces. The pure substance is corrosive to the skin, but 0.1–2.0% solutions in water may be used as disinfectants. It is capable in destroying chemical warfare agents like the skin toxin bis(2-chloroethyl) sulfide (S-Lost) and phosphonic acid-based nerve agents [73].

9.7.5 After inhalation

It may be mandatory to wear respiratory protection before being able to rescue injured people in case of intoxication with toxic gases, vapors or aerosols. If toxic substances were inhaled, first make sure the patient is transferred to a clean air area. Bring the patient who is not unconscious into a comfortable position. In case of shortness of breath, offer medical oxygen, if available. In case of inhalation of irritating gases or substances that can provoke lung edema, offer a corticosteroid aerosol if this was instructed by medical experts.

9.7.6 After ingestion

Ingestion of toxic substances should not occur at workplaces. However, such an incident cannot be excluded to 100%. In case of ingested acids or bases, wash mouth with small portions of drinking water; do not swallow the water but spit it out. After rinsing the mouth, a few small swallows of drinking water may be taken to wash the esophagus, but do not do so if consciousness is impaired or the patient suffers from nausea. Do not induce vomiting. Reflushing the esophagus with the swallowed acid or base may cause rupture of large blood vessels in the esophagus. Also, after having swallowed organic solvents, tensides or in case of impaired consciousness, do not induce vomiting. Liquids may enter the airways and lungs and would create a life-threatening lung inflammation and/or lung edema. In case of doubt, ask the poison center for advice.

Never induce vomiting in case of
- having swallowed acids,
- having swallowed bases,
- having swallowed solvents,
- having swallowed tensides,
- impaired consciousness.

As long as the toxin is in the digestive tract, it is not (yet) in the circulation. Besides vomiting, gastric irrigation may remove toxins form the stomach, but this exercise is to be left to medical experts. Once the toxin has passed the stomach, gastric irrigation is a useless torture. In case of full consciousness of the patient, and if advised by poison centers or medical experts, active medical charcoal dispersed in a glass of drinking water or high molecular weight polyethylene glycol may be swallowed to absorb the toxin in the gut. This needs to be followed by giving a laxative.

9.7.7 Purging the body from toxins

Toxins that have arrived the circulation may be purged by exhalation and/or by urinary excretion. Exhalation depends on the vapor pressure of the substance.

Urinary excretion depends on the water solubility of the toxin and the amount of urine released per time. Acidic toxins are easily excreted if the urine is alkaline and vice versa, and medical experts may influence the urine pH via selected infusions. They may also apply diuretics in combination with infusions so a huge amount of urine can be excreted in short time.

Too large molecules or too lipophilic molecules cannot sufficiently be excreted via the urine. Hemodialysis in specialized centers may be a useful way to clean the blood from toxins (this is also done on a regular scale with uremic patients whose kidneys are insufficient).

9.7.8 Antidotes

Specific antidotes are less commonly available for chemical substances as the general population may imagine. First aid and emergency medicine in case of intoxications are mostly confined to maintaining important body functions, avoiding ongoing toxin uptake by removing the toxin from the body surface and – if possible – purging out the toxin. Nevertheless, more or less specific substances do exist which may support emergency medicine. A selection is summarized below (Tab. 78). In general, the application of antidotes is reserved to medical experts. However, depending on the training and consultation with the company medical doctor, or if advised by poison centers, some of these substances may be applied by the first aid responder or paramedics of the company. Observe national and regional legislation!

9.7.9 Clean up and deactivation

After having taken care for injured persons and clarifying dangers from fire and product release, clean up procedures are the next logical step. Clean up can be done by appropriately trained employees, potentially supported by emergency experts like fire brigades. PPE needs to be fit for that job, and some substances urge to use respiratory protection. Clean up procedures need to be exercised so in case of an incident everybody knows what to do and how to do.

Liquid spillages can be handled in different ways. Flushing away with plenty of water is acceptable only if it is guaranteed that the material can not harm the environment, and this is hardly the case for most substances. Beware! Even a substance that was tested but is not classified as critical to aquatic organisms and that is readily biodegradable can cause mass death of fish: due to the ready biodegradability it

Tab. 78: List of antidotes and their potential field of application (see also Chapter 11).

Antidote/agent	To be applied for/against	Remarks
Sodium thiosulfate	Cyanides	Reserve for sulfur for the enzyme rhodanese which transforms cyanide ions into the less critical thio-isocyanate.
4-Dimethylamino-phenol (DMAP)	Cyanides	Oxidizes a part of the hemoglobin (Hb) to MetHb. Fe(III) in MetHb causes a redistribution of CN^- from cytochrome c to MetHb, so the enzyme cytochrome c which is critical for the inner cellular breathing is reactivated. DMAP is counterproductive in case of fire-gas suffocated victims as a critical part of the Hb may already be occupied by CO.
Hydroxocobalamin (vitamin B_{12})	Cyanides	Releases OH^- while binding CN^-; no need to generate MetHb, and therefore preferably used in case of fire-gas suffocated victims if an intoxication with HCN cannot be excluded.
Oxygen	CO intoxication; shortness of breath	As CO competes with O_2 for the binding to Hb, increasing the external availability of O_2 shifts the equilibrium.
Toluidine blue	Methemoglobinemia, caused by nitrite, aromatic amino or nitro compounds	The oxidized and reduced form of toluidine blue is in equilibrium with Hb and MetHb.
Atropine	Overactivity of neurotransmitter acetyl choline (ACh). Organophosphates, -phosphonates and carbamates that block the acetylcholine esterase (AChE).	Atropine is blocking the ACh receptor without activating it. Doing so, the overflow of ACh in the synaptic gap due to a block of the AChE cannot cause activation of the receiving neuron which would otherwise cause convulsions and cramps. In military and civil defense services available as self-injection set. Can cause intoxication with Atropine.

Tab. 78 (continued)

Antidote/agent	To be applied for/against	Remarks
Obidoxime	Reactivation of AChE in case of intoxication with organophosphonates and organophosphates	Needs to be applied quickly before the AChE-phosphate adduct ages.
Chelators like 2,3-dimercaptopropanesulfonate, Ethylene diamine tetra-acetate (EDTA)	Heavy metals	Chelator for mercury, cadmium and others; care must be taken that essential metals are not extracted from the body
Defoaming silicone (p. e. Sab simplex (R))	Ingestion of surface-active agents	Polysiloxane defoaming agent causes a collapse of foam micelles.
Calcium gluconate	HF and fluorides	Causes precipitation and inhibition of F^- as CaF_2. Prevent/control solution from becoming contaminated with mold.
Chloramine T	Substrates that can be deactivated by oxidation and/or chlorination	Can be held ready as skin decontaminant if appropriate, especially for poorly water-soluble, grease-like compounds.
Dexamethasone or other anti-inflammatory corticosteroid aerosols	Inhalation of irritative gases	Used to fight the development of glottis edema or lung edema, as both may be fatal
Polypropylene glycol Mn > 4,000	Taking up grease-like, poorly water-soluble substances	Decontamination of skin
Copper sulfate solution	White/yellow phosphorus	If advised by poison center: 1% solution for skin

may cause drop of oxygen concentrations in receiving waters. If the substance is not readily biodegradable, it should not just be flushed away.

Absorbents are useful to bind liquids so they can be shoveled into buckets and subjected to further treatment. A few questions need to be addressed in using absorbents:
- The absorbent should not react heavily with the spilled liquid.
  - Wood dust or peat must not be used for spilled concentrated nitric acid or perchloric acid or other strong oxidizers.
- The absorbent may increase the surface for substance evaporation. For flammable liquids, it can act like a wick, so flammability may increase.

– Absorbed material is not necessarily deactivated. Hazardous properties of the spilled liquid may have been transferred to the wetted absorbent.

Beware of slow, gas-generating reactions! If these cannot be excluded, buckets or drums with wetted absorbent should not be tightly sealed. The wetted absorbent is either subject for waste incineration plants which are certified for handling dangerous waste, or deactivation processes have to follow on-site.

Deactivation reactions need to be tested and checked upfront for the specific substance/mixture under consideration. At this place, only a few general remarks can be given (Tab. 79). A textbook dedicated to the destruction of hazardous chemicals on lab scale was published by Lunn and Sansone [74]. Deactivation reactions need to progress under ambient conditions; this is typically the case if activation energies are not too high, and if the reaction is exothermic. Naïve upscaling of laboratory reactions may become dangerous as increase in temperatures accelerates the reaction which accelerates the heat formation, and so on. Dilution with water can

Tab. 79: Examples for hazardous materials and potential deactivators (no warranty!).

Hazardous material	Deactivator	Remarks
Acids	Weak bases, like sodium hydrogen carbonate, sodium carbonate as aqueous solutions	Release of carbon dioxide
Bases	Medium acids, for example diluted acetic acid or formic acid. Weak acids like diluted sodium dihydrogen phosphate	In closed areas, acid vapors may be a nuisance or even irritating to the mucous membranes
Oxidizers	Solution of sodium sulfite; diluted ethanol (for chromium (VI), permanganate)	
Dissolved heavy metals	Bases of buffered bases, sodium carbonate, to precipitate hydroxides and carbonates	Depending on type of heavy metal, pH needs to be controlled to optimize precipitation.
Activated carbonyl compounds like isocyanates, carbonic acid chlorides or anhydrides (including phosgene)	2-Aminoethanol; diluted ammonia solution in water/ ethanol mixtures	Beware of flammability of the deactivator. In confined spaces, ammonia vapors may casue respiratory irritation.
Esters of phosphonic and phosphoric acid, thioethers and thiols (some chemical warfare agents fall into these groups)	Chloramine T, strong alkaline potassium permanganate solution	Hydrolysis of the esters and oxidation of break-down products. Reaction may become violent.

buffer temperature increase, but this measure increases the amount of material where proper disposal has to be sought for. Deactivation reactions on the field are hardly stoichiometric, and the final mixture is likely to have still some hazardous properties. A proper deactivation requires controlled reactions in some kind of reaction vessels, and decisions on whether the products are harmless or not depends on results of product analysis. A few examples of hazardous materials and potential deactivators are given in Table 79.

9.8 Exercises

(1) Which groups of workers deserve special attention and protection?
(2) Before work with hazardous substances can be taken up, what has to be ensured?
(3) Is it allowed to show the content of safety data sheets to a naïve worker? Your supplier has indicated it is confidential and your company has signed a secrecy agreement.
(4) Which measures have to be taken if OELs were exceeded?
(5) What special measures have to be taken if employees have to handle carcinogenic or mutagenic substances?
(6) What is biomonitoring and what are biological guidance values, what are biomarkers (human biomonitoring values)? Name examples.
(7) What is the benefit of biomonitoring against air monitoring? What is the benefit of air monitoring against biomonitoring?
(8) Concerning airborne exposure, should measurement methods be given the preference over computational methods?
(9) Concerning personal equipment at the workplace, what are the obligations of the employer?
(10) What is the ranking of measures to be taken to ensure that occupational exposure levels are met?
(11) What kind of precautions are to be taken by an employer to ensure workplace safety and to limit potential damage, irrespective of the presence of hazardous substances? Could you name at least five?
(12) Name two substances whose use is prohibited by the carcinogen directive.
(13) What limitations are to be borne in mind when working with filter masks?
(14) After a strong acid is splashed over the hand and arm, what should be done (and what should not be done)?
(15) When certain substances had been swallowed, induction of vomiting shall be avoided. Why?
(16) When working with protective gloves, what do you have to think about?

(17) At a workplace a paint is brushed on large surfaces, the paint emits a solvent with a vapor pressure of 800 Pa at 298 K and a molecular weight of 96 g/mol. The OEL is 9,200 mg/m³. Can the OEL be exceeded at 298 K?

(18) Is it allowed to let nonadult apprentices work with a corrosive substance?

(19) After a spillage with a material that has an OEL, the liquid was bound with an adsorbent. What are the next steps?

10 Storage of hazardous substances and mixtures

The storage of substances and mixtures is subject to Regulation (EU) No. 1272/2008. However, entries to be found there are very general:

- The label must be legible if the containment is stored upright.
- Package needs to be designed in such a way that there is no product release.
- Products for the general public containing acute toxic substances category 1–3, corrosive substances or substances with specific target organ toxicity (STOT) category 1 substances must be equipped with child-resistant fastenings. Tactile warnings and pictograms have to be applied for products falling into acute toxicity categories 1–4, eye damage or skin corrosion category 1, STOT category 1, substances which are carcinogenic, mutagenic or toxic to reproduction (CMR) category 2 or flammability category 1 or 2 or presenting an aspiration hazard.

If the product is a dangerous good according to regulations, package must fulfill the defined requirements, which are detailed in terms of quality and testing.

10.1 Control of major-accident hazards involving dangerous substances

This directive 96/82/EC and its successor 2012/18/EU is colloquially termed "Seveso" directive, as it was triggered by a major incident in a chemical production plant in the North Italian town Seveso which had severe impacts on the neighborhood. The aim of the directive is to control major chemical plants handling substances with a certain hazard to avoid risks for the neighborhood in case of incidents. Many details concerning the code of practice are delegated to the EU member states. Therefore, national implementations of this directive have to be observed. Here, only a very brief insight is provided.

Annex I of the directive lists hazardous properties which trigger lower tier requirements (that is, the substance and installation falls under this directive at all) and higher tier requirements, if the storage and production capacity of the site exceeds certain levels. For example, for substances falling into acute toxic category 1, the lower tier is 5 tons, the higher tier requirements are to be applied at 20 tons and above; for substances hazardous to the aquatic environment category 1, the tiers are 100 and 200 tons and for category 2 they are 200 and 500 tons. Part 2 of annex I lists specific substances with their tier 1 and 2. If tier 1 is arrived, the operator of the installation has to notify his establishment to the authorities.

Annex II defines minimum data to be included in the safety report according to article 10. This safety report is required if upper tier levels are met. Between others,

https://doi.org/10.1515/9783110618952-010

this includes a description of the site, geographical and meteorological conditions, activities on site, inventory of dangerous substances and their properties, risk analysis (what can happen) and prevention and measures of protection and intervention to limit consequences of an incident.

Annex III addresses the safety management system and organization of the installation to prevent major incidents, including operational control, audits and reviews.

Annex IV defines data and information to be included in emergency plans according to article 12, for internal and external emergency plans. These plans are required if upper tiers are met.

Annex V covers information of the public as required by article 14. Between others, it deals with the properties of the substance(s), the potential risks and counter-measures, and how the public will be warned.

Annex VI defines criteria of a major incident that has to be reported to the commission.

10.2 German TRGS 510 (technical rules for hazardous substances): storage of hazardous substances in nonstationary containers

As example for a detailed technical guidance for the storage of hazardous substances in nonstationary containers, the German TRGS 510 (technical rules for hazardous substances) shall be mentioned. This document describes "best available technology"; deviations are possible but need to be justified. However, this TRGS 510 is quite useful to describe the sense behind storage conditions. See also: https://www.baua.de/EN/Service/Legislative-texts-and-technical-rules/Rules/TRGS/TRGS-510.html.

10.2.1 Scope of application, risk assessment and basic principles

The TRGS shall be followed, if certain amounts of material are exceeded, depending on the labelling (Tab. 80).

As for workplaces where hazardous products are handled, a risk assessment has to be performed, covering the amount, properties and physical state of the stored goods, activities in the storage area, potential joint reactivity of stored goods, size, location and surrounding of the storage area. Activities need to cover transport into and out of store and handling of spillage. The potential of the formation of explosive atmospheres and countermeasures need to be addressed.

Tab. 80: Cut-off limits for TRGS 510 (examples).

Product class	Permitted storage limit outside warehouse	Additional protective measures requested at
Acute toxic categories 1–3; CMR category 1A, 1B; STOT category 1	50 kg	≥200 kg
Extremely (and highly) flammable liquids	10 kg (20 kg)	≥200 kg
Flammable liquids	100 kg	≥1,000 kg
Flammable solids, pyrophoric, self-heating and self-reactive products	–	≥200 kg
Oxidizers category 1	1 kg	≥5 kg
Pressurized gas in containers	2.5 L	≥2.5 L, etc.

Materials shall be stored in closed containment only. The area has to be kept clean. A register must name the identity of the hazardous substance or mixture, the classification, the amount stored and the area of storage where the substance is located.

For ammonium nitrate storage a separate TRGS exists (TRGS 511), and for organic peroxides reference is made to the German Hazardous Substances Act, annex III. TRGS 509 addresses immobile storage containments and their charge/discharge.

10.2.2 General protective measures

Aisles, corridors, passages, stairs, emergency exits, social rooms and sanitary rooms are no storage areas and must be kept free from hazardous products.

Ignition sources are to be avoided in the storage area. Overheating has to be prevented. Light sources must not affect the stored material (heating up!).

Containment must not be of such a form that it may be mixed up with containment for food or beverages. Outside the storage at workplaces, only that number and amount of hazardous products should be held ready that is needed for the daily task; at the workplace, the maximum size for flammable liquids is 2.5 L for glass bottles and 10 L for nonbreakable bottles.

10.2.3 Store organization

The employer needs to provide safety instruction to the employees on a regular basis. He has to care for working suits and their cleaning and maintenance. Sanitary and first-aid installations have to be provided, and personal protective equipment as needed. Emergency escape routes have to be identified. If substances

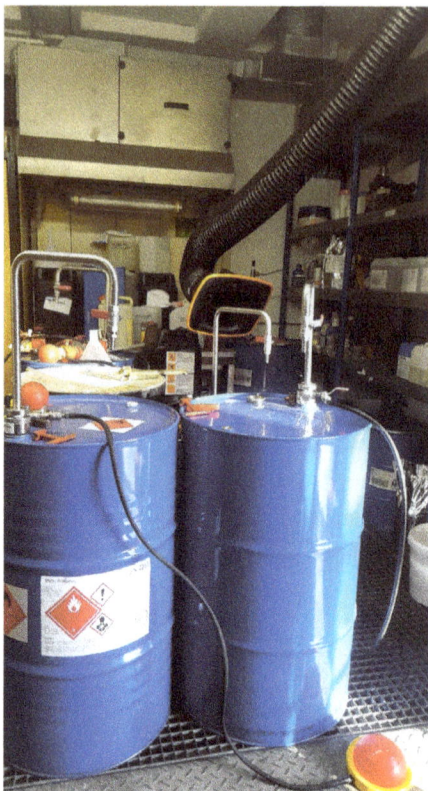

Fig. 139: Store for flammable liquids.

with H330 are stored, employees must be equipped with appropriate breathing apparatus ("escape masks").

If deemed necessary, an alarm system has to be installed. A regular control of the packaging and containment systems (intact and legible) and the ventilation has to be run and recorded. Smoking, drinking and eating is forbidden in the warehouse.

Installations must fulfil requirements for static stability. Containment shall be stored in such a way that contents are always easily identifiable, and leakages can be discovered immediately.

Products of acute toxicity categories 1–3 and CMR substances category 1A and 1B should be kept under seal so only authorized people have access.

Hazardous substances and mixtures must not be stored together with food, beverages, pharmaceuticals and cosmetics and comparable products. Small containments and bottles of hazardous chemicals shall be stored in safety cabinets with 90 min fire-resistant doors.

Fig. 140: Safety cabinets for chemicals.

10.2.4 Additional measures

If more than 200 kg of acute toxic substances category 1–3, STOT category 1, CMR category 1, oxidizers category 1 or extremely or highly flammable gases and liquids are stored, the following has to be assured:
- 30 min fire-resistant walls
- Resistant roofing
- Fire sprinklers and smoke exhaustive system
- Extended escape routes
- Sufficient access routes for fire brigade
- Impermeable containment
- No gutters/floor drains, if these may cause threats
- Hold-back containment at least of volume of the stored goods

For storage areas for oxidizers, the following has to be assured (category 1: from 5 kg onward; excerpts):
- Noncombustible binders must be hold ready in case of spillage
- No storage of combustion engines
- No storage of flammable material
- Impermeable floor and no gutters
- Building only one floor high
  - Unless absence of danger can be demonstrated
- Fire-resistant walls and roofing (at least 90 min) for separation from others storage areas
  - Unless there is at least 10 m distance between wall/roof to oxidizing material

Chemicals of acute toxicity categories 1–3 have to be stored locked up, so only authorized personnel have access. The storage area shall have at least 5 m distance to building entrances. Large areas of 800 m² and more need to be equipped with loudspeaker alarm systems to warn the surrounding in case of incident. At 10 tons or more or at least if 20 tons or more are exceeded, automatic fire alarm systems shall be in place.

Stores for pressurized or liquefied gas and aerosol cans need to fulfil the following requirements (excerpts):
- Protect against temperatures above 50 °C.
- Protect against direct sunlight.
- Stores of 60 m² size or more:
  - Not below social rooms
  - At least two exits

Requirements for the storage of pressurized gas containers are (excerpts)
- gas containers need to be fixed with chains and valves shall be covered by protection caps
- no gutters in floor and no entrance gaps to deeper floor
- in working area: maximum of 50 bottles
  - In safety cabinets
  - Air exchange at least two times per hour
- Gases with H330: install permanent working gas monitors connected to an alarm system.

Requirements for the storage of flammable chemicals (excerpt):
- Walls, ceiling and doors must be of noncombustible material.
- Containment tubs must be of noncombustible, resistant material.
- Drains, openings, ducts and entrances to lower level spaces must be protected against the ingression of vapors and fumes (note: these are typically heavier than air and creep along the floors to ignition sources, see Fig. 9).

- Avoid ignition sources (sparks in non-ex switches; openings to chimneys).
- Storage area/warehouse must not be located next to living rooms or bedrooms. They should also not be located next to social rooms if the storage capacity is 10 tons or more.
- Containment must be capable to hold back the volume of the largest container in the store.
- On the basement of residential buildings, a maximum of 10 kg extremely or highly flammable liquid, and a maximum of 20 kg over all flammable liquids may be stored, provided the container material is nonfragile.
- It is not allowed to store flammable liquids in dwellings.
- Maximum quantities permitted in retails sales rooms are dependent on the space (m²) and flammability categories. They range from 10 to 600 kg. In case of separated areas, safety cabinets, fire alarm systems and/or sprinklers, higher amounts may be acceptable.
- Containers with flammable material shall not be presented near the exits of the sales area.
- Aerosol dispensers and pressurized cartridges in sales area shall not exceed the daily turnover. The area occupied shall not exceed 20 m². They shall not be placed near the exits.
- Sales and storage areas must have fire extinguishers of at least 6 kg capacity for fire classes A, B and C.
- Filled aerosol dispensers and pressurized cartridges shall not be presented in the shop window.

Safety cabinets shall be used for storage only, not for any loading, unloading or re-loading activities. Flammable material shall not be stored together with self-reactive or pyrophoric substances. Containment stored in the cabinet shall be free of external contamination. Ventilation shall ensure that no explosive or toxic atmospheres can buildup.

10.2.5 Joint storage and separation

Substances and mixtures that can react with each other heavily should not be stored together. It is easy to grasp that strong oxidizers in contact with flammable material may produce an explosion. Explosive materials or highly flammable materials would help to spread toxic substances in case of an incident. Substances requiring different fire extinguishing media should not be stored together. In the TRGS 510, and in the German BG RCI M062 leaflet, a table provides a quick overview what may be stored together and what not (Fig. 141).

Separate storage means either storage in different warehouses or within the same warehouse in separated areas. Areas are separated by distance or by noncombustible

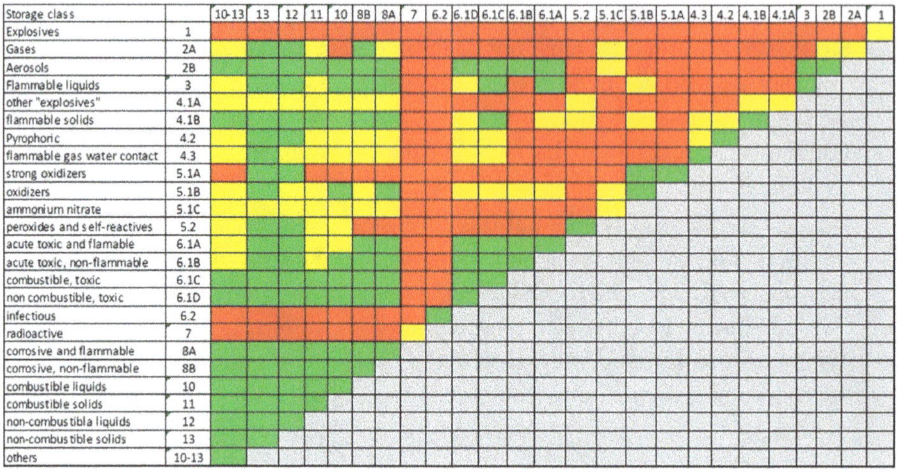

Storage class		10-13	13	12	11	10	8B	8A	7	6.2	6.1D	6.1C	6.1B	6.1A	5.2	5.1C	5.1B	5.1A	4.3	4.2	4.1B	4.1A	3	2B	2A	1
Explosives	1																									
Gases	2A																									
Aerosols	2B																									
Flammable liquids	3																									
other "explosives"	4.1A																									
flammable solids	4.1B																									
Pyrophoric	4.2																									
flammable gas water contact	4.3																									
strong oxidizers	5.1A																									
oxidizers	5.1B																									
ammonium nitrate	5.1C																									
peroxides and self-reactives	5.2																									
acute toxic and flamable	6.1A																									
acute toxic, non-flammable	6.1B																									
combustible, toxic	6.1C																									
non combustible, toxic	6.1D																									
infectious	6.2																									
radioactive	7																									
corrosive and flammable	8A																									
corrosive, non-flammable	8B																									
combustible liquids	10																									
combustible solids	11																									
non-combustibla liquids	12																									
non-combustible solids	13																									
others	10-13																									

Fig. 141: Combined storage: possible (green), possible under restrictions (yellow) and impossible (red). No warranty!

material or storing in separated containment areas (p.e., separate safety cabinets). The fire barrier between two sections withstands at least 90 min.

Deviation of the joint/separated storage rules is acceptable if the total amount of hazardous chemicals does not exceed 400 kg, from which a maximum of 200 kg belongs to one storage class, and risk assessment concludes on low risk.

10.3 Exercises

(1) Name general rules that are valid for the storage of chemicals outside the store.
(2) Name general rules for the storage of materials inside stores.
(3) Which substances should be stored locked up with restricted access?
(4) If the "skull and crossbones" substances (acute toxicity categories 1–3) and many "category 1" substances exceed amounts of 200 kg in the store, which additional requirements should be matched by the storage area?
(5) If you have 50 kg or more of an oxidizer, what precautions have to be taken for the storage?
(6) What safety precautions have to be taken for pressurized gas bottles, in and outside stores?
(7) What requirements must a storage area fulfil where 5,000 kg of highly flammable liquids are stored?

(8) With the help of the annex and Fig. 141, decide which of the following chemicals can be stored together? (a) 1,2-Dichlorobenzene (H302), (b) 1-dimethylaminopropane-2-ol (H226;H302;H314), (c) butanone (H225), (d): di-methylformamide (H360D, H312,; (e) *N,N*-dimethylaniline (H351, H301, H311, H331) and (f) hydrogen peroxide 50–75%.

11 Properties of selected hazardous substances

11.1 Introduction

A listing and description of hazardous substances in a textbook of limited size is always somewhat arbitrary. It is impossible to list all the knowledge that is available. A selection has to be made. This chapter is a compilation of information the author has in mind due to reading about and working with hazardous substances for meanwhile more than 30 years, and by using several sources, to name some:

- Greim H, Snyder R: Toxicology and Risk Assessment. A comprehensive introduction. John Wiley & Sons, Ltd. 2008. ISBN: 978-0-470-86893-5.
- Marquardt H, Schaefer SG, McClellan R, Welsch F: Toxicology. Academic Press San Diego, London, Boston, New York, Sydney, Tokyo, Toronto 1999. ISBN: 0-12-473270-4.
- Hazardous Substance Data Bank: https://toxnet.nlm.nih.gov/newtoxnet/hsdb.htm;
- GESTIS Stoffdatenbank: http://gestis.itrust.de/nxt/gateway.dll/gestis_de/000000.xml?f=templates$fn=default.htm$vid=gestisdeu:sdbdeu$3.0;
- The MAK Collection for Occupational Health and Safety: http://onlinelibrary.wiley.com/book/10.1002/3527600418/topics;
- The ECHA website [51]
- Reichl F-X: Taschenatlas der Toxikologie. Thieme Stuttgart 1997. ISBN: 3-13-108971-8.

Substances included in this chapter are either deemed as relevant in the workplace in general, have a special history in terms of incidences or intoxications, may frequently be found in the nonindustrial area (consumer, professional user), may be relevant due to misuse and/or provide a detailed insight in toxicological mechanisms.

Information provided in this document is based on good faith. No warranty can be accepted for correct reproduction of data, especially in terms of first aid and emergencies.

11.2 Heavy metals

Heavy metals are classics in acute and chronic intoxication. This is owed to the fact that some of these substances are in use since centuries, and toxicological potency combined with limited or even nonexisting workplace safety and hygiene caused numerous human health cases. The most important heavy metals are listed in the following sections.

https://doi.org/10.1515/9783110618952-011

11.2.1 Mercury

The use of mercury longs back into ancient times. Mercury sulfide was used as a pigment. The liquid elemental mercury was used to extract gold or silver from poor ores, and the noble metals were released by evaporating the mercury in open pans over wood fires. This process was also applied for plating of metal articles with thin layers of gold. Mercury salts such as $HgCl_2$ were used as disinfectants and preservatives in textiles, especially hats, and the permanent exposure caused neurological disorders (*mad as a hatter*). The mercury compound thiomersal (sodium-2-(ethylmercurithio)benzoate) is a very powerful biocide against bacteria and mold fungi, and still in use in some pharmaceuticals. As a liquid metal of high density, elemental mercury can be used in measuring devices like thermometers and barometers and in electrical switches.

Entry no. 18 and 18a of Annex XVII of Regulation (EC) No. 1907/2006 (REACh) describes restrictions for the use of mercury and mercury compounds. Mercury compounds must not be used as antifouling agents for the hulls of boats, for the preservation of wood or textiles. Measuring devices containing mercury shall no longer be placed on the market.

At ambient conditions, this liquid metal has a low but not negligible vapor pressure. Uptake by inhalation is the main route for intoxication with elemental mercury, as the retention in the lungs is high. The metal may pass cell membranes. After oral ingestion, the resorption of elemental mercury and Hg(I) is low, and critical acute intoxications are unlikely; Hg(II) can be resorbed around 10% in the digestive tracts, and organic mercury compounds show an absorption rate of around 90%.

In the body, Hg(0) is oxidized to Hg(II) by the enzyme catalase. As Hg(II) can cross cell membranes much less easier than Hg(0), this process results in tissue accumulation during chronic exposure, namely in the brain but also in liver, kidney and testes. Hg(II) is excreted via urine and feces; the excretion of the lipophilic methylmercury depends on its transformation rate to Hg(II).

Mercury forms complexes with sulfur and nitrogen in protein side chains; this can block the activity of enzymes, which is the main interpretation of its mode of toxic action. After intoxication, medical doctors may decide to apply complexing agents to phase out the metal (see Tab. 78).

GHS-symbols (Fig. 142) and hazard phrases for $HgCl_2$:

Fig. 142: GHS symbols for mercury chloride.

- H341: Suspected of causing genetic defects.
- H361fd: Suspected of damaging fertility. Suspected of damaging the unborn child.
- H300: Fatal if swallowed.
- H310: Fatal in contact with skin.
- H372: Causes damage to organs [kidneys, central nervous system (CNS)] through prolonged or repeated exposure.
- H314: Causes severe skin burns and eye damage.

11.2.1.1 Acute toxicity

Inhalation of Hg(0) can result in breathing difficulties, headache, fever, bronchitis and pneumonia. Oral uptake of Hg(II) compounds causes erosion of mucous membranes, nausea and subsequently vomiting, stained with blood. Diarrhea may develop with loss of proteins and minerals, resulting in collapse with first polyuria, then oliguria and finally anuria (kidney failure). Organic mercury compounds trigger predominantly the effects of the CNS: nausea and paresthesia, followed by limb numbness uncoordinated movements (ataxia), gait disturbance, problems with speaking (finding/forming the words) and hearing loss. After methylmercury intoxication, the latency period may take days to years! (Minamata catastrophe!).

11.2.1.2 Chronic toxicity

After chronic inhalation of mercury vapors, kidney failure, inflammations in the mouth and an over excitability (erythrism), sleeplessness and tremor of the hands may occur. Loss of teeth was mentioned in cases of repeated, heavy overexposure. Methylmercury retards the mental and physical development of children. Methylmercury causes kidney tumors in rodents.

Mercury is regarded as a "quasi" persistent substance. Elemental mercury is emitted into the air and is oxidized by radicals (OH*). This reaction leads to Hg_2^{2+} and Hg^{2+} compounds which precipitate with rain. Presence of sulfides, namely, in anoxic sediments results in the formation of HgS. This can be reoxidized microbially to soluble $HgSO_4$ in aerobic sediments, but the more critical reaction is the methylation to methylmercury compounds. These lipophilic compounds can be enriched in biota. In the 1950s, in Minamata, Japan, mercury salts emitted from industrial sites were deposited in aquatic sediments. In this anaerobic environment, mercury was transformed to methyl mercury ions and elemental mercury by microorganisms. As methylmercury is more lipophilic than mercury (II) ions, it was accumulated in fish and subsequently was taken up by the population consuming contaminated fish. In the body, mercury is metabolized to mercury (II) ions and excreted; in the environment, it may partly become reduced to elemental mercury, and the circulation starts again.

11.2.2 Cadmium

Cadmium (Cd) has many similarities with zinc and is frequently a companion in zinc and lead ores. Technical uses are/were CdS as yellow pigment, Ni–Cd rechargeable batteries or Cd plating of steels as anticorrosion protection. Cd carboxylates like cadmium (II) stearate were used as UV stabilizers in polyvinyl chloride (PVC) and as catalysts in polymer production.

Entry no. 21 of Annex XVII of Regulation (EC) No. 1907/2006 (REACh) restricts the use of Cd and its compounds in polymers and paints, brazing fillers and jewelry (content has to be below 0.01%). By the way of derogation, certain articles made from recovered PVC may contain up to 0.1% Cd. Cd must not be used for metal plating if these metal parts come into contact with consumer products with intended or predictable skin or food contact (*for more details, look up the regulation*).

Cd can be taken up as CdO smoke (welding; cigarettes), or via the food chain, as cereals can take up Cd from soil. The intestinal resorption is below 10%, and Cd probably makes use of free Ca- and Fe-transporters. The organs with highest levels are the kidneys, and with a biological half-life of up to 30 years, a steady state is achieved at around 50 years of age. Cd has a high affinity to sulfur in proteins, and zinc-metallo-thioneins are induced by Cd exposure and act as sink for Cd. However, when the metallothionein capacity of the kidney is exhausted, kidney failure can follow after ongoing exposure. After intoxication, medical doctors may decide to apply complexing agents to phase out the metal (see Tab. 78).

GHS symbols (Fig. 143) and hazard statements for $CdCl_2$:

Fig. 143: GHS symbols for cadmium chloride.

- H350: May cause cancer.
- H340: May cause genetic defects.
- H360FD: May damage fertility. May damage the unborn child.
- H330: Fatal if inhaled.
- H301: Toxic if swallowed.
- H372: Causes damage to kidneys through prolonged or repeated exposure.
- H410: Very toxic to aquatic life with long-lasting effects.

11.2.2.1 Acute toxicity

Acute toxicity may be seen at workplaces with heavy overexposure against airborne Cd due to welding processes: cough, headache, dizziness, fever and/or pain behind the sternum, followed by pneumonia and/or lung edema with a delay of up to 24 h. Oral uptake may cause nausea, vomiting and abdominal pain.

11.2.2.2 Chronic toxicity

Intoxication via food chain caused the Itai-Itai-disease in Japan in the 1950s. Waste water containing dissolved Cd was used for watering of rice fields. The metal was taken up by the plants, and contaminated rice grains resulted in a mass intoxication. Cd may block the vitamin D3 synthesis, and together with kidney dysfunction leading to reduced Ca- and phosphate-retention, osteomalacia results in bone malformation. Repeated inhalation affects the olfactory epithelium, and hypoosmia and anosmia may result. Permanent competition with the Fe-transporter in the intestines may end in anemia. Cd-induced kidney damage is progressive even after cessation of exposure. Chronic inhalation induces lung tumors in animal experiments; chronic oral uptake was not associated with cancer. Cd in drinking water was teratogenic in mice.

11.2.3 Lead

Lead is used by man since ancient times: lead shields for roofing, lead pipes for water, lead weights, pigments (carbonate, sulfate), etc. Lead acetate is sweet and was probably used in ancient Rome to sweet-up wine; pottery and metal bowls containing lead gave reason for intoxication if acidic food or beverages was stored over longer periods of time. Today uses span rechargeable batteries, shielding material against X-rays, solder alloys and ammunition.

Entry no. 16 and 17 of Annex XVII of Regulation (EC) No. 1907/2006 (REACh) names lead carbonates and lead sulfates, respectively. The use as pigment is no longer permitted. However, for the restauration of historic artwork, national agencies can grant derogations.

Lead is taken up as inhaled particles with high retention; orally taken up lead compounds are resorbed between 10 and 50%; higher values are possible in case of Ca- and/or Fe-deficiency of the body. Lead in blood is bound to the erythrocytes, and it distributes to liver and kidney; bones are a deep storage of lead, and it shows a biphasic half-life in the body: 20 days and 20 years, the latter resulting from Pb remobilized from bones. Healing processes after braking bones or pregnancy increase the Pb remobilization. After intoxication, medical doctors may decide to apply complexing agents to phase out the metal (see Tab. 78).

GHS symbols (Fig. 144) and hazard statements for PbCl$_2$:

Fig. 144: GHS symbols for lead chloride.

- H360Df: May damage the unborn child. Suspected of damaging fertility.
- H302 + H332: Harmful if swallowed or inhaled.
- H373: May cause damage to kidneys through prolonged or repeated exposure.
- H410: Very toxic to aquatic life with long-lasting effects.

11.2.3.1 Acute toxicity

Acute intoxication may block the heme synthesis in the liver. Heme precursors pile up in the hepatocytes and are subsequently released into the blood where they cause paralysis to nerve cells, which leads to painful intestinal colic (*lead-colic*), obstipation or diarrhea, nausea and vomiting. Spasms of the blood capillaries cause a pale skin. Acute lead encephalopathy shows symptoms like disorientation, insomnia, vomiting, apathy and stupor or hyperactivity and aggressiveness. Disturbances of motor activity and sensation, seizures and Parkinsonism may persist for some time.

11.2.3.2 Chronic toxicity

A typical symptom is anemia, abdominal pain, obstipation. Encephalopathy causes impaired memory, lack of concentration, ataxia with psychomotor impairment, mental disorders (hallucination, excitation, aggressiveness). Lead causes retarded/impaired mental and physical development of children. Kidney effects may lead to proteinuria and glycosuria.

11.2.4 Arsenic

Arsenic is/was used in pigments, pesticides, electronic industry and in metallurgy. Arsenic (III) oxide (As$_2$O$_3$) was used as an assault toxin; it could be mixed into food, and symptoms were similar to infectious diseases. With the upcoming of the "Marsh-test," cases of murder became convictable.

Entry no. 19 of Annex XVII of Regulation (EC) No. 1907/2006 (REACh) lists arsenic and its compounds. Arsenic compounds must not be used as antifouling agents in aquatic applications. They shall not be used in the preservation of wood, and wood preserved by arsenic compounds shall not be placed on the market. Copper-chromium-arsenate (CCA) is still permitted as wood preservative for certain, listed and controlled industrial applications, like structural timber in buildings, bridge-work, avalanche control, to name a few.

As_2O_3 particles in air and AsH_3 are taken up via inhalation. Insoluble arsenic compounds are hardly intestinally absorbed and, therefore, show low toxicity. Of arsenic oxides, less than 10% are absorbed in the intestines. Resorbed arsenic is first distributed to liver and kidneys, but then it is spread throughout the body. It is excreted via urine, hair and nails.

Mimicking phosphorous and binding to sulfur in enzymes like lipoic acid, arsenic disturbs glycolysis and gluconeogenesis, which result in impairment of the production of energy carriers in cells. Organo-chloro-arsenates were used as warfare agents in World War I; skin contact caused long-lasting blisters and swears. After intoxication, medical doctors may decide to apply complexing agents to phase out the metal (see Tab. 78).

GHS symbols (Fig. 145) and hazard statements for As_2O_3:

Fig. 145: GHS symbols for arsenic (III) oxide.

- H350: May cause cancer.
- H300: Fatal if swallowed.
- H314: Causes severe skin burns and eye damage.
- H410: Very toxic to aquatic life with long-lasting effects.

11.2.4.1 Acute toxicity

After oral ingestion of As_2O_3, persons react with abdominal pain, nausea, vomiting and heavy diarrhea, which results in finally fatal electrolyte and water loss. Numbness in limbs and cramps are a result of peripheral neuronal toxicity. Arrhythmia, hallucinations, headache and seizures may result in coma.

11.2.4.2 Chronic toxicity
Repeated respiratory overexposure may result in perforated nasal septum, damage of mucous membranes, rhinitis, pharyngitis, laryngitis and bronchitis with respiratory impairment due to lung emphysema. Disturbance of peripheral nerves results in paresthesia and paralysis. Hyperkeratosis and melanosis are typical skin effects. These neuronal and skin effects are observed after oral exposure as well. Further, damage to peripheral capillaries (black foot disease) and damage to the visual nerve may occur. Arsenic causes cancer of the skin and the respiratory tract.

11.2.5 Chromium-VI compounds

Chromium (VI) can be used as an oxidizer in chemical reactions and for the tanning of leather and wood. Elemental chromium is used as plating material to passivate steel equipment, and it is a component in stainless steels. Chromium (Cr) is an essential element, present in some enzymes. As for all strong oxidizers (this also includes nitrates, hypochlorites, chlorates), contact with flammable material like paper, wool or wood may cause fire, contact with highly flammable material may cause explosions.

Entry no. 47 of Annex XVII of Regulation (EC) No. 1907/2006 (REACh) names Cr (VI) compounds. Cement and cement-containing mixtures shall not be placed on the market if they contain 2 mg/kg or more Cr(VI) by dry weight. If reducing agents were added to the cement to keep the Cr(VI) content below 2 mg/kg, a shelf life for the cement has to be defined and communicated. The Cr(VI) content of leather articles with intended or predictable skin contact must not arrive at or exceed 3 mg/kg.

GHS symbols (Fig. 146) and hazard statements for $K_2Cr_2O_7$:

Fig. 146: GHS symbol for potassium dichromate.

- H272: May intensify fire; oxidizer.
- H350: May cause cancer.

- H340: May cause genetic defects.
- H360FD: May damage fertility. May damage the unborn child.
- H330: Fatal if inhaled.
- H301: Toxic if swallowed.
- H312: Harmful in contact with skin.
- H314: Causes severe skin burns and eye damage.
- H334: May cause allergy or asthma symptoms or breathing difficulties if inhaled.
- H317: May cause an allergic skin reaction.
- H410: Very toxic to aquatic life with long-lasting effects.

Chromium (VI) is a strong skin sensitizer and has caused chromate dermatitis in bricklayers due to its presence in cement. It is also a respiratory sensitizer, and inhalation may cause nasal septum perforation, asthma and bronchitis. Further, necrosis of kidney tubular cells may be induced by oral uptake.

Chromium (VI) is mutagenic and causes lung cancer after inhalation. Oral uptake causes stomach ulceration and gastritis.

11.2.6 Thallium and its compounds

The technical use of thallium (Tl) compounds is limited; some electronic equipment intended for low temperatures can be based on Tl alloys. $Tl_2(SO_4)_3$ is used as rodenticide in some countries; to warn users, grains spiked with this compound may be equipped with deep blue of pink dyes. Tl compounds were used as depilation agents, but this usage is disbanded due to the high toxicity of these substances.

Tl^+ is mimicking K^+ and disturbs nerve impulse conduction. Acute intoxication causes loss of hair, keratitis, and after oral uptake, stomach pain, gingivitis, salivation, convulsions, coma are the main symptoms. The visual nerve may be affected, and impaired vision, headache and sensations may come up within a few days. Dementia and depression may persist for a longer time period. Chronic intoxication may result in weakness of the limbs, muscle atrophy and impaired vision. After intoxication, medical doctors may decide to apply complexing agents to phase out the metal (see Tab. 78).

GHS symbols (Fig. 147) and hazard statements for $Tl_2(SO_4)_3$:

Fig. 147: GHS symbols for thallium (III) sulfate.

- H300: Fatal if swallowed.
- H315: Causes skin irritation.
- H372: Causes damage to nervous system through prolonged or repeated exposure.
- H411: Toxic to aquatic life with long-lasting effects.

11.2.7 Nickel compounds

Nickel is a metal used in the production of special steel, metal alloys (money), metal plating, in Ni-Cd, Ni-Fe and NiH batteries and as catalyst.

Entry no. 27 of Annex XVII of Regulation (EC) No. 1907/2006 (REACh) defines a migration limit below 0.2 µg/cm²/week for piercing jewelry. For articles, for intended contact to intact skin, the migration limit is 0.5 µg/cm²/week.

Nickel is a known contact allergen. For that reason, consumer products have to meet a specific migration limit in appropriate migration tests. Acute oral exposure causes stomach pain, nausea, vomiting and diarrhea. Chronic exposure causes cancer in human beings by inhalation. Welding and grinding processes can cause inhalation exposure. Perforation of the nasal septum, ulceration, hypo-osmia and anosmiamay be caused by repeated inhalation of Ni-compounds. After intoxication, medical doctors may decide to apply complexing agents to phase out the metal (see Tab. 78).

GHS symbols (Fig. 148) and hazard statements for $NiCl_2$:

Fig. 148: GHS symbols for nickel (II) chloride.

- H350i: May cause cancer by inhalation.
- H360D: May damage the unborn child.
- H341: Suspected of causing genetic defects.
- H301 + H331: Toxic if swallowed or inhaled.
- H372: Causes damage to the respiratory tract through prolonged or repeated exposure.
- H315: Causes skin irritation.
- H317: May cause an allergic skin reaction.
- H334: May cause allergy or asthma symptoms or breathing difficulties if inhaled.
- H410: Very toxic to aquatic life with long-lasting effects.

11.3 Inorganic, nonmetal compounds

Besides heavy metals, there are other inorganic compounds – in differentiation to the later listed organic compounds – that have relevance in terms of unpredictable and predictable exposure. A few of these are listed in the following sections.

11.3.1 Carbon dioxide

Carbon dioxide (CO_2) is the physiological product of oxygen breathing, end product in biodegradation of organic substances and is produced in the anaerobic fermentation of starch and sugars. Technical uses are cooling substrate (solid), fire extinguishers, gas for sparkling water, beer and lemonade, to name a few.

CO_2 has only limited toxicity; but heavier as air, it may form invisible lakes in cellars and similar places in case of poor ventilation, like fermenters. People entering these lakes become unconsciousness within seconds. This is also the case if people know about the CO_2, plan to hold their breath and dive into the lake to save others. The suffocation of those first responders was proven with cruel high repeatability and high reliability. Rescue teams need to wear self-sustained breathing apparatus. Why holding the breath is impossible can currently only speculated: because hypercritical CO_2 behaves like a nonpolar, organic solvent, it is conceivable that the gas can easily be taken up via the skin. The CO_2 in blood is an irresistible breathing signal in the stem brain, and people cannot resist breathing when the level of CO_2 increases and on the same line the pH value decreases. This irresistible breathing stimulus is fatal if the person still is diving in the CO_2 lake without fresh air supply.

Above 0.1% in air, people start to react with headache, 4–6% induce tinnitus, agitation, tachycardia and dizziness, 10% causes unconsciousness within 20–30 min, 20–30% cause unconsciousness within minutes and retina damage.

Rescue teams need to be equipped with self-sustained breathing apparatus. Suffocated persons outside the endangered area need to be offered oxygen, probably with artificial breathing and reanimation.

11.3.2 Carbon monoxide

Carbon monoxide (CO) is formed during incomplete combustion, but it has also important technical applications (reducing agent in metallurgical processes, hydroformylation reactions, etc.). CO binds 250 times "tighter" to the Fe(II) in the hemoglobin of the red blood cells than oxygen, and this is the mode of toxic action (in an atmosphere with 20% oxygen and $20/250 = 0.08\%$ carbon monoxide, half of the hemoglobin is occupied by carbon monoxide). As carboxyhemoglobin has a nice red color, the skin of victims of intoxication looks healthy. Intoxication may occur at industrial sites were CO is produced or used, but in many cases every year CO intoxications are caused in the private sector: badly supervised heating in houses; bird nests in chimneys; smoldering indoor fire; falling asleep while eggs are going to be charcoaled in the pan; indoor barbecue or running a combustion motor with doors and windows closed. There is no smell, color or sensation of mucous membranes that could warn endangered people.! Smoldering indoor fires have the additional threat that CO–air mixtures are explosive (explosion limits of CO in air: 11–67%). Opening the door to a burning room allows ingression of fresh air, and the following "flash over" is feared by fire fighters. First reactions of intoxication are headache, soon followed by arrhythmia and finally coma. When rescuing people, do not breathe in the affected area (rescue team should wear self-sustained respiratory equipment), avoid sparks and any ignition sources and take care for good ventilation. For unconscious victims, provide oxygen and call the emergency ambulance as soon as possible.

CO exposure during pregnancy causes reduced oxygen supply of the fetus which may lead to deficits in brain development or death. In people surviving intoxication, but also after repeated overexposure, memory deficits may persist.

GHS symbols (Fig. 149) and hazard statements for CO:

Fig. 149: GHS symbols for carbon monoxide in gas cylinders.

- H220: Extremely flammable gas.
- H280: Contains gas under pressure; may explode if heated.
- H331: Toxic if inhaled.
- H360D: May damage the unborn child.
- H372: Causes damage to the nervous system after prolonged or repeated exposure.

11.3.3 Carbon disulfide

Carbon disulfide (CS_2) can be used as a solvent for extractions. Modification of cellulose fibers in the viscose production is done with the aid of CS_2. It is an extremely flammable liquid, and its vapor has broad explosion limits in air. Intoxication may happen via inhalation and via skin contact. After inhalation, symptoms are cough and burning sensations, head pressure, dizziness and tachycardia. The alcohol dehydrogenase is inhibited for several hours. After swallowing, burning sensations in the mouth appear together with the other symptoms. As CS_2 is a neurotoxin causing a so-called "dying back" axonopathy, chronic intoxication may result in paresthesia and paralysis in hands and feet, Parkinson syndromes, hearing loss (damage to hair cells in the cochlea), loss of libido and sexual potency, and damage to the heart and brain vessels.

GHS symbols (Fig. 150) and hazard statements for CS_2:

Fig. 150: GHS symbols for carbon disulfide.

- H225: Highly flammable liquid and vapor.
- H302: Harmful if swallowed.
- H315: Causes skin irritation.
- H319: Causes severe eye irritation.
- H361: Suspected of damaging fertility or the unborn child.
- H372: Causes damage to the nervous system through prolonged or repeated exposure.
- H412: Harmful to aquatic life with long-lasting effects.

11.3.4 Hydrogen sulfide

Hydrogen sulfide (H_2S) is a component formed in anaerobic decomposition of organic material, appears in natural gas and has some industrial uses for the production/precipitation of sulfides. It is produced in desulfurizing processes of crude oil. At natural gas wells, permanent burning torches are meant to incinerate the natural gas in case of an accidental blow out; it is not only the explosive atmosphere of the gas-air-mix, but the toxicity of H_2S that would pose a critical danger to the surrounding of the well. H_2S has a broad explosion range and may react violently in the presence of oxidizers. The smell of rotten eggs is characteristic, and the odor threshold is below the threshold of toxicity, but the olfactory sense adapts quickly. Exposure of 10 ppm causes irritation of the eyes, and higher levels result in respiratory irritation, headache, dizziness, hypotension, tachycardia and also lung edema, which develops after a latency period of several hours. The sulfide ion blocks cytochrome oxidase in the mitochondria of the cells, and "internal asphyxiation" can occur. Chronic exposure may induce cornea opacity, heart impairment and lung damage (increased respiratory resistance). Memory and concentration deficits may persist.

GHS symbols (Fig. 151) and hazard statements for H_2S:

Fig. 151: GHS labels for hydrogen sulfide gas cylinders.

- H220: Extremely flammable gas.
- H280: Contains gas under pressure; may explode if heated.
- H335: May cause respiratory irritation.
- H400: Very toxic to aquatic life.

11.3.5 Chlorine and chlorine-releasing compounds

Chlorine is a pale, yellow gas with typical odor. It is used in large amounts as a base chemical for chlorination of organic compounds (solvents, polymers, etc.) and as disinfectant for drinking water and pools: it can destroy germs of different origin. Chlorine is heavier than air, and it was used as a warfare agent in the World War I. The odor has only limited warning function as its threshold is in the area of acute toxicity. It irritates eyes and the upper respiratory tract, although not very strongly. Irritation of the lung may cause a fatal lung edema after several hours of a symptom-free latency period. Higher concentrations may cause a critical laryngospasm and bronchospasm.

First responders to accidents need to make sure they are not running into danger. Rescue teams need to wear self-sustained breathing apparatus. Injured people should be placed in a position that makes breathing as comfortable as possible. If inhaled, provide oxygen and call the emergency ambulance and the poison center. In any case, people having inhaled chlorine should consult a lung specialist immediately. Medical doctors may let the victim inhale glucocorticoid aerosols to slow down inflammatory processes in the lung (see Tab. 78).

Some substances can release chlorine over extended periods of time at small amounts and may be used as disinfectants. Examples are salts of hypochloric acid and dichloro-isocyanuric acid (1,3-dichloro-1,3,5-triazine-2,4,6-dione). Sodium hypochlorite, NaOCl, and calcium chloride hypochlorite, $CaCl(OCl)$, slowly release chlorine when dispersed in water. Chlorine release is accelerated in the presence of acids. This has led to accidents in the past when in households, hypochlorite sanitary cleaners were mixed with other cleaners containing acids (p. e. citric acid). Hypochlorites are also strong oxidizers. Contact with easily ignitable material can cause fire.

Besides the toxicological effects, chlorine and chlorine-releasing compounds may act as an oxidizer and react heavily with reducing agents.

GHS symbols (Fig. 152) and hazard statements for chlorine:

Fig. 152: GHS symbols for chlorine gas cylinders.

- H270: May intensify fire; oxidizer.
- H280: Contains gas under pressure; may explode if heated.
- H330: Fatal if inhaled.
- H315: Causes skin irritation.
- H319: Causes severe eye irritation.
- H335: May cause respiratory irritation.
- H400: Very toxic to aquatic life.
- EUH071: Corrosive to the respiratory tract.

GHS symbols (Fig. 153) and hazard statements for dichloro isocyanuric acid:

Fig. 153: GHS symbols for dichloro isocyanuric acid.

- H272: May intensify fire; oxidizer.
- H302: Harmful if swallowed.
- H319: Causes severe eye irritation.
- H335: May cause respiratory irritation.
- H410: Very toxic to aquatic life with long-lasting effects.

11.3.6 Potassium chlorate

Potassium chlorate, $KClO_3$, and other chlorates are strong oxidizers and used as such in industry. Mixtures with flammable material are explosive and may be very shock- and/or friction-sensitive. Chlorate is suspected to be the culprit in accidents with homemade fireworks (which are forbidden, anyway); people frustrated in the mixing of a proper gun powder may have got the advice from "friends" to switch from the oxidizer potassium nitrate to the stronger oxidizer potassium chlorate. However, frustration in trials of preparing a decent gun powder makes people careless, which is punished immediately by the very friction-sensitive mixtures of chlorate with sulfur. Chlorates were used as contact herbicide as they cause burns of plant material. The use as total herbicide is prohibited in the European Union, meanwhile.

After oral ingestion, chlorates cause methemoglobin formation and lysis of erythrocytes. Symptoms are pain, nausea, vomiting and cyanosis. With some delay, debris of blood cells in the glomeruli causes oliguria and finally anuria.

GHS symbols (Fig. 154) and hazard statements for potassium chlorate:

Fig. 154: GHS symbols for potassium chlorate.

- H271: May cause fire or explosion; strong oxidizer.
- H302: Harmful if swallowed.
- H332: Harmful if inhaled.
- H411: Toxic to aquatic life with long-lasting effects.

11.3.7 Hydrogen fluoride and sodium fluoride

Hydrogen fluoride (HF) and sodium fluoride (NaF) are important chemicals in the glass and ceramics industry and may also be used in the synthesis of fluorinated organic compounds. On contact with stronger acids like HCl, NaF releases HF; its boiling point is 19.5 °C, and it is very hygroscopic. Cleaners in bricklayer use may contain HF.

HF causes burning sensations and watering of the eyes. Eye damage may be irreversible. After skin contact, severe skin burns with damage of the bone, nausea and vomiting may follow. The fluoride ions precipitate calcium and cause a painful remineralization of the bones next to the contact surface. After inhalation, a hemorrhagic lung edema may result. Chronic exposure leads to reduced lung function. In case of contact, calcium gluconate solutions shall be used to wash the contact area to trap the fluoride ions. Medical doctors will proceed with injections of calcium gluconate solutions into tissue around the contact area.

GHS symbols (Fig. 155) and hazard statements for HF:

Fig. 155: GHS symbols for hydrogen fluoride.

- H330: Fatal if inhaled.
- H310: Fatal by skin contact.
- H300: Fatal if swallowed.
- H314: Causes severe skin burns and eye damage.
- EUH071: Corrosive to the respiratory tract.

GHS symbols (Fig. 156) and hazard statements for NaF:

Fig. 156: GHS symbol for sodium fluoride.

- H301: Toxic if swallowed.
- H315: Causes skin irritation.
- H319: Causes severe eye irritation.

11.3.8 Sulfuric acid

Sulfuric acid is a strong acid used as base chemical in many industrial and professional processes. 32% H_2SO_4 is used as electrolyte in rechargeable lead batteries. Sulfuric acid is hygroscopic and as such destroys textiles even if diluted; the dilution enthalpy is comparatively high and pouring water into sulfuric acid may cause overheating and splashing.

When exposed against sulfuric acid aerosols, eyes and the upper respiratory tract are irritated, causing pain, bleeding and coughing. Glottis edema may be fatal. After repeated exposure, the lung capacity may decline. Teeth erosions were reported for chronically exposed workers.

GHS symbol (Fig. 157) and hazard statements for sulfuric acid:

Fig. 157: GHS symbol for sulfuric acid.

- H290: May be corrosive to metals.
- H314: Causes severe skin burns and eye damage.

11.3.9 Nitric acid

Nitric acid (HNO_3) is a basic industrial chemical used for many different synthetic processes. It is a strong acid and as such corrosive, and it can corrode many metals, including copper and silver due to its oxidizing capability; the nitrates formed show a good water solubility. Above 65%, the HNO_3 becomes fuming: it is emitting nitrogen oxides (see later). Spots on the skin develop a yellow color due to nitration of the aromatic amino acids in proteins. On chronic exposure, HNO_3 can cause dermatitis and obstructive respiratory diseases. Contact with flammable material can cause fire.

GHS symbols (Fig. 158) and hazard statements for concentrated HNO_3:

Fig. 158: GHS symbols for concentrated nitric acid.

- H272: May intensify fire; oxidizer.
- H290: May be corrosive to metals.
- H314: Causes severe skin burns and eye damage.
- H330: Fatal if inhaled.

11.3.10 Nitrogen oxides and sodium nitrite

Nitrogen oxides are rarely used as such, but they may be formed by oxidative processes, thermal decomposition of fertilizers and reaction of (concentrated) nitric acid with metals. NO_2 has a brown color, a chlorine-like odor and is heavier than air; it is an oxidizer. NO is colorless, induces a decrease in blood pressure (headache) and blocks the iron in hemoglobin.

Nitrogen oxides irritate the eyes and the respiratory tract. First reactions are coughing, headache, dizziness and chest tightness. Inflammation of the respiratory tract and lung edema may follow with some hour delay; this is critical as in the meantime the victim may feel comparatively comfortable. Chronic exposure induces reduced lung capacity (probably due to decay of elastic proteins) and lung fibrosis. Opacity of the cornea may be irreversible.

Nitrite salts cause methemoglobinemia on acute intoxication; initial red discoloration of the skin changes into greyish blue, headache, dizziness, decrease in blood pressure (hypotonia), brady- or tachycardia or arrhythmia are further symptoms. Nitrites are mutagenic in in vitro test systems. Medical emergency staff may decide to apply toluidine blue to counter the methemoglobinemia: toluidine blue establishes an equilibrium between hemoglobin and methemoglobin. Medical doctors may let the victim inhale glucocorticoid aerosols to slow down inflammatory processes in the lung (see Tab. 78).

GHS symbols (Fig. 159) and hazard statements for NO_2:

Fig. 159: GHS symbols for nitrogen dioxide gas cylinders.

- H280: Contains gas under pressure; may explode if heated.
- H270: May cause fire or explosion; strong oxidizer.
- H330: Fatal if inhaled.
- H314: Causes severe skin burns and eye damage.
- EUH071: Corrosive to the respiratory tract.

GHS symbols (Fig. 160) and hazard statements for $NaNO_2$:

Fig. 160: GHS symbols for sodium nitrite.

- H272: May intensify fire; oxidizer.
- H301: Toxic if swallowed.
- H319: Causes severe eye irritation.
- H400: Very toxic to aquatic life.

11.3.11 Hydrogen cyanide and potassium cyanide

Hydrogen cyanide (HCN) is an important base chemical in industrial synthesis. It is a complexing agent used in metallurgical processes and a building block in organic synthesis. The boiling point is 26 °C, and it is easily released from its salts by acids.

Bitter almonds release their bound HCN in the gut, and two bitter almonds may be fatal for little children. HCN may be formed in smoldering fires of nitrogen containing material (nitrile polymers, polyurethanes, polyamides, leather, wool). It is flammable and has a broad range of explosion limits.

Vapors may irritate the eyes and cause a cornea edema. First signs are burning sensations on the tongue and scratching in the throat. Shortness of breath, panic, headache, dizziness, arrhythmia, unconsciousness, convulsions and breathlessness may follow, and the skin of the intoxicated victim has a "healthy" color. The cyanide ions block the Fe(III) of the cytochrome oxidase in the mitochondria. Appropriately trained emergency medical experts may apply hydroxocobalamin (vitamin B12), which traps cyanide ions, or *p*-dimethylamino-phenol to transform part of the hemoglobin to methemoglobin which causes a redistribution of cyanide ions from cytochrome c to the methemoglobin. Sulfuration of cyanide to thiocyanate is a pathway of detoxification which is dependent on the sulfur supply; therefore, emergency centers in hospitals may apply i.v. thiosulfate to the patient (see Tab. 78).

GHS symbols (Fig. 161) and hazard statements for HCN:

Fig. 161: GHS symbols for hydrogen cyanide.

- H224: Extremely flammable liquid and vapor.
- H330: Fatal if inhaled.
- H410: Very toxic to aquatic life with long-lasting effects.

11.3.12 Phosphor (white/yellow)

Phosphor is an important base chemical. The white/yellow form is pyrophoric at air contact and has a noticeable volatility. Mixtures with oxidizers are explosive (this is also the case for red phosphor). White phosphor was used in Napalm bombs.

Yellow phosphor is much more toxic than red phosphor. Skin contact causes severe burns. Arrhythmia may occur after swallowing and after skin contact. After swallowing, heavy stomach pains, vomiting and diarrhea are first symptoms. Phosphor inhibits the rough endoplasmic reticulum, stops protein synthesis and excretion of lipids by the liver. After 1–3 days liver failure becomes obvious, and liver coma may end in death. Chronic exposure may result in osteoporosis, cachexia (loss of weight, weakness), diarrhea, convulsions and pain in the limbs. White phosphorous may be formed unintendedly if phosphates are heated together with coal or other reducing agents in absence of air.

First responders need to protect themselves! Wash off stains on the skin with 1% $CuSO_4$ (forms insoluble, less active Cu_3P_2, see Tab. 78). Wrap affected skin with wet towels and keep them wet. Store yellow phosphor under water in a glass bottle, which itself is placed in a tin can.

GHS symbols (Fig. 162) and hazard statements for yellow phosphor:

Fig. 162: GHS symbols for yellow/white phosphor.

- H250: Catches fire spontaneously if exposed to air.
- H330: Fatal if inhaled.
- H300: Fatal if swallowed.
- H314: Causes severe skin burns and eye damage.
- H400: Very toxic to aquatic life.

11.3.13 Phosphine (Phosphane)

Phosphine (PH_3) is very toxic gas that might be used as fumigant for pest control (rodenticide). It is produced in situ by hydrolysis of AlP or Ca_3P_2 in the presence of humidity. It is extremely flammable and forms explosive mixtures with air over a broad concentrations range. The smell resembles that of garlic; it is said that very pure PH_3 would not have an odor.

Symptoms after exposure are chest tightness and pain behind the sternum. Shortly thereafter headache, nausea, stomach pain and vomiting appear. A lung edema develops which is frequently the cause of death. If survived, lung, kidney and heart functions may remain impaired. Subchronic exposure may cause the same symptoms.

GHS symbols (Fig. 163) and hazard statements for phosphine:

Fig. 163: GHS symbols for phosphine gas cylinders.

- H220: Extremely flammable gas.
- H280: Contains gas under pressure; may explode if heated.
- H330: Fatal if inhaled.
- H314: Causes severe skin burns and eye damage.
- H400: Very toxic to aquatic life.

11.3.14 Ammonia

Ammonia is an important base chemical for the production of further inorganic and organic nitrogen containing compounds, fertilizer production and heat exchange medium in cooling installations. It has a typical odor, is stored and transported in gas cylinders, as liquefied gas, or it may be released from ammonium salts in the presence of strong bases.

Ammonia is strongly irritating to the eyes and respiratory tract. Reactions to inhalation exposure are coughing and nausea. An edema of the glottis and inflammation of bronchi may end in life threatening breathing difficulties. Splashes of liquefied ammonia can cause blindness after a latency period of up to 10 days. Medical doctors may let the victim inhale glucocorticoid aerosols to slow down inflammatory processes of the glottis (see Tab. 78).

GHS symbols (Fig. 164) and hazard statements for ammonia:

Fig. 164: GHS symbols for ammonia gas cylinders.

- H221: Flammable gas.
- H280: Contains gas under pressure; may explode if heated.
- H331: Toxic if inhaled.
- H314: Causes severe skin burns and eye damage.
- H400: Very toxic to aquatic life.

11.3.15 Hydrogen peroxide

Hydrogen peroxide is a base chemical, supplied as aqueous solution up to about 32%. It is used as oxidant and bleaching agent. Solution of 3% is used as disinfectant for skin and wounds. Solutions of higher concentration were misused for the synthesis of acetone peroxide for terroristic suicide assaults.

GHS symbols (Fig. 165) and hazard statements for a 35–< 50% hydrogen peroxide solution:

Fig. 165: GHS symbols for 35–< 50% hydrogen peroxide solution.

- H302 + H332: Harmful if swallowed or inhaled.
- H315: Causes skin irritation.
- H318: Causes severe eye damage.
- H335: May cause respiratory irritation.

GHS symbols (Fig. 166) and hazard statements for a 50–< 75% hydrogen peroxide solution:

Fig. 166: GHS symbols for hydrogen peroxide solution, 50–< 75%.

- H272: May intensify fire; oxidizer.
- H302: Harmful if swallowed.
- H314: Causes severe skin burns and eye damage.

- H332: Harmful if inhaled.
- H335: May cause respiratory irritation.

11.3.16 Potassium permanganate

Potassium permanganate is a strong oxidizer mainly used in laboratories. Strongly diluted solutions may be used as disinfectants. Alkaline solutions may be used to destroy other chemicals.

On contact with the skin, it causes skin discoloration and burns. If swallowed, it causes erosion of mucous membranes, and a glottis edema may result. Erosion of blood vessels may result in delayed, fatal internal bleeding.

GHS symbols (Fig. 167) and hazard statements for $KMnO_4$:

Fig. 167: GHS symbols for potassium permanganate.

- H272: May intensify fire; oxidizer.
- H302: Harmful if swallowed.
- H314: Causes severe skin burns and eye damage.
- H410: Very toxic to aquatic life with long-lasting effects.

11.4 Organic solvents

Many organic solvents are immiscible with water (exemption: small alcohols and ketones), have a considerable vapor pressure and are either highly flammable or flammable and lighter than water (no or only few halogens in the molecule), or they are not flammable and heavier than water (typically chlorinated solvents). All solvents are more or less acting as narcotics, and they are frequently labeled with H336 (may cause drowsiness and dizziness). Vapors are heavier than air and may form invisible pools under certain situations (lower level places, poor ventilation) with the danger of asphyxiation (replacement of oxygen), and in case of flammable solvents, explosive mixtures with air can be formed. In contact with hot surfaces, halogenated solvents may release irritant or corrosive gases (HCl!), and in presence of humidity, they can pose problems due to corrosion of materials.

11.4.1 *n*-Hexane

n-Hexane (hexane) is a component of gasoline and is used as a nonpolar solvent, namely for the extraction of vegetable oils from seeds for food production. It is extremely flammable.

Narcosis is seen only after inhalation of high concentrations. Repeated overexposure results in polyneuropathy, starting with paresthesia and weakness in extremities, leading finally to numbness and paralysis; *n*-hexane causes degeneration of neuronal axons and myelin sheaths. This specific toxicity is not seen with *n*-pentane, *n*-heptane or 2-methyl pentane. Hexane-2,5-dione is the toxic metabolite which is assumed to cross-link transport proteins in the neuronal axons, by pyrrole formation with amino groups and subsequent oxidation. In rats, testicular lesions were observed after repeated respiratory exposure.

GHS symbols (Fig. 168) and hazard statements for *n*-hexane:

Fig. 168: GHS symbols for *n*-hexane.

- H225: Highly flammable liquid and vapor.
- H304: May be fatal if swallowed and enters the airways.
- H315: Causes skin irritation.
- H336: May cause drowsiness or dizziness.
- H361f: Suspected of damaging fertility.
- H373: May cause damage to the nervous system through prolonged or repeated exposure.
- H411: Toxic to aquatic life with long-lasting effects.

11.4.2 Dichloromethane

Dichloromethane is a nonflammable solvent with low boiling point of about 40 °C. It was widely used as a paint stripper, but according to entry no. 59 of the Annex XVII of Regulation (EC) No. 1907/2006, dichloromethane levels for paint strippers for the general public and professional users must be below 0.1%. Under certain, defined conditions, professional use may be permitted. The label of dichloromethane or its mixtures with contents of 0.1% or more must be labelled with: *Restricted to industrial use and to professionals approved in certain EU Member States – verify where use is allowed.* Dichloromethane is/was also used as a solvent for caffeine extraction in closed processes.

Dichloromethane does not show the liver toxicity of chloroform (see later). High concentrations may be narcotic and cause an increase in carboxyhemoglobin. In mice, tumors of the liver and the lung were observed; similar studies with rats and hamsters did not cause increased tumor incidences. It is a bacterial mutagen but not in vivo in mammals. Dichloromethane can undergo metabolism by cytochrome P450 (CYP 450) and, if this pathway becomes saturated, by glutathione-S-transferase. The latter results in the formation of chloro-methyl-glutathione, which is suspected to be an alkylating agent responsible for tumor induction, and may be hydrolyzed by delivering formaldehyde (Fig. 170). Dichloromethane is an example how higher dosages can result in qualitatively different metabolic pathways than lower dosages.

GHS symbol (Fig. 169) and hazard statements for dichloromethane:

Fig. 169: GHS symbol for dichloromethane.

– H351: suspected of causing cancer.

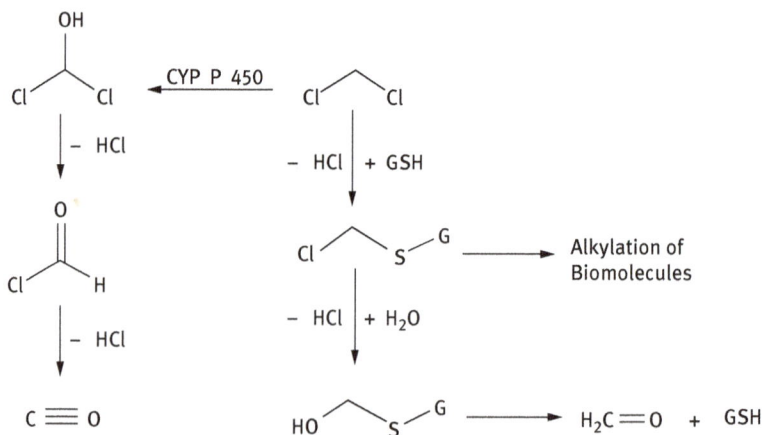

Fig. 170: Metabolism of dichloromethane; if the CYP450 pathway becomes saturated, the glutathione pathway gains importance.

11.4.3 Trichloromethane (chloroform)

Chloroform is a nonflammable solvent with narcotic effects. It was used as a narcotic, but incidences with deaths of patients had led to a discontinuation in the early twentieth century. Like dichloromethane, it was used as a nonflammable solvent for degreasing metal plates. Entry 32 to 38 of the Annex XVII of Regulation (EC) No. 1907/2006 names chloroform and several chlorinated ethanes (trichloro-, tetrachloro-, pentachloroethane and 1,1-dichloroethylene). These substances and mixtures containing 0.1% or more must not be placed on the market for the public or for diffusive applications like cleaning of textiles or large surfaces (for example, degreasing of steel sheets and plates). Such products need to be labeled with "For use in industrial installations only."

After single exposure, narcosis is the most prominent effect of chloroform. It may sensitize the heart muscle so it becomes oversensitive to endogenous stimulants, and after sensitization sudden heart failure may occur. Repeated exposure may cause liver and kidney damage, and at dosages that cause liver damage, liver tumors were observed after chronic exposure in rodents. The toxicity of chloroform to tissue is owed to the metabolite phosgene that is formed after oxidation via CYP 450 (Fig. 171):

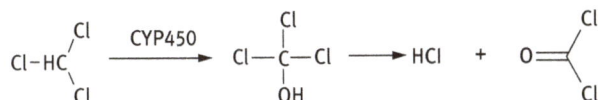

Fig. 171: Transformation of chloroform by CYP 450.

GHS symbols (Fig. 172) and hazard statements for chloroform:

Fig. 172: GHS symbols for chloroform.

- H302: Harmful if swallowed.
- H315: Causes skin irritation.
- H351: Suspected of causing cancer.
- May cause damage to kidneys through prolonged or repeated exposure.

11.4.4 Trichloroethene and tetrachloroethene

Both chlorinated ethenes show a comparatively low acute toxicity and are not flammable. This resulted in a widespread use as solvents and thinners for glues and paints, and also for cleaning and degreasing purposes. Both compounds were used in chemical laundry processes, in daily exposure of craftsmen and painters was high.

Exposure against comparatively high concentrations causes narcosis (misuse for sniffing!).

Liver and kidneys are target organs for chlorinated ethenes. Liver tumors were induced in rodents, and kidney tumors in male rats after chronic exposure to trichloroethylene; the latter was also indicated in worker after prolonged exposure against high concentrations, but not in studies with moderate exposure. It was assumed that at high concentrations, metabolism by P450 monooxygenase becomes exhausted and metabolism via glutathione adducts (starts with an S_N2 reaction at sp^2 carbon!) leads to the formation of the very reactive thioketene in kidneys.

GHS symbols (Fig. 173) and hazard statements for trichloroethane:

Fig. 173: GHS symbols for trichloroethene.

- H350: May cause cancer.
- H341: Suspected of causing genetic defects.
- H315: Causes skin irritation.
- H319: Causes severe eye irritation.
- H336: May cause drowsiness or dizziness.
- H412: Harmful to aquatic life with long-lasting effect.

11.4.5 Methanol

Methanol is a base chemical in industry, used as intermediate and solvent. It is totally miscible with water. Entry 69 of the Annex XVII of Regulation (EC) No. 1907/2006 prohibits the use of methanol in windscreen washing fluids in concentrations at or above 0.6%. Methanol is highly flammable and its vapor forms explosive mixtures with air. In the body, methanol is oxidized to methanol (formaldehyde) and finally formic acid by the enzyme alcohol dehydrogenase and aldehyde dehydrogenase.

Ethanol is a competitor and prevents methanol from transformation to formaldehyde. Therefore, ethanol is an antidote because it is the formaldehyde and its product formic acid which are responsible for the toxic action by cross-linking proteins and acidosis, respectively. The visual nerve is especially sensitive against this acidosis. Overall, acidosis may cause multiple organ failure which finally may be fatal.

Early signs of overexposure are agitation and euphoria, the latter less pronounced against ethanol intoxication, and red flush of the face. Up to 24 h later, headache, nausea, vomiting and impaired vision may occur, and the critical acidosis may be delayed by 2–3 days. Edema in the brain and the lung can result in apnea, and arrhythmia may occur due to impairment of the heart muscle, uremia as a result of kidney damage. Chronic exposure may as well end in impaired vision and additionally in drowsiness, dizziness, tinnitus and paresthesia.

GHS symbols (Fig. 174) and hazard atatements for methanol:

Fig. 174: GHS symbols for methanol.

- H225: Highly flammable liquid and vapor.
- H331: Toxic if inhaled.
- H311: Toxic by skin contact.
- H301: Toxic if swallowed.
- H370: Causes damage to the eyes.

11.4.6 Formaldehyde (methanal)

Formaldehyde is not a solvent (and would, therefore, not fit into this chapter), but it is a metabolite of methanol and for that reason listed at this place. It is a base chemical used in numerous synthetic processes for other chemicals, polymers, glues and wood binders. Formaldehyde may be used as a potent disinfectant and fumigant in pest control. Formaldehyde is marketed as gas, or as aqueous solution (up to 37%).

Due to its high reactivity with biological macromolecules, formaldehyde can cause burns at the site of contact which may appear with some hours delay. Fumes cause severe eye irritation and damage to the olfactory epithelia. In chronic studies, Formaldehyde caused epithelial tumors in the nose of exposed rodents when exposure exceeded the respiratory irritation threshold. Increased incidences of nasal tumors were observed in exposed workers. Formaldehyde is a known skin sensitizer in human beings.

GHS symbols (Fig. 175) and hazard statements for formaldehyde:

Fig. 175: GHS symbols for formaldehyde.

- H301 + H311 + H331: Toxic if swallowed, by skin contact or if inhaled.
- H314: Causes severe skin burns and eye damage.
- H317: May cause an allergic skin reaction.
- H341: Suspected of causing genetic defects.
- H350: May cause cancer.

11.4.7 Benzene

Benzene is one of the most important basic chemicals. As component of the C-6 fraction in petrol refining, it is present in gasoline, and extracting it for "aromatics free" gasoline requires special efforts. It was widely used as a solvent until its carcinogenic potency became apparent. Entry 5 of the Annex XVII of Regulation (EC) No. 1907/2006 names benzene; its content in toys must be below 5 mg/kg, and it shall not be placed on the market as such of in mixtures at 0.1% or more. However, motor fuels may contain higher levels of benzene provided they are compliant with the EU motor fuel directive.

Acute exposure may cause narcosis with headache, nausea, vomiting and dizziness. At high concentrations, convulsions, arrhythmia and respiratory failure may occur. Chronic exposure leads to bone marrow depression and subsequently to anemia, leucopenia and thrombocytopenia. Benzene causes leukemia in human beings. This effect is difficult to be reproduced in animal models. Epoxidation of the aromatic ring creates an electrophilic compound, and follow-up formation of hydroquinone and quinones creates components which may undergo a permanent redox-cycling, releasing reactive oxygen species that can damage the DNA.

GHS symbols (Fig. 176) and hazard statements for benzene:

GHS symbols for benzene.

- H225: Highly flammable liquid and vapor.
- H340: May cause genetic defects.
- H350: May cause cancer.
- H304: May be fatal if swallowed and enters airways.
- H315: Causes skin irritation.
- H319: Causes severe eye irritation.

11.4.8 Toluene

Toluene as well is/was a widespread used organic solvent and is a possible alternative to benzene, as it is not a carcinogen. Entry 48 of the Annex XVII of Regulation (EC) No. 1907/2006 prohibits the marketing and use of liquids that contain 0.1% or more toluene if the product is to be used in adhesives or spray paints for the general public.

Single exposure to high concentrations causes narcosis, headache, dizziness, nausea and euphoria, which is the trigger for misuse of toluene as sniffing agent. Repeated sniffing results in damage in the cerebellum (gait imbalance, abnormal walking, mild tremor). Kidney damage was observed as well after repeated sniffing. Fetuses of addicted mothers had reduced birth weights, and the incidence of still-born was increased. It is not quite clear whether repeated occupational exposure causes impaired hearing.

GHS symbols (Fig. 177) and hazard statements for toluene:

Fig. 177: GHS symbols for toluene.

- H225: Highly flammable liquid and vapor.
- H361d: Suspected of damaging the unborn child.
- H304: May be fatal if swallowed and enters the airways.
- H373: May cause damage to CNS and the kidneys through prolonged or re- peated exposure.
- H315: Causes skin irritation.
- H336: May cause drowsiness or dizziness.

11.4.9 Xylenes

Xylene is available as *ortho-*, *meta-* or *para*-xylene or as mixed isomers. Xylenes are important intermediates for the production of phthalic and terephthalic acid and the subsequent polyesters, and they are used as organic solvents which may re- place the carcinogenic benzene.

Acute exposure may result in headache, dizziness, lack of concentration, numb- ness and sleepiness. Irritation of the respiratory tract and the eyes due to vapor expo- sure may be observed, depending on the isomers. Functional disorders of heart, liver and namely kidney may occur after chronic exposure; neuronal effects were weak- ness in limbs, memory deficits, gait imbalance.

GHS symbols (Fig. 178) and hazard statements for xylene, mixed isomers:

Fig. 178: GHS symbols for xylenes, mixed isomers.

- H226: Flammable liquid and vapor.
- H304: May be fatal if swallowed and enters the airways.
- H312: Harmful in contact with skin.
- H332: Harmful if inhaled.
- H315: Causes skin irritation.
- H319: Causes severe eye irritation.
- H335: May cause respiratory irritation.
- H373: May cause damage to kidneys through prolonged or repeated exposure.

11.5 Intermediates and fine chemicals

11.5.1 Ethane diol (glycol) (and oxalates)

Ethane diol/glycol is used as intermediate, antifreeze additive and solvent. It has a more pronounced toxicity in man than in rodents. It is oxidized to glycol aldehyde, glyoxylic acid, glycol dialdehyde and finally oxalic acid. This process is faster in man than in the rat. Oxalic acid precipitates calcium ions in tissues. The minimum lethal dose in man is 100 mL, and the intoxication progresses via four phases: first the CNS and the digestive tract are affected accompanied with dizziness, nausea, vomiting, convulsions and perhaps coma within 0.5–12 h. Within 12–24 h, the second phase appears with high blood pressure, lung edema, tachycardia and tachypnea. Between 24 and 72 h, in the third phase, oliguria and anuria indicate kidney damage due to precipitated calcium oxalate crystals. In the fourth phase, precipitates in the brain cause impaired vision, hyperreflexia and difficulties in swallowing. Chronic exposure can lead to kidney and circulation impairment.

GHS symbols (Fig. 179) and hazard statements for ethane diol:

Fig. 179: GHS symbols for ethane diol.

– H302: Harmful if swallowed.
– H373: May cause damage to kidneys through prolonged or repeated oral exposure.

11.5.2 Aromatic amines and aromatic nitro compounds

Aromatic amines and aromatic nitro compounds are intermediates which split up in multiple branches of uses and applications, covering nearly every area of products based on organic chemistry like glues, adhesives, colorants, pigments, polymers, to name a few. The use of nitrobenzene as solvent is prohibited in the EU. Uptake of these compounds may occur via inhalation, but skin contact is also critical as these substances penetrate the skin easily. In the body, aromatic amines are oxidized to N-hydroxyl amines (perhaps after N-de-alkylation of alkylated amines), and nitro-compounds are reduced via the nitroso-compounds to N-hydroxyl amines. This common intermediate is the reason for the common mode of action, the formation of methemoglobin. In presence of oxygenated hemoglobin, the hydroxyl amines are oxidized to nitroso-compounds while the hemoglobin is oxidized to methemoglobin. A reductase reduces the nitroso-compound to hydroxyl amine, so these compounds act catalytically (Fig. 181). The methemoglobin is reduced back to functional hemoglobin by the enzyme cytochrome b_5 reductase and the cofactor NADH; further reductase activity can be induced, and the cofactor is NADPH; this reductase activities are not fully developed in new born babies who are, therefore, much more sensitive against methemoglobin forming agents than adult people. Rodents are much less sensitive than human beings because they have a higher activity of methemoglobin reductase. Dogs and especially cats show a sensitivity resembling that of human beings.

The formation of methemoglobin which can no longer transport oxygen causes an oxygen deficit in tissues, and as a consequence pale to blueish discoloration of the skin, headache, drowsiness, dizziness, nausea and agitation, tachycardia and low blood pressure are seen. In critical cases, victims become unconscious and stop breathing. In such situations, emergency personnel may decide to apply redox-active dyes like toluidine blue which supports reduction of methemoglobin. Chronic exposure may show similar symptoms, and also mental depression, anemia and impaired liver and kidney function.

Some hydroxyl amines may become O-acetylated or sulfated, introducing the good leaving groups acetate and sulfate, releasing a nitrenium cation; this nitrenium ion is formed in the cytosol and highly reactive, so it may react with proteins and deactivate them; if it has a sufficient lifetime to enter the nucleolus, it reacts with the DNA bases and may cause mutations which finally transform the cell into a cancer cell (Fig. 180).

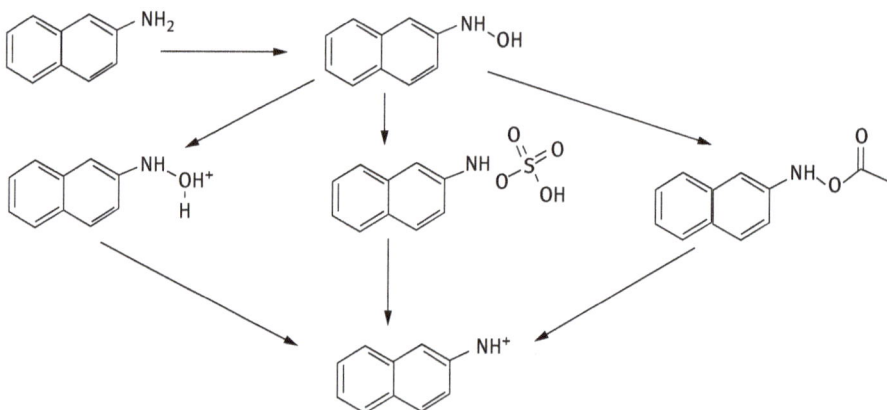

Fig. 180: Activation of primary, aromatic amines to electrophiles.

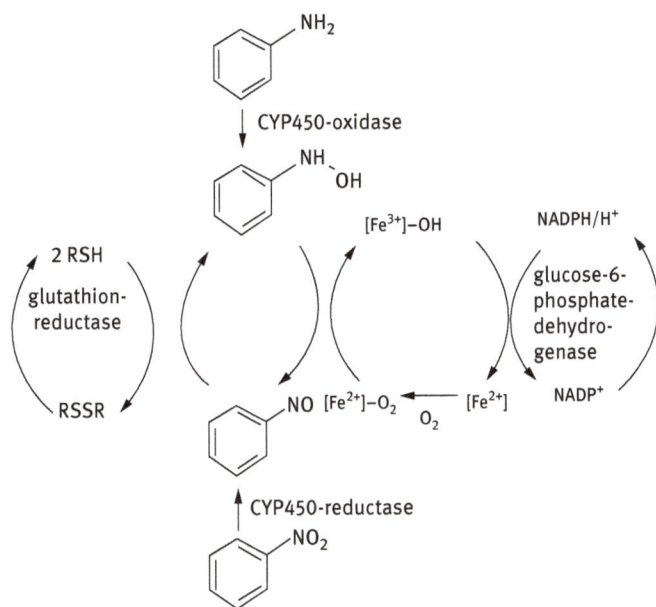

Fig. 181: Catalytic activity of primary aromatic amines, aromatic nitro- and nitroso-compounds in the formation of methemoglobin.

Typical cancer sites of carcinogenic primary aromatic amines are the bladder, but also the liver, kidneys, thyroid gland and lung. Aromatic amino- and nitro-compounds proven to be strong carcinogens in human beings (category 1A) are sketched in Fig. 182.

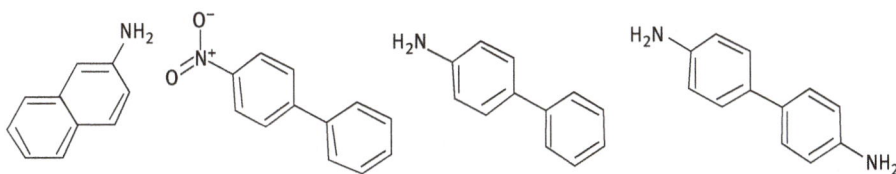

Fig. 182: Proven strong human carcinogens: 2-amino naphthalene, 4-nitro-biphenyl, 4-aminobiphenyl, 4,4'-diamino biphenyl (benzidine).

These substances are strongly restricted concerning production and use, and azo dyes which may liberate these amines (and other carcinogenic, primary aromatic amines) are not allowed for dyeing textiles or other articles with intended skin contact or toys in the EU. It could be demonstrated that rats fed with such azo dyes developed tumors of the same kind as after dosage of the free amines. In the large intestine, these azo dyes are reduced to the corresponding amines. Entry 48 of the Annex XVII of Regulation (EC) No. 1907/2006 restricts the use of such azo colorants for skin contact applications that can release carcinogenic primary aromatic amines; the amines are listed in appendix 8 to the Annex XVII.

Aniline and nitrobenzene caused spleen tumors in mice and rats, but these appeared only at dosages which caused such a high turnover of erythrocytes that the iron released could no longer be picked up by storage proteins. Free iron compounds are a source of reactive oxygen species which cause oxidative DNA damage. Nitrobenzene damaged testicles in some of the rats exposed against 40 ppm, and the fertility was decreased.

GHS symbols (Fig. 183) and hazard statements for aniline:

Fig. 183: GHS symbols for aniline.

- H351: Suspected of causing cancer.
- H341: Suspected of causing genetic defects.
- H331: Toxic if inhaled.
- H311: Toxic by skin contact.
- H301: Toxic if swallowed.
- H372: May cause damage to blood through prolonged or repeated exposure.
- H318: May cause severe eye damage.
- H317: May cause an allergic skin reaction.
- H400: Very toxic to aquatic life.

GHS symbols (Fig. 184) and hazard satements for nitrobenzene:

Fig. 184: GHS symbols for nitrobenzene.

- H351: Suspected of causing cancer.
- H360F: May damage fertility.
- H331: Toxic if inhaled.
- H311: Toxic by skin contact.
- H301: Toxic if swallowed.
- H373: Causes damage to the blood through prolonged or repeated exposure.
- H411: Toxic to aquatic life with long-lasting effects.

11.5.3 Polychlorinated biphenyls, dibenzodioxins, dibenzofuranes

Polychlorinated Biphenyls (PCB) were produced and used as plasticizers, capacitor oils and dielectrics, heat exchange medium, to name a few applications. They comprise 209 congeners. Halogenated dibenzodioxins and halogenated dibenzofurans are byproducts in the production of PCBs, but they are also formed when organic material is overheated in the presence of chlorine. Brominated analogues share many properties with the chlorinated compounds.

The planar PCBs share many toxicological properties with the halogenated dibenzodioxins and dibenzofurans (Fig. 185). On the molecular level, they bind to the Aryl-hydrocarbon receptor (Ah-receptor) in the cytosol of the cell. This complex migrates into the nucleus and induces the transcription of certain genes. Based on their affinity to this Ah-receptor, the dioxins, furans and PCBs could be allocated to toxicity classes. These substances are very persistent and very bioaccumulating.

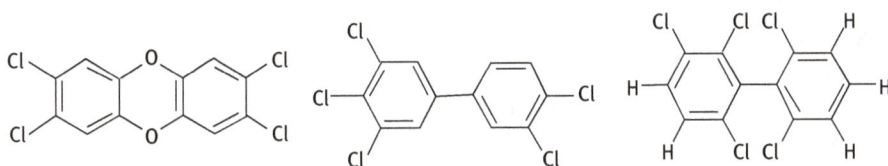

Fig. 185: 2,3,7,8-Tetrachloro dibenzo-*p*-dioxin (TCDD); 3,4,5,3′,4′-pentachlorobiphenyl (a dioxine-like PCB); 2,3,6,2′,6′-pentachlorobiphenyl (a nondioxin-like PCB).

Acute and chronic intoxication causes a longer lasting, so called "chloro"-acne. Dark pigmentation of the skin is another typical effect, which is also induced in fetuses (*cola-babies*). Acute lethal toxicity is very species dependent, with oral LD_{50} of 0.001 mg/kg for guinea pigs, 0.045 mg/kg for the rat, and 5 mg/kg for beagles; the enzyme phosphoenolpyruvate carboxykinase which is required for the gluconeogenesis is inhibited by the dioxins. These substances are also immune suppressants and cause thymus atrophy. Chronic exposure is associated with weight reduction, hyperkeratosis, swollen eyelids, liver damage, erosion and swelling of the gum, numbness in legs and arms. Although not being mutagenic, dioxins are strong tumor promotors. This substance class disturbs fertility and development. In animal experiments, malformations of fetuses, prenatal death and reduced birth weight are observed. PCBs with at least two ortho-chlorines are twisted and resemble the hormones thyroxin and triiodothreonin, which play a role in metabolism and the physical and mental development (Fig. 186). Affected behavioral and cognitive development was observed in children of heavily exposed mothers.

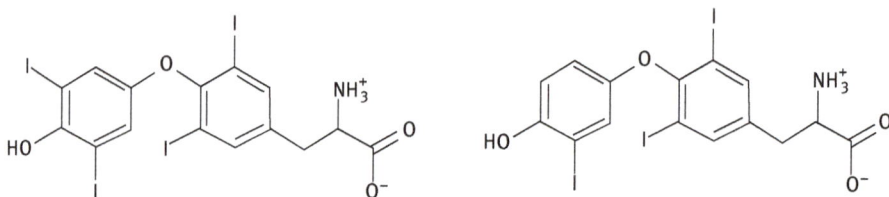

Fig. 186: Thyroid gland hormones thyroxine (tetraiodothreonine) and triiodothreonine.

11.5.4 Phthalic acid, diesters with C8–C10 alcohols

Diesters of phthalic acid (benzene-1,2-dicarboxylic acid) with C4–C10 alcohols are/ were used as low vapor pressure solvents and plasticizers in thermoplastics. Half-life in freshwater sediments is up to 300 days in some experiments which would render these compounds as persistent; however, some tests on biodegradation indicate ready biodegradability. During metabolism, the monoesters are formed which have antiandrogenic effects in mammals. Testicular atrophy and delayed sexual maturation were observed in exposed male rodents. According to entry 51 of the Annex XVII of Regulation (EC) No. 1907/2006, such phthalates must not be found in toys and childcare articles at concentrations of 0.1% or more.

GHS symbol (Fig. 187) and hazard statement for phthalic acid, bis(2-ethylhexyl) ester:
– H360: May damage fertility or the unborn child.

Fig. 187: GHS symbol for bis(2-ethyl-hexyl)phthalate.

11.5.5 Diphenylmethane diisocyanate/methylene diphenyl diisocyanate

Methylene diphenyl diisocyanate (MDI) is used as monomer in the production of flexible foams (furniture), integral skin foams (arm rests, seats, steering wheels), insulation panels, canned foam and adhesives at construction sites. It is a mixture of isomers and homologues (Fig. 188).

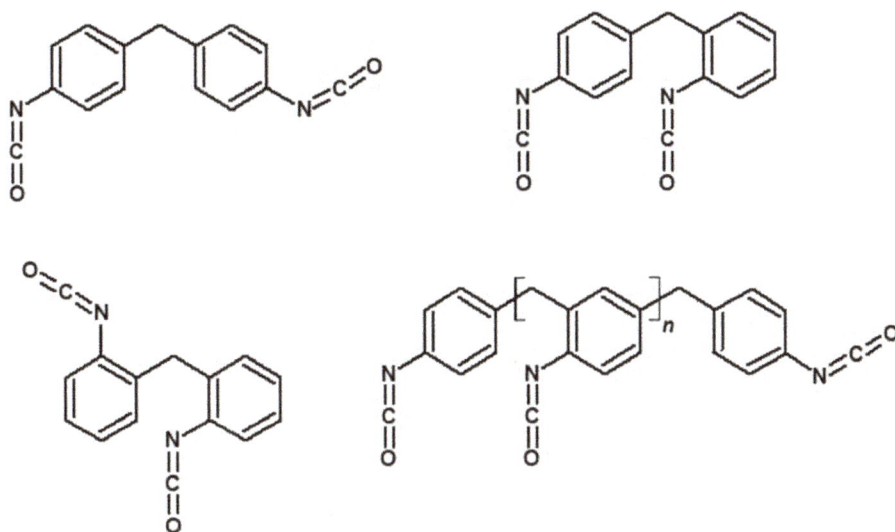

Fig. 188: Methylene-diphenyl-4,4′-diisocyanate, isomers and homologues.

The NCO groups react with OH groups at a rate dependent on catalysts present, rapidly with NH groups and in the presence of water, which hydrolysis an NCO group to an NH_2 group, which then reacts rapidly with another NCO group. This reactivity is the reason for the versatile use of MDI, but it is as well the reason for its toxicity. For the use in glues and one-component foams by the general public, entry 56 of the Annex XVII of Regulation (EC) No. 1907/2006 requests that these products must be equipped with suitable gloves and detailed safety instructions if the content of MDI is 0.1% or more.

MDI is irritating to the respiratory tract, the skin and the eyes. Of practical importance is its capability to introduce respiratory sensitization which may show up by symptoms like coughing, shortness of breath (isocyanate asthma), chest tightness (. . . *as if somebody is tightening an iron clamp around your chest* . . .), fever and flu-like symptoms (isocyanate flu). The symptoms may start immediately at exposure or may appear with some hour delay. Allergic skin reactions are less common, but reports exist. Animal experiments indicate that skin contact may facilitate the acquisition of respiratory allergy. Repeated overexposure may cause a decrement of the lung capacity. The LC_{50} value in rats would call for a classification of toxic if inhaled; however, the artificially fine aerosol can practically not be produced at workplaces, and larger aerosols were demonstrated to be much less critical, so MDI is classified as harmful if inhaled. Rats exposed to irritating concentrations of an MDI aerosol developed lung tumors. At concentrations below the irritation threshold, no tumors were observed.

On skin contact to MDI or fresh can foam, material should be wiped off with a towel, then washed with soap and water. Do not use organic solvents! Residual-cured MDI/foam will stick to the skin, but it is no longer dangerous. It can be removed mechanically or await the natural renewal of the skin surface.

GHS symbols (Fig. 189) and hazard statements for MDI:

Fig. 189: GHS symbols for MDI.

- H351: Suspected of causing cancer.
- H332: Harmful if inhaled.
- H373: May cause damage to the respiratory system through prolonged or repeated exposure.
- H319: Causes severe eye irritation.
- H335: May cause respiratory irritation.
- H315: Causes skin irritation.
- H317: May cause an allergic skin reaction.
- H334: May cause allergy or asthma symptoms or breathing difficulties if inhaled.

11.5.6 Epoxide resins

Epoxide resins used as strong adhesives on construction sites are formed by the reaction of bisphenol A with epichlorohydrin (Fig. 190). As two-component systems, they are mixed with a hardener which may be a polyamine or polythiol compound, and the cross-linking reaction forms the final polymer.

The epoxide systems are strong skin sensitizers which can induce allergic contact eczema. However, the potency is reduced with increasing mean molecular weight. The hardeners may be irritating, corrosive and/or skin sensitizers. Fine tuning of the cured material by cutting or heat-cutting generates dust and/or fumes that may have irritating and sensitizing properties.

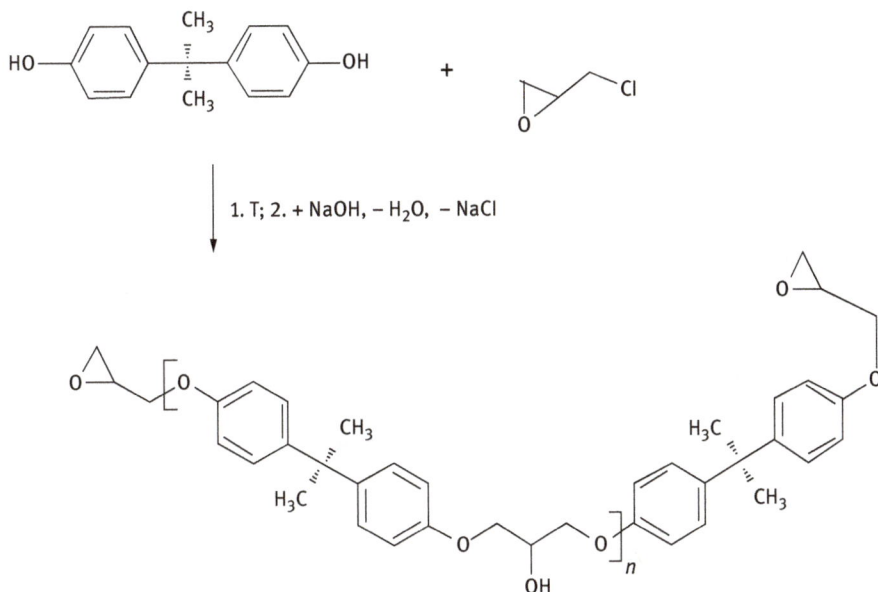

Fig. 190: Synthesis of epoxy resin.

11.5.7 Phenol

Phenol is an important intermediate in the chemical and plastics industry, like phenol-formaldehyde resins and wood binders. Phenol itself as well as its ring-methylated compounds (cresols) are powerful disinfectants. Phenol actually was one of the first disinfectants that was consistently used by the Scottish medical doctor Lister who had amazing success in curing wounds in his time.

Phenol is a weak acid, but it is also cytotoxic and anesthetizing so small area, skin contact is not always recognized immediately. Large surface skin contact can be fatal. Phenol causes methemoglobin, headache, dizziness, nausea, vomiting and the decay of red blood cells results in dark discoloration of urine and kidney failure. Arrhythmia may lead to breakdown of circulation. Chronic overexposure may as well cause dark discoloration of urine, discoloration of the skin, kidney impairment and impairment of the liver which may show up as retardation of blood clotting. Phenol is equivocal in mutagenicity tests but was negative in cancer tests.

GHS symbols (Fig. 191) and hazard statements for phenol:

Fig. 191: GHS symbols for phenol.

- H331: Toxic if inhaled.
- H311: Toxic by skin contact.
- H301: Toxic if swallowed.
- H314: Causes severe skin burns and eye damage.
- H341: Suspected of causing genetic defects.
- H373: May cause damage to blood, liver and kidney through prolonged exposure.

11.5.8 Nonylphenols

Isomers of alkylated phenols with C-9 alkyl chains are summarized as nonylphenol. Their ethoxylates are very potent surface-active agents and were used in many technical applications as detergents and lubricants. However, in the environment, biodegradation processes digested the ethoxy units, leaving the nonylphenol as less biodegradable unit (although not persistent). The nonylphenol interferes with the estrogen receptor, mimicking estrogen activity, and disturbance of sexual development was observed in fish in polluted waters. Entry 46 of the Annex XVII of Regulation (EC) No. 1907/2006 regulates the use of nonylphenol and its ethoxylates. Nonylphenol and the ethoxylates must not be contained in detergents, cosmetics at or above 0.1%, and from 2021 onward, the maximum permitted content in textiles is 0.01% if they are expected to undergo laundry processes.

GHS symbols (Fig. 192) and hazard statements for branched nonylphenols:

Fig. 192: GHS symbols for branched nonylphenols.

- H302: Harmful if swallowed.
- H314: Causes severe skin burns and eye damage.
- H361fd: Suspected of damaging fertility. Suspected of damaging the unborn child.
- H410: Very toxic to aquatic life with long-lasting effects.

11.5.9 Phenols, chlorinated or nitrated

Chlorophenols and nitrophenols are intermediates in the production of pesticides and metabolites of these. One example is the herbicide 2,4-dichloro phenoxy acetic acid, or the fungicide pentachlorophenol, which was used as a wood preservative which is no longer permitted in the EU.

With increasing number of substituents, these phenols become more acidic. In a certain pK_a range, which is the case for nitro-, dinitro- and dichlorophenol, these substances act as uncouplers of the oxidative phosphorylation in the cell, because they induce a breakdown of the proton-gradient at the inner mitochondrion membrane. Instead of producing the energy carrier adenosine triphosphate, the cell generates heat. Cases of fatal intoxication were reported after skin contact due to accidents, of after oral ingestion of so-called slim-pills, which actually contained 2,4-dinitrophenol. This toxic mechanism is also described as "internal asphyxiation."

Symptoms of intoxication are hyperthermia, fever, heat-shock, seizures, corrosion of mucous membranes or skin, unconsciousness, respiratory and circulation failure. Chronic exposure may affect the liver and induce a porphyria, bone marrow depression. Nitrophenols can induce a cataract (turbidity of the lens of the eye) several months after acute intoxication, and after chronic exposure impairment of peripheral nerves showing up as numbness or rheumatic pain.

GHS symbols (Fig. 193) and hazard statements for 2,4-dichlorophenol:

Fig. 193: GHS symbols for 2,4-dichlorophenol.

- H302: Harmful if swallowed.
- H311: Toxic by skin contact.

- H314: Causes severe skin burns and eye damage.
- H411: Toxic for aquatic life with long-lasting effects.

GHS symbols (Fig. 194) and hazard statements for 2,4-dinitrophenol:

Fig. 194: GHS symbols for 2,4-dinitrophenol.

- H301: Toxic if swallowed.
- H311: Toxic by skin contact.
- H331: Toxic if inhaled.
- H373: May cause damage to the nervous system through prolonged or repeated exposure.
- H400: Very toxic to aquatic life.

11.5.10 Organotin compounds

The terminus "organotin compounds" covers aliphatic and/or aromatic substituents bonded directly to tin (IV). Between four and one organic residues may be bound to tin, the remaining valences saturated by inorganic or organic anions. Tributyltin and other triorganotin and organotin compounds were used as biocides (antifouling varnishes) and as UV-stabilizers, dialkyltin compounds were/are used as catalysts, polymerization modifiers or UV stabilizers in polymers. Most of these uses are restricted (see entray no. 20, Annex XVII of regulation (EU) No. 1907/2006). Tributyltin compounds turned out to be endocrine disruptors so their use as antifouling paints was prohibited. Further, they are suppressing the immune system, and this property is also observed with dioctyltin compounds.

GHS symbols (Fig. 195) and hazard statements for dibutyltin dichloride:

Fig. 195: GHS symbols for dibutyltin dichloride.

- H341: Suspected of causing genetic defects.
- H360FD: May damage fertility. May damage the unborn child.
- H330: Fatal if inhaled.
- H301: Toxic if swallowed.
- H312: Harmful by skin contact.
- H372: Causes damage to the immune system on prolonged or repeated exposure.
- H314: Causes severe skin burns and eye damage.
- H410: Very toxic to aquatic life with long-lasting effects.

11.6 Insecticides

Concerning the group of insecticides, only very few examples will be mentioned which have either historical relevance, or they have some current relevance for the consumer. For some insecticides, the toxic mechanisms were identified on a molecular level.

11.6.1 Lindane

Lindane is a powerful insecticide. It is produced by perchlorination of benzene, but only the γ-isomer (15% yield) is active (Fig. 196). The other isomers are less or even nonactive, but partly they are very persistent and very bioaccumulating. The use of lindane, therefore, is restricted in the EU. Member states can grant limited derogations in cases of emergency. It is listed in annex I part A or Regulation (EU) No. 2019/1021; therefore, its manufacture, marketing and use is prohibited, and waste has to undergo special treatment if it contains 50 ppm or more hexachlorocyclohexanes.

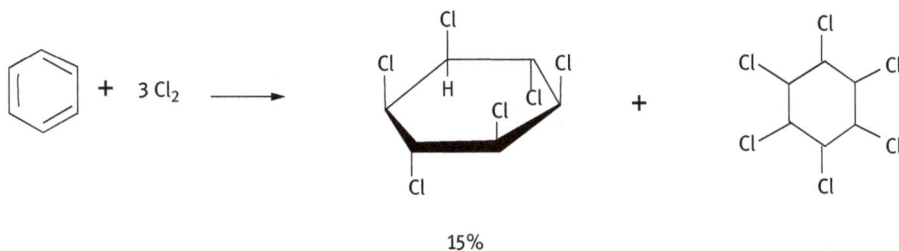

Fig. 196: Lindane synthesis produces a large number of persistent byproducts.

Lindane is much more toxic to insects than to mammals, but it can cross the skin easily, so dermal uptake is an issue.

Lindane and other chlorinated insecticides interrupt the nerve signals, probably predominantly of inhibiting neurons. Symptoms are mental and motor retardation, tonic and clonic convulsions, pulmonary edema and respiratory failure. Animal experiments indicate that lindane may affect the female and male fertility, and the immune system is depressed. Chronic symptoms are memory deficits, sleeplessness and motoric weakness.

GHS symbols (Fig. 197) and hazard statements for hexachlorocyclohexane:

Fig. 197: GHS symbols for hexachlorocyclohexane.

- H301: Toxic if swallowed.
- H332: Toxic if inhaled.
- H312: Harmful by skin contact.
- H373: May cause damage to the nervous system through prolonged or repeated exposure.
- H362: May cause harm to breast-fed children.
- H410: Very toxic to aquatic life with long-lasting effects.

11.6.2 Methiocarb (carbamates)

Methiocarb (N-methyl-(3,5-dimethyl-4-methylthio)phenyl carbamate) is an insecticide belonging to the group of carbamates. Carbamates inhibit the acetylcholine esterase (AChE), in insects much more than in mammals. In that respect, they act via the same mechanism as neurotoxic phosphates (see later). However, the inhibition of AChE by carbamates is much easier reversed than that by organophosphates, so application of other carbamates for the reactivation of the AChE is counterproductive. However, atropine may be given as antidote by medical experts.

GHS symbols (Fig. 198) and hazard statements for methiocarb:

Fig. 198: GHS symbols for methiocarb.

- – H301: Toxic if swallowed.
- – H410: Very toxic to aquatic life with long-lasting effects.

11.6.3 Parathion (organophosphate compounds)

Parathion, (4-nitrophenyl-diethyl-thiophosphate), is a member of the organic phosphates that are neurotoxic. Certain organic phosphates and phosphonates were synthesized as chemical warfare agents. The mechanism of molecular toxicity is similar to that of parathion. By oxidation in cells, parathion loses the sulfur and is transferred to the phosphoric acid ester. The p-nitrophenol residue is replaced by the OH group of the serine side chain in the active center of the enzyme AChE, which is inhibited and suffers slow recycling (Fig. 200). The toxification process is faster, and then detoxification process is slower in insects compared to mammals; that is the reason why parathion can be used as an insecticide. However, as dermal uptake is possible, users need to wear personal protective equipment.

The block of AChE means that the neurotransmitter acetylcholine stays active and leads to tonic and clonic convulsions and finally respiratory arrest. Victims show heavy salivation, tearing headache, vomiting, diarrhea, cramps, blurred vision and loss of sphincter control (urination and defecation). First responders need to take care of self-protection. Artificial breathing might be given by breathing apparatus only (saliva may contain toxin). Medical staff may provide intramuscular injections of atropine (antagonist of acetylcholine) and oximes like obidoxime or pralidoxime which regenerate the AChE as long as its complex with the phosphate is not aged. Survival of the intoxication, or on prolonged overexposure, drowsiness, headache, salivation, tremor, weakness may persist.

GHS symbols (Fig. 199) and hazard statements for parathion:

Fig. 199: GHS symbols for parathion.

- H300: Fatal if swallowed.
- H311: Toxic by skin contact.
- H330 Fatal if inhaled.
- H372: Causes damage to the nervous system through prolonged or repeated exposure.
- H410: Very toxic to aquatic life with long-lasting effects.

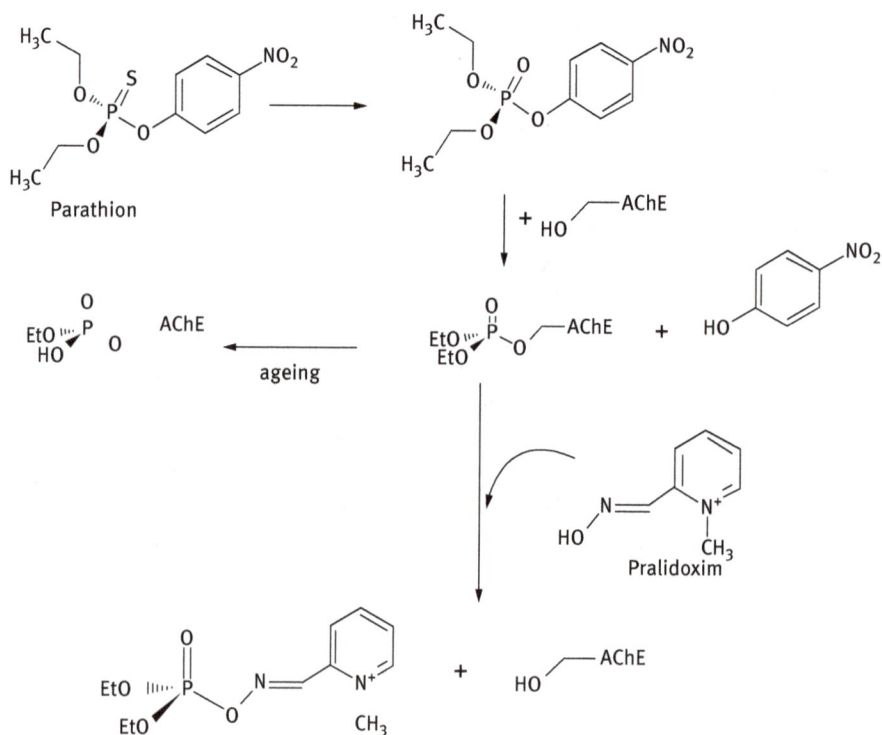

Fig. 200: Inactivation of acetylcholine-esterase (AChE) by parathion and reactivation with pralidoxime.

12 Appendix 1: solutions to the exercises

12.1 Chapter 2

(1) Solids: melting point; liquids: refraction index

(2) Molar mass allows us to calculate concentrations in terms of mass per volume and in terms of mole per volume. For reactions in the environment, the concentration as mole per volume is mostly required to calculate transformations, but data on ecotoxicity, substance thresholds or tolerable limits in environmental media (if existent) are mostly given as mass per volume. The density provides an idea as to where in aqueous media the substance is likely to be found, and the boiling point allows a first rough estimate in how far a vapor phase might be important.

(3) The ideal gas law shall be valid, and we need the molar amounts of substance instead of μg. For test runs 1–5, we have 7.692E-09, 1.539E-08, 1.731E-08, 1.827E-09 and 1.923E-08 mol. The volume collected always was 3 h × 1 L/min = 0.18 m^3. The respective vapor pressure is calculated via the ideal gas law as

$$p = \frac{n \times R \times T}{V} = \frac{n \times 8.314 \times T}{0.18}$$

which results in 0.104, 0.208, 0.234, 0.248 and 0.26 mPas. The air freezed-out seems to be nearly saturated with the compound at a filling height of 10 cm in the test tube. Further experiments are not deemed necessary.

(4) In case of nonspecific measurements care has to be taken that the substance under consideration is as pure as possible. Even in case of 99% purity, 1% trace impurity may dominate the vapor pressure measurement if it is nonspecific.

(5) The molecular mass of $C_{20}H_{40}O_2$ is about 312. 5 g/mol. The content of carbon is 76.8%. A DOC of 2 mg/L means that 2.6 mg of the substance are dissolved in the saturated water. Concerning the COD, the stoichiometric equation is $C_{20}H_{40}O_2 + 29\ O_2 = 20\ CO_2 + 20\ H_2O$. 4.45 mg O_2 is equivalent to 1.392E-04 mole O_2 and according to the reaction equation, the content of the ester in one liter of saturated water is 4.8E-06 mole = 1.5 mg/L. The difference between the DOC method and the COD method needs to discuss the quantification limits of both procedures. For the DOC method, it should be clarified whether inorganic carbon (CO_2, carbonates) was present and addressed in the analysis; inorganic carbon can cause overreporting of DOC. Repetition of both tests should deliver means and standard deviations, and perhaps the difference between both methods is no longer significant. If the measured values are compared to the calculated result for water solubility, the difference is more than factor 10. It needs to be considered that many IT tools for the calculation of physical constants may deliver

https://doi.org/10.1515/9783110618952-012

rather crude estimates. However, looking at the chemical identity of the compound (ester) and its potential synthesis, it cannot be excluded that there is residual ethanol in the product. Ethanol is miscible with water at any ratio at 293 K, and traces in the product may interfere strongly with the DOC and COD analysis. A way forward could be a specific analysis for the target molecule in water, for example HPLC or GC.

(6) A surface-active substance lowers the surface tension of water (at maximum 1% solution below 60 mNm). Surface activity can be measures either as force necessary to pull a ring or a plate in contact with the liquid surface away or as reduction of capillary force.

(7) The Henry constant is the equilibrium distribution of a substance between water and air. It can be expressed as pressure divided by concentration in water, Pa $\times$ m³/mol; if this value is divided by $R \times T$, the dimension-free Henry constant results.

(8) Water solubility 1,800 g/L = 14.63 mol/m³. $H' = 0.01$ Pa/14.63 mol/m³ = 6.8E-04 Pa $\times$ m³ $\times$ mole^{-1}. $H = 2.76$E-07.

(9) The K_{OC} is a distribution constant, normalized for the organic carbon content of the soil. The measured distribution constant, $K = C_{soil}/C_{water}$ is divided by the percentage of organic carbon in the soil, multiplied with 100. The K_{OC} is independent of soil type. However, it neglects the importance of clay minerals in soil; these play a role for the absorption of positively charged molecules.

(10) Azo-m nitro-, nitroso-, azido-, chloro amino- and so on.

(11) Answer:

$$OB = -1,600 \times \frac{\left\{ 2x + \frac{y}{2} - z \right\}}{MW}, \text{for } C_x H_y O_z$$

(a) Dinitrophenol, $C_6H_4O_5N_2$, MW = 184 g/mol; OB = -78; as this is larger than – 200, and suspicious functional groups are present (nitro groups), this substance could be an explosive.

(b) 2-Azo acetic ester ethyl ester, $C_4H_7O_2N_2$, M = 115 g/mol; OB = -132; as this result is larger than – 200 and the molecule contains a suspicious functional group (azo-group), 2-azo acetic acid ethyl ester could be an explosive.

(c) $C_2H_2O_4$, M = 74 g/mol; OB = -21; the OB would indicate explosive properties; however, the molecule does not contain suspicious functional groups.

(12) The flammability describes that grow easy substances can catch fire. For gases, the concentration range in air and the lower concentration limit in mixtures decide the category. For liquids, the flash point (temperature, at which the vapor in equilibrium with the surface spreads a flame) in combination with the boiling point, and for solids the propagation speed of a flame and effect of a barrier discriminates the different categories.

(13) Any ignition source can ignite the liquid and its vapors. Therefore, care has to be taken to avoid open flames, sparks and static electric charge. Vapor–air mixtures may be explosive in a certain concentration range. The vapors are heavier than air and may creep along the floor gutters, entering an area where nobody is aware of the danger.

(14) Oxidizers may react heavily with flammable materials. On contact with flammable material, inflammation may occur or even explosion.

(15) Hydroperoxides can be formed in the presence of oxygen and light by those organic solvents that can build comparatively stable radicals. This is the case for tetrahydrofuran and decalin.

(16) In presence of hydroperoxides, iodide ions form iodine which dyes dissolved starch deep blue. Aqueous Fe(II) is oxidized to Fe(III) which then forms a deep red complex with phenanthroline. Shaking with Fe(II) solutions or sodium bisulfite solutions deactivates the hydroperoxide.

(17) 14 mL of a 0.1 N thiosulfate solution consumed means that 1.4 mmol thiosulfate were consumed. Therefore, 0.7 mmol I_2 were reduced, which means that 0.7 mmol H_2O_2 equivalents were present, that is 23.8 mg H_2O_2 per 10 g solvent or 2.38 mg per 1 g solvent:

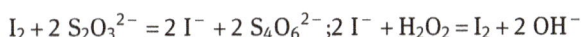

$$I_2 + 2 S_2O_3^{2-} = 2 I^- + 2 S_4O_6^{2-} ; 2 I^- + H_2O_2 = I_2 + 2 OH^-$$

(18) In case of spillages, care must be taken so that metallic containment can resist the contact to the spilled substances. Metal containment has the advantage of thermal stability, easy detection of temperature increase and better efficiency of external cooling, for example with water spray in situations where an unexpected temperature increase occurs. However, metal corrosive substances can result in material failure and formation of hydrogen. Increase in pressure can result in violent rupture of the containment and the hydrogen presents a fire risk.

(19) Using the formulas for the volume and surface of spheres, the original ball starts with a surface of 4.8E-04 m^2. With a radius of 1,000 μm, you have 239 particles to match the weight of 1 g, and their total surface is 3.0E-03 m^2. 100 μm radius makes 2.39E + 05 particles with a total surface of 3.0E-02, and those of 10 μm radius in total 2.38E + 09 particles with a total surface 3 m^2. If the speed of combustion is set = 1 for the original particle, the increase in combustion speed for the 1 g of wood with decreasing radius is listed in Tab. 81:

Tab. 81: Results for exercise 19.

Radius [m]	Surface [m²] (all particles)	Relative speed of combustion
6.2E-03	4.8E-04	1
1.0E-03	3.0E-03	6.25
1.0E-04	3.0E-02	62.5
1.0E-05	3.0E-01	625

12.2 Chapter 3

Exercises

(1) In general, molecules can enter the cell, the easier the smaller, more lipophilic and neutral they are.

(2) The hormones bind to specific receptors on the cell surface. The receptor transmits a response through the cell membrane.

(3) On a molecular level, molecules can disturb the membrane, block or overactivate receptors, block enzymes and, in doing so, inhibit the synthesis of important molecules and interfere with the genetic code.

(4) The toxic substance may itself be a substrate to the enzyme (competitive inhibition) or may undergo allosteric interaction with the enzyme (noncompetitive inhibition) or both (mixed inhibition).

(5) Local toxicity describes the adverse reaction of a chemical at the site of contact. Systemic toxicity describes adverse reaction on a target tissue or a target organ that is reached via distribution of the substance in the organism.

(6) Inhalative uptake of substances increases with respiration rate (volume per time), concentration of the substance in air and with the retention factor.

(7) Substance factors that favor dermal uptake are low vapor pressure, low molecular weight (and molecular size), limited water solubility, medium lipophilicity; dermal uptake increases with contact surface, contact time, circulation in the effected skin (which increases with temperature), number of hair follicles and decreases with the thickness of the stratum corneum.

(8) Organic solvents can accelerate dermal uptake of substances.

(9) HCl will predominantly irritate the eyes and the upper respiratory tract due to its very high water solubility. Cl_2 has limited water solubility; it reaches the lower airways and alveoli. It induces an inflammatory response, and after a lag time free of symptoms a lung edema with shortness of breath can occur.

(10) Propanol - > propanal - > propanoic acid. Ethene - > oxirane - > (ethane-1,2-diol, 2-(glutathionyl)-ethanol). *ortho*-Xylene - > (*o*-methyl-benzyl alcohol); *o*-xylene - > 2,3-dimethylphenol - > 2,3-dimethylphenylsulfate.

(11) 1-Butanol - > butanal or dihydroxy-butane. Styrene - > phenyl-oxirane or ethenyl-phenol.

(12) 1-Naphthol - > napthohydroquinone (dihydroxy naphthalin); 1-naphthol - > 1-naphthylsulfate. 1-Naphthol - > 1,1′-naphthyl-glucuronide.

(13) 1,2-Dihydroxybutane, glutathionyl-hydroxybutane.

(14) Answer:

$$\text{concentration in urine} \left[\frac{mg}{L}\right] \times \text{urine flow} \left[\frac{L}{h}\right] = \text{concentration in plasma} \left[\frac{mg}{L}\right] \times CL$$

$$CL = \frac{30\mu g/L \times 2L}{24\,h \times 1,500\,\mu g/L} = 1.67\,\frac{mL}{h}$$

(15) $D/V = C$; $D = 30$ mg/L, $C = 0.2$ mg/L -> $V = 150$ L.

(16) Allometric scaling addresses the fact that smaller animals have a higher turn-over for substances than larger ones, normalized to body weight. For example, per body weight, they have a higher clearance, higher breathing rate and higher metabolism rate. Plasma and tissue concentrations after a fixed dose are anti-proportional to these parameters. Therefore, a lower dose is resulting in the same tissue concentrations in larger animals. Interspecies-extrapolation: extrapolation from one to another species, which covers toxicokinetics and toxicodynamis. Intraspecies extrapolation addresses the fact that data gained with inbred laboratory animals with small genetic diversity are to be extrapolated to a human population with broad genetic diversity.

(17) Answer:

$$\text{Dose}_{man} = \text{Dose}_{rat} \times (\text{Body weight, rat})^{0.25}/(\text{Body weight, man})^{0.25}$$

$$= 10 \text{ mg/kg} \times \{\,0.5 \text{ kg}/\,70 \text{ kg}\}^{0.25} = 2.9 \text{ mg/kg}.$$

(18) The uptake and elimination rate constants shall be k_1 and k_2, respectively. For A, $k_{1A} = 1.25$ m³/h $\times$ 10 mg/m³ = 12.5 mg/h = I_A, and $k_{2A} = 0.5$/h; $V_A = 6$ L. For B, $k_{1B} = 2.5$ m³/h $\times$ 10 mg/m³ = 25 mg/h = I_B, and $k_{2A} = 0.2$/h; $V_B = 5$ L:

$$C = \frac{I}{V \times k_2} *(1 - \exp[-k_2 t])$$

$C_A = 3.6$ mg/L; $C_B = 13.8$ mg/L.

(19) Organs with a high demand for oxygen are nerve cells, and the brain will react rapidly to a lack of oxygen. Permanent working muscles have a high oxygen demand and they react rapidly in case of oxygen shortage, so the heart is expected to be affected.

(20) The molecule is expected to be lipophilic. Dissolved in water, it is probably resorbed faster in the stomach than it is after dissolution in corn oil.

(21) Dermal resorption is slower than oral (peak later, but still higher), after oral dose there is a first-pass effect; the intestines provide a much greater uptake area than the external skin, but substances taken up are first delivered to the liver. Obviously, significant metabolism takes place in the liver, so a part of the parent compound is transformed. Dermal exposure results in a higher systemic availability of the parent compound than oral exposure, and dermal exposure is more critical for triggering brain effects.

(22) It is not C_{max}, but the area under the curve, that is responsible for the power of the substance effect. B must have a much higher elimination half-life than A.

(23) Clotting factors are synthesized in the liver.

(24) Oligo- or anuria, if the kidney can no longer produce urine; polyuria when the kidney cannot resorb the water. Blood, proteins and/or carbohydrates in the urine in case of kidney damage and/or failure to resorb proteins and saccharides.

(25) Substances can interfere with the myelin sheets so nerve conductance is slowed down; substances can interfere with the ion channels so the ion gradient along the nerve is disturbed (hyper- or hyporeactivity); the axon can be destroyed (axonopathy); the release of neurotransmitters can be blocked, or their activity can be increased due to block of neurotransmitter decay in the synaptic gap, or receptors for neurotransmitters are blocked or overactivated by the toxin.

(26) Look up the text.

(27) Genotoxicity is the overarching expression and means any effect on the DNA. Mutagenicity means the induction of a mutation: a gene is changed, or the reading of a gene is changed, resulting in changed phenotype of the organism; this can also mean a transformation of a cell so it changes its genetic programs.

(28) Look up the text

(29) Predominantly, chromosome breaks and unbalanced segregation of chromosomes. Point mutations may result in micronuclei, but absence of micronuclei is not a proof for the absence of point mutations.

(30) The substance can damage the fetus, which is developmental toxicity, but it also can damage the male and/or female fertility. Damage to fertility or reproduction which is clearly secondary to parental effects (i.e., severe loss of weight of the parent animals, very poor general state) would not result in rating the substance as toxic to reproduction. Effects on development or fertility caused at dosages without other parental effects will most likely result in rating the chemical as toxic to reproduction.

12.3 Chapter 4

(1) The organic carbon of the substance is not only oxidized to carbon dioxide, it is also incorporated in organic molecules of the bacteria. Therefore, 100% CO_2 – production is unlikely and even not achieved if sugar is presented to the bacteria.

(2) Ad a) A DOC decay may occur by adsorption and by evaporation of the parent compound as well as of metabolites, both processes not being biodegradation. For both tests, the analytics should be cross-checked for errors. Ad b) If the inoculants originate from different sources, this may explain the different findings, especially if for test A the activated sludge had been adapted to the substance. DOC decay results may also originate from a test for inherent biodegradability, which has rather favorable conditions for the biodegradation than tests for ready biodegradability.

(3) 65% CO_2 production could result solely from the degradation of the polyethylene glycol chain and would leave the pentachlorophenol moiety undamaged. The decay of the phenol ring should be checked before a conclusion can be drawn that emission to sewage treatment plants (STPs) results in breakdown to harmless products.

(4) For 1-chloro-2-propanol, $M = 94.5$ g/mol. 1 g = 0.0106 mol. Formal oxidation states: $Cl^{(-1)}$-$C^{(-1)}H_2$-$C^{(0)}H(OH)$-$C^{(-III)}H_3$. In CO_2 every C-atom is $(+ IV)$, that is, one molecule delivers 16 e$^-$ which can be taken up by 4 O_2.

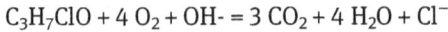

$$C_3H_7ClO + 4\ O_2 + OH^- = 3\ CO_2 + 4\ H_2O + Cl^-$$

1 g substances = 0.0106 mol chloropropanol; this consumes the fourfold amount of di-oxygen, which is 0.0423 mol $O_2 = 1.35$ g O_2.

(5) Three-hour value: pure absorption, no biodegradation. 3 day is lag time, as at this point 10% biodegradation was achieved (25–15%). The 10-day window reaches out to day 13. At day 13, (87–15%) = 72% DOC–decay. This is more than 70%, and the substance can be rated as readily biodegradable.

(6) k (January) = 628 day^{-1} = > $t_{0,5}$ = 1.1×10^{-3} day; k (June) = 3,870 day^{-1} = > $t_{0.5}$ = 1.79×10^{-4} day.

(7) Half-life = $t_{0.5}$ = ln 2/k. $k = k_{OH} \times$ [OH*]. Therefore, $t_{0.5,\ June}$ = 12.7 days, $t_{0.5,\ January}$ = 382 days.

(8) First, the half-lives have to be transformed to first-order rate constants: $t_{0.5}$ = ln 2/k. 3.285 h, 1.59 h and 0.03 h are equivalent to (pseudo-) first-order rate constants of 0.211 h^{-1}, 0.436 h^{-1} and 23.325 h^{-1}. With

$$k = k_a \times [H+] + k_n \times [H_2O] + k_b \times [OH^-],$$

the following expressions can be set up:

$$0.211 = 10{-}4\ k_a + 55.56 \times k_n + 10{-}10\ k_b$$
$$0.436 = 10{-}7\ k_a + 55.56 \times k_n + 10{-}7\ k_b$$
$$23.325 = 10{-}9\ k_a + 55.56 k_n + 10{-}5\ k_b$$

The solution of these three equations delivers k_a = 59.7 L/(mol × min), k_n = 3.69E-03 L/ (mol × min) and k_b = 2.31E + 06 L/(mol × min).

(9) The higher the LOG K_{OW}, the higher the concentration of a substance in fat tissues and the lower the concentration in surrounding water. Such a substance may accumulate in living organisms simply due to the physical distribution. However, only if the substance is resistant against metabolism, the LOG K_{OW} is a suitable substitute for experimental BCF.

(10) Due to the use of the substance, ecosystems should not be endangered. No species shall disappear because of substance use. However, it is impossible to test all potentially exposed species. Therefore, ecotoxicity tests are run with selected species, and assessment factors are allocated hoping that the predicted no effect concentration is safe for all species.

(11) In fish testing, it is easily recognized whether an exposed fish is dead. In daphnia testing, affected daphnia do not move; it is difficult to say whether they are really dead. Therefore, the concentration that paralyzes 50% of the exposed daphnia is taken as endpoint. In algae testing, the inhibition of growth is measured and, therefore, delivers an average inhibiting concentration.

(12) Freshwater and seawater, sediments in freshwater and seawater and soil.

(13) Activated sludge in STP consists to a large extent of bacteria which can digest organic pollutants. A substance which is toxic to the bacteria results in less oxygen consumption and, therefore, less biodegradation. This bears the danger that pollutants enter the receiving water bodies of the STP.

(14) Look up the text.

(15) To be able to derive a PNEC for freshwater organisms, the minimum requirement is the availability of acute toxicity data for fish, daphnia and algae.

(16) The fish, as it is more sensitive in acute testing than daphnia. For algae, we already have a chronic data point, the IC_{10}.

12.4 Chapter 5

(1) Increasing dosages should increase the likelihood of damage and/or the extent of damage.

(2) A dichotomous response is a yes/no answer: cancer, death, ... A quantal answer is allocated to categories, for example, "no effect," "weak effect," "medium effect," "strong effect." A continuous response has a continuous measuring scale, for example, "% body weight reduction," "liver weight," "enzyme activity."

(3) Answer (Tab. 82):

Tab. 82: Logit analysis of the acute toxicity data.

Dose (mg/L)	ln dose	N_j	R_j (dead)	% Mortality	w_i	Logit	Probit
3,2	1.1632	10	0	0	0.25	−3.7136	2,674
7	1.9459	10	4	40	0.25	−0.3857	4,747
9	2.1972	10	8	80	0.25	1.2993	5,842
10	2.3026	10	10	100	0.25	3.7136	7,326
Logit = 0	LN(LC50) =	1.863	LC50 =	6.44 mg/L			
Probit = 5	LN(LC50) =	1.861	LC50 =	6.43 mg/L			

(4) If the substance is dosed above the water solubility limit, a linear uptake with concentration via the gills is no longer necessarily the case. At 10 mg/L, droplets of the liquid may have been swallowed by the fish, creating a high peak exposure; this could explain the up-levelled Logit and Probit qualitatively.

(5) $EC_{50} = 0.90$ ppm [0.61;1.33]. Note that in this case, there is no need to incorporate the weight of the dose groups. Details:

Dose (ppm)	Number of daphnia exposed	Number of immobile daphnia	LN(dose)	Logit	Error (Y-(A+$B \times X$))	(Error)^2
0.40	20	2	-0.9163	-2.1972	0.16231629	0.02634658
0.80	20	7	-0.2231	-0.6190	-0.27378405	0.07495771
1.20	20	14	0.1823	0.8473	0.01427142	0.00020367
2.40	20	19	0.8755	2.9444	0.09712683	0.00943362
		Sum:				0.1109
		Average:	-0.0204	0.2439		
		SD:	0.6495	1.8948		

By linear regression: $y = A + B \times X$

A:	0.3032	SD(E):	0.23552237
B:	2.906	SD(P):	0.26390314
ED(50):	0.90	$t(n-2;0.975)$:	4.303
Logit(LC50, lower 95% CI)	-1.1356		
Logit(LC50, upper 95% CI)	1.1356		
ED50, lower 95% CI	0.6104		
ED50, upper 95% CI	1.3305		

(6) $RR = 5$; in the exposed population, it is five times more likely to get cancer than in the control population. $AR = 0.8$, 80% of cancer cases in the exposed population are attributable to the exposure. $ER = 8\%$; exposure increases the cancer risk by 8%.

(7) Excess risk: $32-2\% = 30\%$. 10 [mg/kg]/30% = $T25/25\%$ ⟺ $T25 = 8.3$ mg/kg. To extrapolate to human beings, an allometric scaling factor of 4 has to be applied, and a further factor of 250,000 to extrapolate a risk of 0.25 to 1.00E-06. That is, MEL = 8.3 mg/kg/4/250,000 = 8 ng/kg.

(8) Answer:

$$R + 2L = RL_2; \quad K_D = \frac{[R] \times [L]^2}{[RL_2]}; \quad P = \frac{[RL_2]}{[R] + [RL_2]} \rightarrow P = \frac{[L]^2}{K_D + [L]^2}$$

$$1 - P = \frac{K_D}{K_D + [L]^2} \rightarrow \frac{P}{1 - P} = \frac{[L]^2}{K_D} \rightarrow \text{Logit} = \text{LN}\left\{\frac{P}{1 - P}\right\} = 2 \times \text{LN}[L] - \text{LN}[K_D]$$

12.5 Chapter 6

(1) Answer:

$100/\text{ATE}_{mix} = 6/1.5 + 11/0.6 + 40/1.1 = 49$

$\quad\text{ATE}_{mix} = 1.7$ mg/L/4 h

This value has to be matched against cut-offs for dusts and mists: $=>$ Category 4 (1 mg/L/4 h $<$ LC50 $<$ 5 mg/L/4 h)

(2) Answer:

$100/\text{ATE}_{mix} = \Sigma(C_i/\text{ATE}_i)$

Oral toxicity: $\text{ATE}_{mix} = 100$: {10/500 + 50/500} = 833 mg/kg; - > H302, GHS07.

Dermal toxicity: $\text{ATE}_{mix} = 100$: {10/100 + 50/500} = 500 mg/kg. - > H312, GHS07.

(3) Answer:

$(100-40)/\text{ATE}_{mix} = \Sigma (C_i/\text{ATE}_i)$

Oral: $\text{ATE}_{mix} = 60/$ {10/500 + 50/500} = 500 mg/kg; 300 $<$ ATE_{mix} ≤ 2,000: H302:

Dermal: $\text{ATE}_{mix} = 60/$ {10/300 + 50/1,100} = 761 mg/kg. 200 $<$ ATE_{mix} ≤ 1,000: H311, GHS06:

Addition on the label: 40% of the mixture consists of components with unknown toxicity.

(4) Oral: $\text{ATE}_{mix} = 100/$ {12/500} = 4,166 $>$ 2,000, no classification.

Dermal $\text{ATE}_{mix} = 100/$ {12/1,100 + 50/1,980} = 2,765, $>$ 2,000, no classification.

(5) Answer: 1% Cat. 1 $<$ 3% - $>$ not cat. 1. But 10 $\times$ 1% + 2% = 12% ≥ 10% - $>$ cat. 2; H319: causes severe eye irritation, GHS07.

(6) Skin corrosion cat. 1A? 40%/70%, 1, - $>$ not corrosive cat. 1A.

Skin corrosion cat. 1B? 40%/50% + 2%/3% = 1.25 $>$ 1 - $>$ skin corrosive, cat. 1B, H314, GHS05.

Classification as corrosive covers the effect of severe damage to the eyes.

(7) As A is below its lowest SCL of 5%, and because it is not acting additively, it does not contribute neither to corrosion nor irritation. Calculations now concentrate on component B.

Skin corrosion: 2% cat. 1B, $<$ 3% - $>$ not corrosive;

Skin irritation: 10 $\times$ 2% = 20% $>$ 10% - $>$ irritating to the skin, H315, GHS07.

Severe eye damage: 2% $<$ 3% - $>$ no classification as causing severe eye damage.

Severe eye irritation: 2% $>$ 1% - $>$ causes severe eye irritation, H319, GHS07.

(8) Note that component B is bioaccumulative and C is not rapidly degradable; A is not bioaccumulative and rapidly degradable.

$$\frac{100}{\text{NOEC}_{mix}} = \frac{[A]}{\text{NOAEC}_A} + \frac{[B]}{0.1\times\text{NOAEC}_B} + \frac{[C]}{0.1\times\text{NOAEC}_C}$$

For fish: $\dfrac{100}{\text{NOAEC}_{mix}} = \dfrac{20}{0.02} + \dfrac{40}{0.1\times 0.8} + \dfrac{40}{0.1\times 0.25} \rightarrow \text{NOAEC}_{mix, fish} = 0.0328\,\text{mg/L}$

For daphnia: $\dfrac{100}{NOAEC_{mix}} = \dfrac{20}{2} + \dfrac{40}{0.1 \times 0.6} + \dfrac{40}{0.1 \times 1.2} \rightarrow NOAEC_{mix, daphnia} = 0.099 \, mg/L$

For algae: $\dfrac{100}{NOAEC_{mix}} = \dfrac{20}{1} + \dfrac{40}{0.1 \times 4} + \dfrac{40}{0.1 \times 0.1} \rightarrow NOAEC_{mix, algae} = 0.024 \, mg/L$

The lowest result is $NOAEC_{mix, algae} = 0.024$ mg/L; looking up the cut-off concentrations for chronic data for rapidly degradable, nonbioaccumulative substances (Table 65), the mixture is H411: Toxic to aquatic organisms with long lasting effects. Symbol: GHS09.

(9) The content of cobalt sulfate is below the general concentration limit of 0.1% which would be applicable according to its classification. However, the specific concentration limit mentioned in the annex is 0.01%, although it is mentioned in combination with cancer. To be on the safe site, we assume that this limit is applicable for all endpoints. For aquatic toxicity, this is the case anyway as there is a factor $M = 10$ to be used.

Acute oral toxicity: $\dfrac{100}{ATE_{mix}} = \dfrac{10}{100} + \dfrac{0.01}{500} \leftrightarrow ATE_{mix} = 1,000$ category 4, H302

Acute dermal toxicity: $\dfrac{100}{ATE_{mix}} = \dfrac{10}{300} \leftrightarrow ATE_{mix} = 3,000$ no category

Acute respiratory toxicity: $\dfrac{100}{ATE_{mix}} = \dfrac{10}{3} \leftrightarrow ATE_{mix} = 23.7 > 20$ no category

Skin sensitization: 0.05% < 0.1%, no skin sensitizer, but substance to be named on the label and in the safety data sheet as 10% of the C&L limit are exceeded.

Respiratory sensitization: 0.05% < 0.1%, no respiratory sensitizer, but substance to be named on the label and in the safety data sheet as 10% of the C&L limit are exceeded.

Mutagen category 2: 0.05% < 1.0%, not mutagenic.

Carcinogen category 1B: 0.05% > 0.01%; carcinogen cat. 1B, H350 (overwrites H351)

Toxic to reproduction 1B: 0.05% < 0.3%; not to be classified as toxic to reproduction.

Acute aquatic toxicity category 1: 0.05% × 10 (M-factor) = 0.5% < 25%; not acute toxic category 1 to aquatic organisms.

Chronic aquatic toxicity category 1: 0.05% × 10 (M-factor) = 0.5% < 25%: no.

Chronic aquatic toxicity category 2: 0.05% × 10 × 10 + 10% = 15% < 25%: no.

Chronic aquatic toxicity category 3: 0.05% × 10 × 10 × 10 + 10 × 10% + 10% ≥ 25%; H312: Harmful to aquatic organisms with long lasting effects.

(10) Including the H-phrases, the mixture is composited as

5% chloroacetic acid {H301, H311, H314 (cat. 1B), H335 (SCL = 5%), H400}, 5% 1-chloro-4-nitrobenzene {H301, H331, H341, H351, H373, H411}, 81% butanone

{H225, H319, H336}, 2% 4-tert-butylphenol {H315, H318, H361f}, 2% 2-methyl-5-tert-butylthiophenol {H226, H304, H315, H317, H319, H336, H361d, H373, H400, H411} and 5% water. Go through the different endpoints step by step: Flammability: flash point to be measured (or make use of the calculation method, if possible).

$$\text{Acute oral: } \frac{100}{\text{ATE}_{\text{mix}}} = \frac{5}{100} + \frac{5}{100} = \frac{1}{10}; \rightarrow \text{LD}_{50} = 1,000 \rightarrow \text{Cat. 4, H302Acute}$$

$$\text{dermal: } \frac{100}{\text{ATE}_{\text{mix}}} = \frac{5}{300} = ; \rightarrow \text{LD}_{50} = 3,000 \rightarrow \text{no classification}$$

$$\text{Acute inhalation, vapor: } \frac{100}{\text{ATE}_{\text{mix}}} = \frac{5}{3}; \rightarrow \text{LC}_{50} = 30 \rightarrow \text{no classification}$$

H304: 2% components with H304, < 10%, → no classification.

Corrosion/irritation: 5% 1B ≥ 5% → 1B, H314; this makes H315, 318 and 319 obsolete.

CMR: C and M category 2: 2% ≥ 1%, → H341, H351

Skin Sensitization: category 1: 2% ≥ 1%, → H317

STOT SE 3, H336: 81% ≥ 20%, → H336 (also EUH 066).

STOT SE, H335: 5% ≥ 5% (SCL!), → H335

H361f: 2% < 3%, → no classification

H361d: 2% < 3%, → no classification

STOT RE 2: H373 = 2% < 10%, → no classification

Acute aquatic toxicity, cat 1: 5% + 2% = 7% < 25%; → no classification

Chronic aquatic toxicity, cat 1: 2% < 25%; → no classification

Chronic aquatic toxicity, cat. 2: 10 × 2% (H410) + 5% (H411) ≥ 25%, → H411

Hazard phrases to be put on the label: H302, H314 (cat. 1B), H317, H335, H336, H341, H351, H411; (H225/H226: to be tested)

GHS labels: (02 (depends on test results); 05, 07, 08, 09)

Substances to be named: all but 4-tert-burtyl-phenol, as they contribute significantly to the C&L

(11) Answer:

H225: measure flash point (or make use of the calculation method, if possible).

$$\text{Acute oral toxicity: } \frac{100}{\text{LD50}_{\text{mix}}} = \frac{22}{100} + \frac{40}{500} + \frac{9}{500} + \frac{1}{5} = \frac{259}{500}; \rightarrow \text{LD50} = 193 \text{ Cat 3,}$$
H301

$$\text{Acute dermal toxicity: } \frac{100}{\text{LD50}_{\text{mix}}} = \frac{22}{300} + \frac{1}{5} = \frac{82}{300}; \rightarrow \text{LD50} = 366, \text{ Cat 3, H311}$$

$$\text{Acute inhalation toxicity (mist): } \frac{100}{\text{LC50}_{\text{mix}}} = \frac{22}{0.05} + \frac{1}{0.05} = \frac{23}{0.05}; \rightarrow \text{LC50} = 0.217$$
Cat 2, H330

Skin corrosion, H314 (1B): 8% < 25%, → no classification

Skin irritation, H315: 8%/10% < 1, → no classification.

Eye damage, H318: 2% cat. 1B with limit of 25 % for cat.1; cat.1 skin substance counts 10 – fold:

(8% / 25%) × 10 = 3.2 % > 3 %; → H318 causes serious eye damage. This makes H319 obsolete.

Skin sensitization: 22% ≥ 1%, → H317

H335? 20%/25% < 1, → no classification

H351? 1% ≥ 1%, → H351

Acute aquatic toxicity, cat. 1: 22% < 25%, → no classification

Label:

On the label, the following hazard statements have to be mentioned: H301, H311, H330, H317, H318, H351; precautionary statements have to be selected accordingly. Symbols on the label: GHS 06 and 08; and 07 (sensitizer! Therefore, GHS 07 is replaced neither by GHS08 nor by GHS06!). Whether or not GHS02 and H225 ff has to be added depends on test results. Substance to be named with identifiers (EC-No.): THF, H_3PO_4, dichlorphos, bis(2-chloroethylether). Name and address of the supplier needs to be mentioned and the content of product.

(12) With that LD50, the substance is acute toxic category 3, H301: Toxic if swallowed, GHS06. As the cramps were the cause for muscle paralysis resulting in death, classification as STOT SE is not indicated.

(13) This dermal LD50 is classified as acute toxic category 4, H312: Harmful by skin contact, GHS08. The degeneration of the retina is a severe organ failure, and at that dose, classification as STOT SE category 2 is appropriate, H371: May cause damage to the eyes by skin contact, GHS08.

(14) With a subchronic dermal LOAEC of 15 mg/kg which is below 20 mg/kg, the substance is classified as STOT RE category 1. SCL = LOAEC/$GV_{1 \text{ or } 2}$; SCL_1 = 15/20 = 0.75 = 75%; this has to be rounded down to 50%. SCL_2 = 15/200 = 7.5%, which has to be rounded down to 5%. However, both values are higher than the GCL, so the GCLs have to be used in classification and labeling of mixtures with this substance.

(15) $Y = A + B \times X$, Y = effect, X = dose.

$$B = \frac{20 - 4}{80 - 40} = 0.4; \rightarrow A = 20 - 0.4 \times 80 = -12$$

$$Y = 10 = -12 + 0.4 \times X \rightarrow X = 55$$

The ED_{10} is 55 mg/kg. The substance is a reproduction toxin category 1 of medium potency and the SCL is 0.3%.

(16) First a mixture of A, B and C is calculated:

$$\frac{\sum^{c_i} + \sum^{c_j}}{NOEC_{mix}} \sum \frac{C_i}{NOAEC_i} + \sum \frac{C_j}{0.1 \times NOAEC_j} \leftrightarrow \frac{80}{NOAEC_{mix}} = \frac{10}{0.2} + \frac{30}{0.1 \times 0.9} + \frac{50}{0.1 \times 8} \leftrightarrow$$

$NOAEC_{mix}$ = 0.179 mg/L, that is, chronic category 3.

Complete mixture: chronic aquatic toxicity category 2: 20% components are of chronic aquatic toxicity category 2, < 25%: not category 2. Therefore, the mixture is to be classified and labeled as chronic aquatic toxicity category 3.

12.6 Chapter 7

(1) If a product is a dangerous good and a dangerous substance, the outer (transport) packaging must carry labels according to dangerous goods regulations.
(2) If the outer packaging is the transport package, and if the inner packaging carries the labels according to the CLP regulation, labeling as hazardous good is sufficient for the outer packaging. If the transport packaging is also the "use" packaging, labels have to be added according to dangerous goods regulations and according to the CLP regulation.
(3) The local fire brigade handling the incidence, ambulances and hospitals, but also individual companies can call the ICE centers and ask for support. Support can be granted on three levels: level 1 is instructions by experts for the substance via phone. Level 2 is delegating experts to the scene of incident. On level 3, specific equipment and trained experts handling the equipment will be sent to the incident.
(4) The class 6 for dangerous goods does also cover contagious material, which is not addressed by the hazardous substance regulations, but by other specific regulations concerning handling and use. For dangerous goods, the focus is on accidental release and single exposure, whereas the hazardous substance regulations also cover chronic exposure and repeated dose toxicity.
(5) Flammable or self-reactive solid; flammable gas; very toxic flammable liquid; toxic and corrosive gas; flammable or self-heating solid; toxic oxidizer; corrosive oxidizer; highly oxidizing gas; toxic, very flammable liquid; very flammable solid, reacts with water.

12.7 Chapter 8

(1) Yes, every company bringing a substance on the market needs to register as soon as 1 t/a is arrived/exceeded.
(2) This exercise addresses the question of polymers. The polymer definition is not matched by product B (most abundant molecule < 50%) and C (at least 3 monomers + one starter molecule in one chain; three ethylene oxides may be evenly allocated to the three OH groups of the glycerol), whereas A fulfills the polymer definition. As the monomers are registered, A does not need to be registered.

(3) If the content is 0.1% or more, I need to inform the agency. The agency may subsequently request a registration.

(4) Ad a): The catalyst is an internal isolated, nontransported intermediate and to be registered as such. Ad b): the agency needs to be informed about the program, and information about the program have to be submitted, that is timing, tasks and workforces involved, location and all available data on the catalyst. Material exclusively exported outside the EU is not subject to registration, and the 500 kg for the EU market are below the cut-off of 1 t/year. Ad c): A registration is required, but due to use as intermediate, requirements are limited, provided strictly controlled conditions can be guaranteed. If 1,000 t/year are arrived at or exceeded, a data set according to annex VII has to be generated. Without strictly controlled conditions, full registration is required. Ad d) Annex VII data have to be generated and to be included in the registration dossier. In strictly controlled conditions cannot be guaranteed, a full registration is necessary.

(5) As the solvent in the printer cartridge is intended to be released under use, I need to register that solvent as soon as I have an annual import of cartridges that is equivalent to 1 t of solvent or more. My supplier can run the registration via his EU office, or via a service provider who then is the only representative.

(6) This is the case only if the use of the chemicals is covered by the registration. This is the case if the PROCs and ERCs that are applicable to my processes are covered in the extended safety data sheets of my suppliers.

(7) As far as tests with vertebrate animals are concerned, the regulation forbids to perform unnecessary animal testing. I have to inform the ECHA upfront, need to submit a C&L proposal and ask the agency whether data for acute toxicity and sensitization in mammals are available.

(8) If the mixture contains 0.1% of: a substance of very high concern; a mutagen or carcinogen category 2, or a substance toxic to reproduction; a very persistent and very bioaccumulative substance; or 1% of another hazardous substance; or it contains a substance with a community occupational exposure limit, or a sensitizer is present at least at 10% of its concentration limit, a SDS has to be supplied, at least upon request.

(9) Authorization concerns substances which are listed in annex XIV. Due to their hazardous properties, the final aim is their complete disappearance from the market, if possible. For that reason, their marketing and use needs to be endorsed by the agency. The user needs to demonstrate that there is no technical viable alternative for the substance for a specific use, and he has to demonstrate that risk management methods have been installed to minimize and control exposure of human beings and environment as far as possible. Authorization can be granted for limited time periods, and before the deadline, the user needs to apply for an extension of the authorization.

(10) Intermediates are exempt from authorization. Of course, all substances that do not fall under REACh will not be subject to REACh authorization, that is, radioactive substances, pharmaceuticals, pesticides, as long as they are covered by their specific legislation.

(11) As pharmaceuticals are exempt from REACh, only the 800 to which are used as solvent need to be registered under REACh.

(12) Annex XVII of the REACh regulation lists substances and their specific restrictions for the marketing, as such, in mixtures and in articles. These restrictions can specifically permit or prohibit certain uses, conditions of uses, and specific concentration limits in dependence on the substance use can be set.

(13) Regulations concerning persistent organic pollutants, ozone depleting substances and greenhouse gases go beyond marketing. The regulations define sunset dates for substances which are still permitted and cut-downs can be defined per substance and per application. Beyond that, member states are committed to survey the market, update and report inventories and take care for substances being in the market and take provisions to remove substances from the market. For example, installations that contain ozone depending substances or greenhouse gases are subject to defined maintenance, and destruction has to be organized in such a way that no substances are released but destroyed by authorized methods and authorized companies. For POPs, environmental monitoring programs have to be run.

(14) Biocides are intended to be applied in more or less dispersive uses. The general population is likely to be exposed, and target-organisms in the environment are killed on purpose. Therefore, the use of biocides is to be authorized. In addition, it has to be demonstrated that the biocide is fit for its intended use.

(15) Plant protection products are intended for dispersive use, and the general population and ecosystems are going to be exposed against the active agents and their metabolites. For that reason, any risks need appropriate control and must be outweighed by benefits for the population. The product must be fit for purpose, that is, protecting the plant against target organisms while not harming nontarget organisms beyond an acceptable extent. PPP are subject to an authorization that needs to be renewed in regular time intervals. The performance of the product in the market must be monitored by the notifier, and must register undesired effects and take measures to control them as necessary. Organisms can be notified as plant protection products.

(16) Yes, if due to likely damage to health or environment immediate action is required. The national body must inform the ECHA immediately, and ECHA will follow up the case.

12.8 Chapter 9

(1) Young people, pregnant and breastfeeding workers.

(2) The workplace has to be evaluated against any possible risks to health and risks have to be minimized. Focusing on the hazardous substances, the safety evaluation has to cover questions as whether a substance can be replaced by a less risky substance (substitution), how exposure can be minimized, to check airborne and dermal exposure, observe occupational exposure limits. Workers must be trained in precautionary measures, self-protection, appropriate handling, they must understand the hazards of the substance, appropriate personal protective equipment, hygiene measures, what to do in case of accidents (first aid, firefighting, dealing with spillages) and proper disposal.

(3) It is not only allowed, but it is subject to prosecution to withhold information from safety data sheets from workers.

(4) All concerned workers and the occupational physician have to be informed. Records have to be stored for a certain period of time. The origin for the overexposure has to be identified, remediate actions to be taken and checked for effectiveness before regular work can be taken up again.

(5) Workers need to be informed about risks before they are asked to resume the concerned tasks. Records have to be taken, to be stored and made retrievable for several decades, listing the identity of exposed people, time of exposure, tasks performed and the identity of the substances. Any available exposure data (air monitoring and biomonitoring) have to be added to these records. Workers have to be offered appropriate medical surveillance.

(6) Biomonitoring is a measurement at/in the body to check systemically availability of substances, their metabolites or their effect on the body. For example, specimen like urine or blood may be analyzed for the respective hazardous substance or its breakdown products. Biomonitoring may also address changed body parameters, for example, blood pressure, methemoglobinemia, micronuclei in peripheral lymphocytes. Human biomonitoring guidance values can either define what levels of biomarkers deemed as safe, or whether the level of biomarkers calls for a reduction of exposure.

(7) Biomonitoring gives a value for the internal exposure. At a certain airborne concentration, the uptake depends on the breathing rate of the person and the location of the emission source. If metabolites are monitored, the individual differences in metabolic capacity are covered as well. Depending on the half-life of the biomarker, biomonitoring can provide an idea on exposure which happened some time ago. For certain substances/metabolites, biomonitoring can disclose whether there was overexposure or not, although there was no air monitoring. Biomarkers can be detected after dermal exposure, which is missed out by air monitoring. Biomonitoring integrates exposure over a certain time. It does not disclose at which precise location the exposure had

occurred, and whether there was a short peak exposure followed by a long period of low exposure, or whether the exposure was more uniform.

(8) Meanwhile, computational methods are quite powerful and the risk of false-
negatives is small. Computational methods like ART or ECETOC TRA deliver
data for a huge variety of different situations on short notice, which is physically not feasible with real-life measurements where computational results indicate that an OEL can be approached or even exceeded. Confirmation/cross-
check by certified air monitoring (personal/area monitoring) shall be sought.
This is also recommendable for substances with critical health effects.

(9) The employer must make available any clothing that is specifically required
for performing the work, for example, safety shoes, lab coats, goggles, gloves.
He has to take care for timely exchange of dirty clothing and has to organize
appropriate laundry. Protective equipment must be fit for purpose.

(10) Processes have to be organized in such a way that hazardous substances are
enclosed or segregated from the worker as far as technically feasible. Even
open handling is unavoidable, point ventilation and general ventilation have
to be installed so the OELs are met; personal protective equipment and handling tools are to be used to minimize dermal exposure. Exposure time per
work shift can be limited as the next measure. Personal respiratory protection
is the last option, and the special burden to the worker has to be accounted
for in organizing the working process.

(11) Workers must receive a safety-training specific for their workplace before they
take up work, and regularly thereafter. Precautions against fire, depending on
fire risk, indicated escape routes (which must be hold free), no material storage in aisles and on stairs, fire extinguishers and fire detectors, emergency
plan for external fire brigade. First-aid equipment and employees should be
trained as first-aid responders on a regular basis. General separation of working area and leisure area, installations for personal hygiene. Handholds at
stairs, ceilings at elevated platforms and floor must be nonslippery. Sufficient
and appropriate light to perform work (not too dark, not too dazzling) micro-
climate must be ok for the kind of work to be done (temperature, humidity, air
flow). Workplace safety needs to be evaluated by either internal or external
experts. Shortcomings need to be reported and a plan for timely remediation
has to be taken.

(12) Benzidine, 4-aminobiphenyl, 2-naphthylamine, 4-nitrobiphenyl

(13) The filter must be suitable for the gas under consideration. For particles, a particle filter is required. Filter masks do not allow for unlimited exposure, but
OELs can be exceeded only up to a certain factor. The surrounding air needs
to contain sufficient oxygen. Filters present a respiratory resistance, and working while carrying filter masks is much more fatiguing than working without
filters.

(14) Take off soaked clothing immediately, but first-aid responders need to take on gloves. Wash skin with tap water for several minutes. Do not wash with bases in an attempt to neutralize! Call ambulance and poison center and take further first-aid methods as necessary.

(15) Vomiting can be dangerous after having swallowed organic solvents or tensides, as these may enter the lung inducing a potentially fatal lung inflammation. After having swallowed acids or bases, vomiting brings the esophagus again in contact with corrosive material and large blood vessels may rupture.

(16) There is no universal glove material that protects against all kinds of chemicals. Even with the same material and same thickness, the quality may be different. Suppliers have to be asked for break-through times against typical chemicals. When the glove is disposed, it needs either to be decontaminated or to be disposed properly. Never touch items with a gloved hand that will be touched by other with unprotected hands (door handle, telephone!).

(17) Worst-case assumption: no air exchange, so the solvent can achieve the saturated vapor concentration. With help of the ideal gas law, $p/(R \times T) = n/V$, the saturated gas concentration is 31,000 mg/m^3 which is more than the OEL.

(18) Only if this is necessary to achieve the goals of the education, and if the work is closely supervised by an experienced worker.

(19) The soaked adsorbent has to be shoveled into buckets of resistant material, and further treatment depends on the specific properties of the spilled material (buckets to be closed or not? Further special treatment?). The floor may need further decontamination. Normal work can be taken up only if overexposure is no longer anticipated.

12.9 Chapter 10

(1) Aisles, staircases and escape routes are no storage areas. Access should be restricted. Containment must not be of such form and size that it could be mixed up with containment for beverages. The workplace is not a store. The maximum containment size at the workplace for flammable products is 2.5 L in breakable bottles and 10 L in nonbreakable containments.

(2) Storage areas should have an impermeable floor. Containments must be stored in such a way that labels can always be read easily, and any leakage should be easily detectable.

(3) Substances of acute toxicity categories 1–3; substances of carcinogenic or mutagenic category 1 or toxic to reproduction category 1; corrosive or eye damaging substances category 1.

(4) Fire resistant walls and roofing, fire sprinklers, smoke exhaust system, access routes for the fire brigade, not gutters or drains in the floor, hold back containment.

(5) Do not store together with flammable material; no combustion engines! Floor must be impermeable, no gutters. Storage building shall be only one floor high. Walls and roofing must be fire resistant.

(6) Gas bottles need to be fixed so they cannot tumble over. Valves not in operation have to be covered by caps against mechanical shock. At working areas, the bottles preferably have to be placed in safety cabinets. Air exchange at least 2 - per hour. For gases labeled with H330, fatal if inhaled, a permanent working air monitoring system must be installed.

(7) Walls, ceiling and floor must be of noncombustible material and fire resistant. There has to be a separation from areas where toxic or oxidizing material is stored. No drains or gutters that would allow the ingression of vapors to lower levels. Any ignition source has to be avoided. There must be no social rooms directly next to the store (strictly if the amount of flammable liquid exceeds 10 t).

(8) Answer:

	A	b	c	d	e	f
A	+	+	+	+	+	+
B	+	+	+	+	+	−
C	+	+	+	+	+	−
D	+	+	+	+	+	+
E	+	+	+	+	+	−
F	+	−	−	−	−	+

Appendix 2

Note: This is a kind of an exercise copy, a surrogate-table of Annex VI, Part 3 of Regulation (EC) No. 1272/2008 for training purposes. For the current valid classification and labeling, look up the latest publication of that regulation.

https://doi.org/10.1515/9783110618952-013

International chemical identification	CAS No	Classification		Labeling			Specific concentration limits, M-factors	Notes
		Hazard class and category code(s)	Hazard statement code(s)	Pictogram	Hazard statement code(s)	Suppl. hazard statement code (s)		
1-Chloro-4-nitrobenzene	100-00-5	Carc. 2 Muta. 2 Acute Tox. 3 Acute Tox. 3 STOT RE 2 Aquatic Chronic 2	H351 H341 H331 H301 H373 H411	GHS06 GHS08 GHS09	H351 H341 H331 H301 H373 H411			
1-Dimethylamino-propan-2-ol	108-16-7	Flam. Liq. 3 Acute Tox. 4 Skin Corr. 1B	H226 H302 H314	GHS02 GHS05 GHS07	H226 H302 H314			
1,2-Dichlorobenzene	95-50-1	Acute Tox. 4 Eye Irrit. 2 STOT SE 3 Skin Irrit. 2 Aquatic Acute 1 Aquatic Chronic 1	H302 H319 H335 H315 H400 H410	GHS 07 GHS 09	H302 H319 H335 H315 H410			
2,6-Dimethylheptan-4-one; di-isobutyl ketone	108-83-8	Flam. Liq. 2 STOT SE 3	H226 H335	GHS02 GHS07	H226 H335		STOT SE 3; H335: C ≥ 10%	
2-Hexyloxyethanol;	112-25-4	Acute Tox. 4 Acute Tox. 4 Skin Corr. 1B	H312 H302 H314	GHS 05 GHS 07	H312 H302 H314			

Name	CAS	Classification	H-code	GHS	H-code	Note
2-Methyl-5-*tert*-butylthiophenol	–	Flam. Liq. 3 Repr. 2 STOT RE 2 Asp. Tox. 1 Eye Irrit. 2 Skin Irrit. 2 Skin Sens. 1 STOT SE 3 Aquatic Acute 1 Aquatic Chronic 1	H226 H361d H373 H304 H319 H315 H317 H336 H400 H410	GHS02 GHS08 GHS07 GHS09	H226 H361d H373 H304 H319 H315 H317 H336 H410	
2-Methylaziridin; propylene imine	75-55-8	Flam. Liq. 2 Carc. 1B Acute Tox. 2 Acute Tox. 1 Acute Tox. 2 Eye Dam. 1 Aquatic Chronic 2	H225 H350 H330 H310 H300 H318 H411	GHS02 GHS06 GHS08 GHS05 GHS09	H225 H350 H330 H310 H300 H318 H411	Carc 1B; H350: C ≥ 0,01%
2-Phenylpropene; α-methylstyrene	98-83-9	Flam. Liq. 3 Eye Irrit. 2 STOT SE 3 Aquatic Chronic 2	H226 H319 H335 H411	GHS02 GHS07 GHS09	H226 H319 H335 H411	STOT SE 3; H335: C ≥ 25%
4-Tert-butylphenol	98-54-4	Repr. 2 Skin Irrit. 2 Eye Dam. 1	H361f H315 H318	GHS08 GHS05	H361f H315 H318	

(continued)

(continued)

International chemical identification	CAS No	Classification		Labeling			Specific concentration limits, M-factors	Notes
		Hazard class and category code(s)	Hazard statement code(s)	Pictogram	Hazard statement code(s)	Suppl. hazard statement code (s)		
Acetic acid . . .%	64-19-7	Flam. Liq. 3 Skin Corr. 1A	H226 H314	GHS 02 GHS 05	H314		Skin Corr. 1A; H314: $C \geq 90\%$ Skin Corr. 1B; H314: $25\% \leq C < 90\%$ Skin Irrit. 2; H315: $10\% \leq C < 25\%$ Eye Dam. 1; H318: $C \geq 25\%$ Eye Irrit. 2; H319: $10\% \leq C < 25\%$	B
Ammonia (%)	1336-21-6	Skin Corr. 1B Aquatic Acute 1 STOT SE 3	H314 H400 H335	GHS05 GHS09	H314 H400 H335		STOT SE 3; H335: $C \geq 5\%$	
Ammonia, anhydrous	7664-41-7	Flam. Gas 2 Press. Gas Acute Tox. 3 Skin Corr. 1B Aquatic Acute 1	H221 H331 H314 H400	GHS 04 GHS 06 GHS 05 GHS 09	H221 H331 H314 H400			
Ammonium chloride	10043-52-4	Eye irrit. 2	H319	GHS07	H319			

Name	CAS	Classification	Hazard statements	Pictograms	Hazard statements	Supplemental
Benzene	71-43-2	Flam. Liq. 2 Carc. 1A Muta. 1B STOT RE 1 Asp. Tox. 1 Eye Irrit. 2 Skin Irrit. 2	H225 H350 H340 H372 H304 H319 H315	GHS02 GHS08 GHS07	H225 H350 H340 H372 H304 H319 H315	
Bis(2-chloroethyl) ether	111-44-4	Carc. 2 Acute Tox. 2 * Acute Tox. 1 Acute Tox. 2 *	H351 H330 H310 H300	GHS08 GHS06	H351 H330 H310 H300	
Butane-1-ol	71-36-3	Flam. Liq. 3 Acute Tox. 4 Skin Irrit. 2 Eye Dam. 1 STOT SE 3 STOT SE 3	H226 H302 H315 H318 H335 H336	GHS 02 GHS 05 GHS 07	H226 H302 H315 H318 H335 H336	
Butanone; ethyl methyl ketone	78-93-3	Flam. Liq. 2 Eye Irrit. 2 STOT SE 3	H225 H319 H336	GHS02 GHS07	H225 H319 H336	EUH066
Calcium cyanide	592-01-8	Acute Tox. 2 Aquatic Acute 1 Aquatic Chronic 1	H300 H400 H410	GHS06 GHS09	EUH032	

(continued)

(continued)

International chemical identification	CAS No	Classification		Labeling			Specific concentration limits, M-factors	Notes
		Hazard class and category code(s)	Hazard statement code(s)	Pictogram	Hazard statement code(s)	Suppl. hazard statement code (s)		
Chloroacetaldehyde	107-20-0	Carc. 2	H351	GHS06 GHS08 GHS05 GHS09	H351		STOT SE 3; H335:	
		Acute Tox. 2	H330		H330		C ≥ 5%	
		Acute Tox. 3	H311		H311			
		Acute Tox. 3	H301		H301			
		Skin Corr. 1B	H314		H314			
		Aquatic Acute 1	H400		H400			
		STOT SE 3	H335		H335			
Chloroacetic acid	79-11-8	Acute Tox. 3	H331	GHS06 GHS05 GHS09	H331		STOT SE 3; H335:	
		Acute Tox. 3	H311		H311		C ≥ 5%	
		Acute Tox. 3	H301		H301			
		Skin Corr. 1B	H314		H314			
		Aquatic Acute 1	H400		H400			
		STOT SE 3	H335		H335			
Cobalt sulfate	10124-43-3	Carc. 1B	H350i	GHS08 GHS07 GHS09	H350i		Carc. 1B; H350i:	
		Muta. 2	H341		H341		C ≥ 0.01%	
		Repr. 1B	H360F		H360F		M = 10	
		Acute Tox. 4	H302		H302			
		Resp. Sens. 1	H334		H334			
		Skin Sens. 1	H317		H317			
		Aquatic Acute 1	H400		H410			
		Aquatic Chronic 1	H410					

Substance	CAS No.	Classification		Pictograms		Notes
Di-ammonium-tetrachloroplatinate	13820-41-2	Acute Tox. 3 Skin Irrit. 2 Eye Dam. 1 Resp. Sens. 1 Skin Sens. 1	H301 H315 H318 H334 H317	GHS06 GHS05 GHS08	H301 H315 H318 H334 H317	
2,2-Dichlorovinyl dimethyl phosphate	62-73-7	Acute Tox. 2 * Acute Tox. 3 * Acute Tox. 3 * Skin Sens. 1 Aquatic Acute 1	H330 H311 H301 H317 H400	GHS06 GHS09	H330 H311 H301 H317 H400	
Diisopropyl ether; [1] dipropyl ether; [2]	108-20-3; [1] 111-43-3; [2]	Flam. Liq. 2 STOT SE 3	H225 H336	GHS02 GHS07	H225 H336	EUH019 EUH066 C
Dimethylformamide	68-12-2	Repr. 1B Acute Tox. 4 Acute Tox. 4 Eye Irrit. 2	H360D H332 H312 H319	GHS 07 GHS 08	H360D H332 H312 H319	
Divanadium pentoxide	1314-62-1	Muta. 2 Repr. 2 STOT RE 1 Acute Tox. 4 Acute Tox. 4 STOT SE 3 Aquatic Chronic 2	H341 H361d H372 H332 H302 H335 H411	GHS08 GHS07 GHS09	H341 H361d H372 H332 H302 H335 H411	

(continued)

(continued)

International chemical identification	CAS No	Classification		Labeling			Specific concentration limits, M-factors	Notes
		Hazard class and category code(s)	Hazard statement code(s)	Pictogram	Hazard statement code(s)	Suppl. hazard statement code (s)		
Ethane-1,2-diol; ethylene glycol	107-21-1	Acute Tox. 4 *	H302	GHS07	H302			
Ethylbenzene	100-41-4	Flam. Liq. 2 Acute Tox. 4 STOT RE 2 Asp. Tox. 1	H225 H332 H373 (hearing organs) H304	GHS02 GHS07 GHS08	H225 H332 H373 (hearing organs) H304			
Hydrochloric acid (%)	7647-01-0	Skin Corr. 1B STOT SE 3	H314 H335	GHS 05 GHS 07	H314 H335		Skin Corr. 1B; H314: $C \geq 25\%$ Skin Irrit. 2; H315: $10\% \leq C < 25\%$ Eye Dam. 1; H318: $C \geq 25\%$ Eye Irrit. 2; H319: $10\% \leq C < 25\%$ STOT SE 3; H335: $C \geq 10\%$	B

Name	CAS	Classification	H-codes	Pictograms	H-codes	Notes	Concentration limits
Methanol	67-56-1	Flam. Liq. 2 Acute Tox. 3 Acute Tox. 3 Acute Tox. 3 STOT SE 1	H225 H301 H311 H331 H370	GHS 2, GHS 6, GHS 8			STOT SE 1; H370: $C \geq 10\%$ STOT SE 2; H371: $3\% \leq C < 10\%$
N,N-Dimethylaniline	121-69-7	Carc. 2 Acute Tox. 3 Acute Tox. 3 Acute Tox.3 Aquatic Chronic 1	H351 H331 H311 H301 H411	GHS06 GHS08 GHS09	H351 H331 H311 H301 H411		
n-Butyl acetate	123-86-4	Flam. Liq. 3 STOT SE 3	H226 H336	GHS 02 GHS 07	H226 H336	EUH066	
o-Xylene; [1] m-xylene; [2] p-xylene; [3] xylene; [4]	95-47-6 106-42-3 108-38-3 1330-20-7	Flam. Liq. 3 Acute Tox. 4 Acute Tox. 4 Skin Irrit. 2	H226 H312 H332 H315	GHS02 GHS07	H226 H312 H332 H315	C	
p-Benzoquinone	106-51-4	Acute Tox. 3 Acute tox. 3 Eye Irrit. 2 STOT SE 3 Skin Irrit. 2 Aquatic Acute 1	H331 H301 H319 H335 H315 H400	GHS06 GHS09	H331 H301 H319 H335 H315 H400	$M = 10$	
Phosphoric acid . . .%	7664-38-2	Skin Corr. 1B	H314	GHS05	H314	B	Skin Corr. 1B; H314: $C \geq 25\%$ Skin Irrit. 2; H315: $10\% \leq C < 25\%$

(continued)

(continued)

International chemical identification	CAS No	Classification		Labeling			Specific concentration limits, M-factors	Notes
		Hazard class and category code(s)	Hazard statement code(s)	Pictogram	Hazard statement code(s)	Suppl. hazard statement code(s)		
Piperazine	110-85-0	Repr. 2 Skin Corr. 1B Resp. Sens. 1 Skin Sens. 1	H361 fd H314 H334 H317	GHS05 GHS08	H361 fd H314 H334 H317			
Piperazine hydrochloride; [1] piperazine di-hydrochloride; [2]	6094-40-2 [1] 142-64-3 [2]	Repr. 2 Eye Irrit. 2 Skin Irrit. 2 Resp. Sens. 1 Skin Sens. 1 Aquatic Chronic 3	H361fd H319 H315 H334 H317 H412	GHS08	H361fd H319 H315 H334 H317 H412			
Potassium hydroxide	1310-58-3	Acute Tox. 4 Skin Corr. 1A	H302 H314	GHS05 GHS07	H302 H314		Skin Corr. 1A; H314: C ≥ 5% Skin Corr. 1B; H314: 2% ≤ C < 5% Skin Irrit. 2; H315: 0.5% ≤ C < 2% Eye Irrit. 2; H319: 0.5% ≤ C < 2%	
Propane-2-ol, isopropanol	67-63-0	Flam. Liq. 2 Eye Irrit. 2 STOT SE 3	H225 H319 H336	GHS 02 GHS 07	H225 H319 H336			

	CAS	Classification	H-codes	GHS	H-codes	Notes
Solvent naphtha (petroleum), light arom.	64742-95-6	Carc. 1B Muta. 1B Asp. Tox. 1	H350 H340 H304	GHS08	H350 H340 H304	M2
Styrene	100-42-5	Flam. Liq. 3 Repr. 2 STOT RE 2 Skin Irrit. 2 Eye Irrit. 2	H226 H361d H372 (hearing organs) H315 H319	GHS02 GHS08 GHS07	H226 H361d H372 (hearing organs) H315 H319	D
Tetrahydrofuran	109-99-9	Flam. Liq. 2 Eye Irrit. 2 STOT SE 3	H225 H319 H335	GHS02 GHS07	H225 H319 H335	Eye Irrit. 2; H319: $C \geq 25\%$ STOT SE 3; H335: $C \geq 25\%$
Tert-butylmethylether (MTBE)	1634-04-4	Flam. Liq. 2 Skin Irrit. 2	H225 H315	GHS02 GHS07	H225 H315	
Toluene	108-88-3	Flam. Liq. 2 Skin Irrit. 2 Repr. 2 STOT SE 3 STOT RE 2 Asp. Tox. 1	H225 H315 H361d H336 H373 H304	GHS 02 GHS 07 GHS 08	H225 H315 H361d H336 H373 H304	
Tributyl phosphate	126-73-8	Carc. 2 Acute Tox. 4 Skin Irrit. 2	H351 H302 H315	GHS08 GHS07	H351 H302 H315	

(continued)

(continued)

International chemical identification	CAS No	Classification		Labeling		Suppl. hazard statement code (s)	Specific concentration limits, M-factors	Notes
		Hazard class and category code(s)	Hazard statement code(s)	Pictogram	Hazard statement code(s)			
Triethyl phosphate	78-40-0	Acute Tox. 4	H302	GHS 07	H302			
Triethylamine	121-44-8	Flam. Liq. 2	H225	GHS 02	H225		STOT SE 3, H335:	
		Acute Tox. 4	H332	GHS 05	H332		C ≥ 1%	
		Acute Tox. 4	H312	GHS 07	H312			
		Acute Tox. 4	H302		H302			
		Skin Corr. 1A	H314		H314			
Triethylenetetramine	112-24-3	Acute Tox. 4	H312	GHS 05	H312			
		Skin Corr. 1B	H314	GHS 07	H314			
		Skin Sens. 1	H317		H317			
		Aquatic Chronic 3	H412		H412			
Triphenyl phosphite	101-02-0	Eye Irrit. 2	H319	GHS07	H319		Skin Irrit. 2; H315:	
		Skin Irrit. 2	H315	GHS09	H315		C ≥ 5%	
		Aquatic Acute 1	H400		H410		Eye Irrit. 2; H319:	
		Aquatic Chronic 1	H410				C ≥ 5%	

References

[1] P. Rasmussen and J. Gemehling, "Flash points of flammable liquid mixtures using UNIFAC," *Ind. Eng. Fundament*, Bd. 21, pp. 186–188, 1982.

[2] R. S. Helmut Greim, Toxicology and Risk Assessment: A Comprehensive Introduction, Hoboken: Wiley, 2019.

[3] L. Robinson, A Practical Guide to Toxicology and Human Health Risk Assessment, Hoboken: Wiley, 2019.

[4] C. Lipinski, F. Lombardo, B. Dominy and P. Feeney, "Experimental and computational approaches to estimate solubility and permeability in drug discovery and development settings," *Adv. Drug Delivery Rev*, 23, pp. 3–25, 1997.

[5] USEPA, "EPISUITE (TM) Estimation Program Interface," United States Environmental Protection Agency, 2012. [Online]. Available: https://www.epa.gov/tsca-screening-tools/epi-suitetm-estimation-program-interface. [Zugriff am 22 August 2019].

[6] ECHA, "Guidance on information requirements and chemicals safety assessment, version 2.1, chapter R.8," European Chemicals Agency, P.O. Box 400, FI-00121 Helsinki, Finland, 2012. [Online]. Available: https://www.echa.europa.eu/documents/10162/13632/information_re quirements_r8_en.pdf/e153243a-03f0-44c5-8808-88af66223258. [Zugriff am 06 09 2019].

[7] M. Mayersohn and M. Gibaldi, "Mathematical methods in pharmacokinetics. II. solution of the two compartment open model," *Am J Pharmaceutical Education*, Bd. 35, pp. 19–28, 1971.

[8] T. Cahill, I. Cousins and D. Mackay, "Development and application of a generalized physiologically based pharmacokinetic model for multiple environmental contaminants," *Environ. Toxicol. Chem.*, Bd. 22, pp. 26–34, 22 2003.

[9] C. C. f. E. M. a. Chemistry, "PBPK beta version 1.0," Trent University, 2003. [Online]. Available: http://www.trentu.ca/academic/aminss/envmodel/models/PBPK.html. [Zugriff am 02 09 2019].

[10] F. Joneneelen and W. Ten Berge, "CEFIC Long Range Research Initiative," European Chemical Industry Council, 2017. [Online]. Available: http://cefic-lri.org/toolbox/induschemfate/. [Zugriff am 02 09 2019].

[11] F. Jongeneelen and W. Ten Berge, "A generic, cross-chemical predictive PBTK model with multiple entry routes running as application in MS Excel; design of the model and comparison of predictions with experimental results," *Ann. Occup. Hyg.*, Bd. 55, Nr. 8, pp. 861–864, 2011.

[12] G. Agricola, De Re Metallica Libri XII; Zwölf Bücher vom Gerg- und Hüttenwesen, Wiesbaden: marixverlag; ISBN-13: 978-3-86539-097-4., 2006 (Reprint).

[13] OECD, "OECD Test Guidelines for the Chemicals," Organization for economic cooperation and development, 2019. [Online]. Available: http://www.oecd.org/chemicalsafety/testing/oecdg uidelinesforthetestingofchemicals.htm. [Zugriff am 16 08 2019].

[14] EMEA, "European Agency for Evaluating of Medicinal Products," October 1998. [Online]. Available: https://www.ema.europa.eu/en/documents/mrl-report/atropine-summary-report-committee-veterinary-medicinal-products_en.pdf. [Zugriff am 16 08 2019].

[15] D. S. Stefan Berger, Classics in Spectroscopy, Weinheim: Wiley-VCH; ISBN: 978-3-527-32516-0, 2009.

[16] HSDB, "Hazardous Substance Data Bank," United States National Library of Medicine, [Online]. Available: https://toxnet.nlm.nih.gov/cgi-bin/sis/search2/f?./temp/~iwmRay:1:ani mal. [Zugriff am 16 08 2019].

[17] B. K. Thomas Efferth, "Chemical Carcinogenesis: Genotoxic and Non-Genotoxic Mechanisms," in *Toxicology and Risk Assessment: A Comprehensive Introduction*, Hoboken, Wiley, 2019, pp. 162.

https://doi.org/10.1515/9783110618952-014

[18] OECD, "The OECD QSAR Toolbox," OECD, 19 February 2019. [Online]. Available: http://www.oecd.org/chemicalsafety/risk-assessment/oecd-qsar-toolbox.htm. [Zugriff am 15 October 2019].

[19] T. Schupp, "Read across for the derivation of indoor air guidance values supported by PBTK modelling," *EXCLI J.*, Bd. 17, pp. 1069–1078, 2018.

[20] G. Patlewicz, N. Jeliazkova, R. Safford, A. Worth and B. Aleksiev, "An evaluation of the implementation of the Cramer classification scheme in the Toxtree software," *SAR QSAR Environ. Res.*, Bd. 19, Nr. 5–6, pp. 495–524, 2008.

[21] I. Ltd, "Toxtree 3.1.0," Ideaconsult Ltd GPL, May 01, 2018. [Online]. Available: http://toxtree. sourceforge.net/download.html. [Zugriff am October 15, 2019].

[22] ECHA, "Guidance on information requirements and chemical safety assessment. Chapter R.6: QSARs and grouping of chemicals," May 2008. [Online]. Available: https://www.echa.eu ropa.eu/documents/10162/13632/information_requirements_r6_en.pdf/77f49f81-b76d-40ab-8513-4f3a533b6ac9. [Zugriff am 15 October 2019].

[23] T. Schupp, H. Allmendinger, C. Boegi, B. Bossuyt, S. Shen, B. Tury and R. J. West, "The environmental behavior of methylene-4,4'-dianiline," *Rev. Environ. Contam. Toxicol.*, 246 pp. 91–132, 2018.

[24] W. Mabey and T. Mill, "Critical review of hydrolysis of organic compounds in water under environmental conditions," *J. Phys. Chem. Refer. Data*, Bd. 7, pp. 383–415, 1978.

[25] R. Frank and W. Klöppfer, "Spectral solar photon irradiance in Central Europe and the adjacent north sea," *Chemosphere*, Bd. 17, Nr. 5, pp. 985–994., 1988.

[26] R. Zepp, J. Hoigné und H. Bader, "Nitrate-induced photooxidation of trace organic chemicals in water," *Environ. Sci. Technol.*, Bd. 21, Nr. 5, pp. 443–450, 1987.

[27] ECHA, "Guidance on Information Requirements and Chemical Safety assessment, Chapter R.11: PBT / vPvB assessment," June 2017. [Online]. Available: https://www.echa.europa.eu/ documents/10162/13632/information_requirements_r11_en.pdf/a8cce23f-a65a-46d2-ac68-92fee1f9e54f. [Zugriff am August 22, 2019].

[28] ECHA, "Guidance on information requirements and chemical safety assessment, chapter R.10: characterization of dose [concentration]-response for environment," European Chemicals Agency, 2008. [Online]. Available: https://www.echa.europa.eu/documents/ 10162/13632/information_requirements_r10_en.pdf/bb902be7-a503-4ab7-9036-d866b8ddce69. [Zugriff am 07 09 2019].

[29] ECHA, "Guidance on information requirements and chemical safety assessment, chapter R.16: environmental exposure assessment, version 3.0," European Chemicals Agency. P.O. Box 400, FI-00121 Helsinki, Finland, February 2017. [Online]. Available: https://www. echa.europa.eu/documents/10162/13632/information_requirements_r16_en.pdf/b9f0f406-ff 5f-4315-908e-e5f83115d6af. [Zugriff am 07 September 2019].

[30] J. Lijzen and M. Rikken, "EC (2004) European Union System for the Evaluation of Substances 2.0 (EUSES 2.0). Prepared for the European Chemicals Bureau by the National Institute of Public Health and the Environment (RIVM), Bilthoven, The Netherlands (RIVM Report no. 601900005)," January 2004. [Online]. Available: https://www.rivm.nl/bibliotheek/rap porten/601900005.pdf. [Zugriff am 07 September 2019].

[31] EUSES, "The European Union System for the Evaluation of Substances," The European Commission's science and knowledge service, June 06, 2016. [Online]. Available: https://ec. europa.eu/jrc/en/scientific-tool/european-union-system-evaluation-substances. [Zugriff am 07 September 2019].

[32] T. C. C. f. E. M. a. Chemistry, "Level III Model," The Canadian Centre for Environmental Modelling and Chemistry, March 2002. [Online]. Available: https://www.trentu.ca/academic/ aminss/envmodel/models/VBL3.html. [Zugriff am 07 September 2019].

[33] D. Mackay, S. Paterson und W. Shiu, "Generic models for evaluating the regional fate of chemicals," *Chemosphere*, Bd. 24, Nr. 6, pp. 695–717, 1992.

[34] ECHA, "Acetophenone," September 13, 2019. [Online]. Available: https://www.echa.europa.eu/de/web/guest/registration-dossier/-/registered-dossier/14683. [Zugriff am 13 09 2019].

[35] T. C. C. f. E. M. a. Chemistry, "Level I Model," The Canadian Centre for Environmental Modelling and Chemistry, September 2004. [Online]. Available: https://www.trentu.ca/academic/aminss/envmodel/models/L1300.html. [Zugriff am 13 September 2019].

[36] M. Brinkmann, C. Schlechtriem, M. Reininghaus, K. Eichbaum, S. Buchinger, G. Reifferscheid, H. Hollert und T. Reuss, "Cross-species extrapolation of uptake and disposition of neutral organic chemicals in fish using a multispecies physiologically-based toxicokinetic model framework," *Environ. Sci. Technol.*, Bd. 50, pp. 1914–1923, 2016.

[37] A. Franco und S. Trapp, "A multimedia activity model for ionizable compounds: validation study with 2,4-dichlorophenoxyacetic acid, aniline, and trimethoprim," *Environ. Toxicol. Chem.*, Bd. 29, Nr. 4, pp. 789–799, 2010.

[38] USEPA, "Benchmark Dose Tools," United States Environmental Protection Agency, 2019. [Online]. Available: https://www.epa.gov/bmds. [Zugriff am 13 September 2019].

[39] "The R-project," The R Foundation, July 5, 2019. [Online]. Available: https://www.r-project.org/. [Zugriff am 25 10 2019].

[40] RIVM, "PROAST," RIVM: National institute for public health and the environment, ministry of health, welfare and sport, September 09, 2019. [Online]. Available: https://www.rivm.nl/en/proast. [Zugriff am 13 September 2019].

[41] R. Conolly and W. Lutz, "Nonmonotonic dose-response relationships: mechanistic basis, kinetic modeling, and implications for risk assessment," *Toxicol. Sci.*, Bd. 77, pp. 151–157, 2004.

[42] L. Vandenberg, T. Colborn, T. Hayes, J. Heindel, D. Jacobs, D.-H. Lee, T. Shioda, A. Soto, F. Vom Saal, W. Welshons, R. Zoellr and J. Myers, "Hormones and endocrine-disrupting chemicals: low-dose effects and nonmonotonic dose responses," *Endocr. Rev.*, Bd. 33, Nr. 3, pp. 378–455, 2012.

[43] M. Spassova, "Statistical approach to identify threshold and point of departure in dose–response data," *Risk Anal.*, Bd. 39, Nr. 4, pp. 940–956, 2019.

[44] ECHA, "Understanding CLP," European chemicals agency, 2019. [online]. available: https://www.echa.europa.eu/web/guest/regulations/clp/understanding-clp. [Zugriff am 26 September 2019].

[45] U. Nations, "Globally Harmonized System odf Classification and Labelling of Chemicals; forth, revised edition," 2011. [Online]. Available: https://www.unece.org/fileadmin/DAM/trans/danger/publi/ghs/ghs_rev04/English/ST-SG-AC10-30-Rev4e.pdf. [Zugriff am 02 05 2019].

[46] ECHA, "Guidance on labelling and packaging in accordance with Regulation (EC) No 1272/2008," March 2019. [Online]. Available: https://www.echa.europa.eu/documents/10162/23036412/clp_labelling_en.pdf/89628d94-573a-4024-86cc-0b4052a74d65. [Zugriff am 14 September 2019].

[47] ECHA, "Guidance on the Application of the CLP Criteria," July 2017. [Online]. Available: https://www.echa.europa.eu/documents/10162/23036412/clp_en.pdf/58b5dc6d-ac2a-4910-9702-e9e1f5051cc5. [Zugriff am 14 September 2019].

[48] UNECE, "About the ADR," United Nations Economic Commission for Europe, 2019. [Online]. Available: https://www.unece.org/trans/danger/publi/adr/adr_e.html. [Zugriff am 21 September 2019].

[49] U. S. N. L. o. Medicine, "Hazardous Substances Data Bank," 2019. [Online]. Available: https://toxnet.nlm.nih.gov/cgi-bin/sis/htmlgen?HSDB. [Zugriff am 08 08 2019].

[50] IFA, "GESTIS Substance Database; Institute for Occupational Safety and Health of the German Social Accident Insurance," 2019. [Online]. Available: http://gestis-en.itrust.de/nxt/gateway. dll/gestis_en/000000.xml?f=templates$fn=default.htm$vid=gestiseng:sdbeng$3.0. [Zugriff am 08 08 2019].

[51] EChA, "Information on Chemicals," European Chemicals Agency, January 2020. [Online]. Available: https://echa.europa.eu/information-on-chemicals. [Zugriff am 15 February 2020].

[52] J. Young, M. How, A. Walker and W. Worth, "Classification as corrosive or irritant to skin of preparations containing acidic or alkaline substances, without testing on animals," *Toxicol. in Vitro*, Bd. 2, Nr. 1, pp. 19–26, 1988.

[53] S. E. Manahan, Environmental Chemistry, Boca Raton: CRC Press Taylor & Francis Group; ISBN: 978-1-4200-59205, 2010, p. 368 ff.

[54] ECE-ITE, "European agreement concerning the international carriage of Dangerous Goods by road," January 2019. [Online]. Available: https://www.unece.org/fileadmin/DAM/trans/publi cations/ADR_2019_vol1_1818953_E.pdf. [Zugriff am 21. September 2019].

[55] UN, "Part II: Classification," 2009. [Online]. Available: http://www.unece.org/fileadmin/ DAM/trans/danger/publi/unrec/rev16/English/02E_Part2.pdf. [Zugriff am 20 September 2019].

[56] CEFIC, "ICE (International Chemical Environment) guidance," European Chemical Industry Council, 24 July 2018. [Online]. Available: https://cefic.org/guidance/transport-and-logistics /ice-international-chemical-environment/. [Zugriff am 21 September 2019].

[57] EChA, "Guidance for identification and naming of substances," May 2017. [Online]. Available: https://echa.europa.eu/documents/10162/23036412/substance_id_en.pdf/ee696bad-49f6- 4fec-b8b7-2c3706113c7d. [Zugriff am 13 February 2020].

[58] ECHA, "Guidance on Information Requirements and Chemical Safety Assessment, chapter R.12: Use description," December 2015. [Online]. Available: https://echa.europa.eu/docu ments/10162/13632/information_requirements_r12_en.pdf/ea8fa5a6-6ba1-47f4-9e47- c7216e180197. [Zugriff am 21 September 2019].

[59] N. C. Wells, The Atmosphere and Ocean: A Physical Introduction, Third Edition, John Wiley & Sons, Ltd, 2012.

[60] RAC, "Committee for Risk Assessment,"" European Chemicals Agency, October 2019. [Online]. Available: https://echa.europa.eu/about-us/who-we-are/committee-for-risk- assessment. [Zugriff am 8 October 2019].

[61] W. Fransman, J. Schinkel, T. Meijster, J. van Hemmen, E. Tielemans and H. Goede, "Development and Evaluation of an Exposure Control Efficacy Library (ECEL)," *Ann. Occup. Hyg.*, Bd. 52, Nr. 7, pp. 567–575, 2008.

[62] W. Fransman, J. Cherrie, M. van Tongeren, M. Tischer, J. Schinkel, H. Marquart, N. Warren, S. Spankie, H. Kromhout and E. Tielemans, "Development of a mechanistic model for the Advanced REACH Tool (ART), Version 1.5," TNO Quality of Life, Utrechtsweg 48, 3700 AJ Zeist, The Netherlands, 2013.

[63] E. C. Agency, "CHESAR 3.4," European Chemicals Agency, 20 11 2018. [Online]. Available: https://chesar.echa.europa.eu/. [Zugriff am 12 10 2019].

[64] ECETOC, "Targeted Risk Assessment version 3.1," ECETOC, 9 2 2018. [Online]. Available: http://www.ecetoc.org/tools/targeted-risk-assessment-tra/. [Zugriff am 12 10 2019].

[65] ECETOC, "ECETOC TRA version 3: Background and rationale for the improvements. Technical Report No. 114," 07 2012. [Online]. Available: http://www.ecetoc.org/publication/tr-114-ecetoc- tra-version-3-background-and-rationale-for-the-improvements/. [Zugriff am 12 10 2019].

[66] A. f. Gefahrstoffe, "Ermitteln und Beurteilen der Gefährdungen bei Tätigkeiten mit Gefahrstoffen: Inhalative Exposition," February 15, 2017. [Online]. Available: https://www.

baua.de/DE/Angebote/Rechtstexte-und-Technische-Regeln/Regelwerk/TRGS/pdf/TRGS-402. pdf?__blob=publicationFile&v=4. [Zugriff am 12 10 2019].

[67] D. G. Unfallversicherung, "DGUV-Regel 112-190: Benutzung von Atemschutzgeräten," December 2011. [Online]. Available: https://publikationen.dguv.de/regelwerk/regelwerk-nach-fachbereich/persoenliche-schutzausru-stungen/atemschutz/1011/benutzung-von-atemschutzgeraeten. [Zugriff am 12 October 2019].

[68] BAuA, "TRGS 900," Ausschuss fuer Gefahrstoffe, August 08, 2019. [Online]. Available: https://www.baua.de/DE/Angebote/Rechtstexte-und-Technische-Regeln/Regelwerk/TRGS/ pdf/TRGS-900.pdf?__blob=publicationFile&v=13. [Zugriff am 17 February 2020].

[69] X. Zhao, H. Yoshioka, T. Noguchi, S. Fujimoto, Y. Tanaike, T. Hayakawa, Y. Hase and T. Naruse, "Fundamental study of gas toxicity with respect to fire stages," *Fire Sci. and Tech.*, Bd. 36, Nr. 1, pp. 11–24, 2017.

[70] J. Pauluhn, G. Kimmerle, T. Maertins, F. Prager and W. Pump, "Toxicity of combustions gases from plastics: relevance and limitations of results obtained in animal experiments," *J. Fire Sci.*, Bd. 12, pp. 63–104, 1994.

[71] C. Fuilla, P. Ménage, M. Imbert, H. Julien and F. Baud, "Cyanide compounds in fire gas toxicity and the role of hydroxycobalamin," in *The Management of Mass Burn Casualties and Fire Disasters*, Dordrecht, Springer, 1992, pp. 268–270.

[72] R. Wester, X. Hui, T. Landry and H. Maibach, "In vivo skin decontamination of methylene bisphenyl isocyanate (MDI): Soap and water ineffective compared to polypropylene glycol, polyglycol- based cleanser, and corn oil," *Toxicol. Sci.*, Bd. 48, Nr. 1, pp. 1–4, 1999.

[73] S. Talmage, A. Watson, V. M. N. Hauschild and J. King, "Chemical warfare agent degradation and decontamination," *Curr. Org. Chem.*, Bd. 11, pp. 285–298, 2007.

[74] G. Lunn and E. Sansone, Destruction of hazardous chemicals in the laboratory. 3rd edition, Hoboken, New Jersey.: John Wiley & Sons, 2012.

Index

www.ingramcontent.com/pod-product-compliance
Lightning Source LLC
Chambersburg PA
CBHW080706220326
41598CB00033B/5330